Green Energy and Technology

Climate change, environmental impact and the limited natural resources urge scientific research and novel technical solutions. The monograph series Green Energy and Technology serves as a publishing platform for scientific and technological approaches to "green"—i.e. environmentally friendly and sustainable—technologies. While a focus lies on energy and power supply, it also covers "green" solutions in industrial engineering and engineering design. Green Energy and Technology addresses researchers, advanced students, technical consultants as well as decision makers in industries and politics. Hence, the level of presentation spans from instructional to highly technical.

Indexed in Scopus.

Indexed in Ei Compendex.

Sachin Kumar · Suresh Sundaramurthy ·
Deepak Kumar · Anuj K. Chandel
Editors

Clean Energy Transition-via-Biomass Resource Utilization

A Way to Mitigate Climate Change

Springer

Editors
Sachin Kumar
Biochemical Conversion Division
Sardar Swaran Singh National Institute of Bio-Energy
Kapurthala, Punjab, India

Suresh Sundaramurthy
Department of Chemical Engineering
Maulana Azad National Institute of Technology
Bhopal, Madhya Pradesh, India

Deepak Kumar
College of Environmental Science and Forestry
State University of New York
New York, NY, USA

Anuj K. Chandel
Department of Biotechnology
Engineering School of Lorena
University of São Paulo (USP)
Lorena, São Paulo, Brazil

ISSN 1865-3529 ISSN 1865-3537 (electronic)
Green Energy and Technology
ISBN 978-981-97-6323-8 ISBN 978-981-97-6321-4 (eBook)
https://doi.org/10.1007/978-981-97-6321-4

This Springer imprint is published by the registered company Springer Nature Singapore Pte Ltd.
The registered company address is: 152 Beach Road, #21-01/04 Gateway East, Singapore 189721, Singapore

Foreword

The bioresources have tremendous potential to mitigate global warming. Also, biomass is expected to play a multifunctional role including food production, source of energy and fodder, biodiversity conservation, and yield of goods and services to society. Clean energy transition via utilizing biomass resources has been projected as an important climate change mitigation strategy. A vital characteristic of biomass is its localized nature; therefore, bioenergy utilization should follow decentralized planning.

This book consists of 12 chapters that highlight different biomass resource utilization for clean energy transition objectives. A few of them are described here. Sustainable utilization of biomass resources focuses on the potential of biomass to mitigate climate change, enhance energy security, and promote rural development. Algal feedstocks are the key component of the 3G biorefinery that strengthens the bio-renewables industry.

Policies for systematizing biomass collection, handling, storage, and transport to have a robust supply chain, improved product yields, and quality would drastically bring down the huge investment needed for separation and further upgradation and in turn, boost the competitiveness of the processes.

Among the sources used for biomaterials and related products, the use of vegetable oil seeds as a biomass resource is of great prominence due to the availability of a wide range of sources and varieties of plant-based seeds. Utilization of castor seed oil and its products (an agricultural-based feedstock) as a biomass source for a wide range of oleochemical products for potential applications ranging from bio-lubricants to biomaterials. Management, processing, and value addition of fish waste in the sense of supplying different agricultural leftovers as a co-digesting medium to a floating drum biogas plant. The slurry from the bio-digester has been analyzed with respect to N, P, and K content and treated properly for utilization as manure for eco-farming. Perspectives on industrial symbiosis, biomass banking, and a government-funded co-operative also covered the view of bioenergy society to promote indigenous biofuel production and improve a nation's gross domestic product value.

One of the chapters focuses on the mechanisms of production of biohydrogen by microalgae, microalgal cultivation methods, types of reactors used for cultivation of microalgae, parameters influencing the production of biohydrogen by microalgae, metabolic limitation and bottlenecks and economic evaluation for the biohydrogen produced by microalgae.

Biofuels such as ethanol and biodiesel produced from different feedstocks as promoted in the National Biofuel Policy 2018 of India are critically discussed and economic feasibility is of great importance considering the application of the non-edible oil of *Jatropha curcas* process at an industrial scale. Furthermore, it explores the cost-effective and sustainable technologies associated with this sector. These factors, when combined, could have a positive impact on the local or regional economy availability and cost of biomass feedstocks, the efficiency and cost of conversion technologies, and the market demand for biomass-derived energy.

UN Sustainable Development Goals associated with biomass utilization, emphasizing the importance of policy frameworks for the social inclusion of minorities. It highlights the need for transparent and accountable reporting mechanisms to ensure the credibility of biomass utilization practices, promoting informed decision-making and responsible resource management. The integration of social accounting principles into the assessment of biomass resource utilization provides a holistic understanding of its impacts, beyond traditional economic and environmental metrics. By recognizing the social dimensions and involving stakeholders, the credibility of biomass applications can be significantly enhanced and can mobilize a renewable biomass-based economy, which is the need of the day.

Satinder Kaur Brar, Ph.D.
Professor and James and Joanne Love
Chair in Environmental Engineering
Department of Civil Engineering
Lassonde School of Engineering
York University
Toronto, ON, Canada

Preface

Nowadays energy is a big crisis in all sectors for which there is a big change to store and optimize its uses. Energy usage in industrial, domestic, and transportation accounts for an overwhelming proportion of greenhouse gas emissions. The cost of fossil fuels is increasing and fluctuating due to depleting resources and growing concerns for climate change that require an immediate solution. Globally, the determination to achieve 'net carbon-neutral' targets by 2050, requires new clean energy sources and efficient energy systems. Green Energy and Technology focus on energy and power supply, it also covers 'green' solutions in industrial engineering and engineering design. The Clean Energy Transition-via-Biomass Resource Utilization will uplift the focus on Green Energy and Technology which addresses researchers, advanced students, technical consultants as well as decision-makers in industries and politics.

This book consists of 12 chapters that highlight different biomass resource utilization for clean energy transition objectives. This book highlights the clean energy transition via sustainable utilization of biomass resources, viz. forestry, agriculture, agroforestry, grassland, and seaweeds to climate change mitigation. Bioresources have tremendous potential to mitigate global warming. Also, biomass is expected to play a multifunctional role including food production, source of energy and fodder, biodiversity conservation, and yield of goods and services to society. This book was elaborated on considering the biological and physiochemical characteristics of these bioresources—the major criteria on which the choice of application(s) could be reserved for each biomass. Sustainable utilization of biomass resources focuses on the potential of biomass to mitigate climate change, enhance energy security, and promote rural development. Algal feedstocks are the key component of the 3G biorefinery that strengthens the bio-renewables industry. Policies for systematizing biomass collection, handling, storage, and transport to have a robust supply chain, improved product yields, and quality would drastically bring down the huge investment needed for separation and further upgradation and in turn, boost the competitiveness of the processes. Among the sources used for biomaterials and related products, the use of vegetable oil seeds as a biomass resource is of great prominence due to the availability of a wide range of sources and varieties of plant-based seeds.

Utilization of castor seed oil and its products (an agricultural-based feedstock) as a biomass source for a wide range of oleochemical products for potential applications ranging from bio-lubricants to biomaterials. Fish waste management, processing, and value addition in the sense of feeding to a floating drum biogas plant with various agro-residues as a co-digesting medium. The slurry from the bio-digester has been analyzed with respect to N, P, and K content and treated properly for utilization as manure for eco-farming. Perspectives on industrial symbiosis, biomass banking, and a government-funded co-operative also covered the view of bioenergy society to promote indigenous biofuel production and improve a nation's gross domestic product value. One of the chapters focuses on the mechanisms of production of biohydrogen by microalgae, microalgal cultivation methods, types of reactors used for cultivation of microalgae, parameters influencing the production of biohydrogen by microalgae, metabolic limitation and bottlenecks, and economic evaluation for the biohydrogen produced by microalgae.

Presently, the global economy is based on the fossil fuel economy, which is referencing the global hydrocarbon industry. The energy used mostly comes from fossil fuels like crude oil, natural gas, and coal. Problems of using these fossil fuels are the emission of greenhouse gases (GHGs), price fluctuation, issues with supply/distribution network, etc., these problems can be minimized by adopting clean energy or alternative clean fuels such as Hydrogen. The hydrogen economy relies on the use of hydrogen as a low-carbon commercial fuel in the transportation sector, alternative industrial feedstock, power generation, energy storage, etc. A few strategies were covered that could help to reduce greenhouse gas emissions and achieve the net carbon neural target are:

- Increase or optimise the energy efficiency of existing/new energy systems.
- Biofuels with Carbon Capture, Utilization, and Storage.
- Shift to clean energy as H_2 from conventional energy systems.

This book covers energy transition for various industries and techno-economic analysis. This book content will be informative to all industrial and research professionals along with academicians.

We sincerely hope that our contribution throughout this book will be a valuable asset to researchers, instructors, decision-makers, practising professionals, senior undergraduate and graduate students, and others interested in clean energy production and storage using renewable and low-cost bioresources. The book could even be used as a reference book for researchers teaching courses dealing with biomass valorisation. All the chapters were contributed by professionals from academia and government laboratories from various countries. The editors thank Springer for believing in our book project. We gratefully acknowledge all the authors who have contributed to this book for sharing their views and research findings with the scientific and professional communities through our book. The views or opinions expressed in each chapter of this book are those of the authors.

Kapurthala, India — Dr. Sachin Kumar
Bhopal, India — Dr. Suresh Sundaramurthy
New York, USA — Dr. Deepak Kumar
Lorena, Brazil — Dr. Anuj K. Chandel

Contents

About the Editors

Dr. Sachin Kumar is a Deputy Director in the Biochemical Conversion Division at the Sardar Swaran Singh National Institute of Bio-Energy (Formerly Sardar Swaran Singh National Institute of Renewable Energy), Kapurthala, India. He obtained his Ph.D. in Chemical Engineering from the Indian Institute of Technology Roorkee, India. He has more than 12 years of research experience in Biochemical Conversion of Biomass to Biofuels including lignocellulosic ethanol, biogas, biohydrogen, etc. During his career at SSS-NIBE, he has developed state-of-the-art laboratory facilities for Biochemical Conversion of Biomass to Biofuels. His research interests are in biofuels, including bioethanol, biogas, biobutanol, biohydrogen, algal biomass, bioprocess engineering, enzyme technology, metabolic engineering, etc.

Dr. Suresh Sundaramurthy is the Faculty & Head, Department of Chemical Engineering, Maulana Azad National Institute of Technology (MANIT) Bhopal, India. He has received Ph.D. in Chemical Engineering from the Indian Institute of Technology Roorkee, India. Prof. Suresh was the recipient of Postdoctoral Fellow awarded by IUSSTF, Govt. of India, and at that time, he was associated at The City University of New York, USA. He was also the recipient of Visiting Faculty and Researcher awarded by Govt. of India. During this award, he was deputed to the Asian Institute of Technology Thailand and the International Centre for Materials Science (JNCASR) Bengaluru, India. He has also worked at Pondicherry University and Indian Institute of Technology Kanpur. He has supervised 09 Ph.D. students, 60 M.Tech. students, more than 145 B.Tech./project students, and a few students along with co-supervisors. His total number of research journal publications are more than 120 in the areas of Environmental Biotechnology, New sorbents & Catalysts, Reactor Design, Waste-to-Energy, Nanoengineered materials, biofuels and intermediates from CO_2 and Biomass Decarbonization, Energy Transition, Technical Textiles, Gas separation, Air pollution abatement, Sensing of toxic gases, and Process Safety. He has more than 75 publications in international and national conference proceedings and few with co-authored. He has undertaken 23 R&D and consultancy projects. He has published 01 Indian patent and 01 technology synchronized and many more in progress, 04 textbooks, and more than 33 book chapters, and 06 edited books and proceedings.

In recognition of his research contributions, he has received a number of awards and honors including the National Innovation-Rashtrapati Bhavan, office of the President of India, and DST-Young Scientist Award; Visiting Research Fellowship; and Prof. R C Singh Medal and IEI Young Engineers Award.

Dr. Deepak Kumar is an Assistant Professor in the Chemical Engineering Department at the State University of New York College of Environmental Science and Forestry (SUNY-ESF). Dr. Kumar received his Ph.D. in Biological and Ecological Engineering from Oregon State University. His research focuses on the development of sustainable bioprocess technologies for the production of high-value bioproducts, bioplastics, and biofuel from agro-food processing waste and forest residues using integrated experimental and modeling approaches. He has authored more than 90 peer-reviewed research articles and 15 book chapters, with over 3200 citations and an H-index of 32. He has received several awards, including the Bioenergy Society of Singapore (BESS) Achievement Award 2016. He is listed as one of the top 2% cited scientists in the field of energy by the research conducted by Stanford University in collaboration with Elsevier-Scopus for the years 2022 and 2023.

Dr. Anuj Kumar Chandel is a Professor in the Department of Biotechnology, Engineering School of Lorena, University of São Paulo, Brazil. He has over 22 years of experience working on various aspects of industrial biotechnology (production of industrial enzymes, biofuels, and bio-based products) and membrane-based separation of fats, proteins, and viruses. He obtained his Ph.D. in 2009 and then did three post-doctoral fellowships at Stellenbosch University, South Africa, Engineering School of Lorena, University of São Paulo, Brazil, and University of Arkansas, Fayetteville, USA. Dr. Chandel has worked in three biotech industries for 7 years at large-scale operations. His primary research interest is to develop sustainable processes for the bioconversion of lignocellulosic biomass into renewable fuels and biochemicals by bridging the gap between research laboratories and industries. Dr. Chandel has published more than 116 peer-reviewed scientific papers, edited 15 books, and wrote 56 book chapters. He has one Brazilian patent to his credit.

Sustainable Utilization of Biomass Resources

V. Varalakksmi, S. Sudalai, and A. Arumugam

Abstract The chapter explores the sustainable utilization of biomass resources, focusing on the potential of biomass to mitigate climate change, enhance energy security, and promote rural development. It highlights the global need to transition away from fossil fuels and reduce greenhouse gas emissions. The chapter also discusses various biomass feedstock forms and their applications, including bioenergy production, biofuels, biochemicals, and bioproducts. It emphasizes the importance of sustainable practices like crop rotation, agroforestry, and waste-to-resource conversion. It also highlights the role of policy frameworks and regulatory incentives in promoting sustainable biomass utilization. The chapter also addresses challenges such as resource availability, logistical constraints, and competition for land use. Strategies for mitigating these include efficient supply chain management and integrating biomass into circular economies. The chapter offers valuable insights for policymakers, researchers, industry stakeholders, and environmentalists seeking to harness biomass's full potential while safeguarding our planet's natural resources and ecosystems.

Keywords Sustainable utilization · Biomass · Climate mitigation · Bioenergy · Bioproducts · Policies

1 Introduction

In an era characterized by the dire consequences of unchecked industrialization and fossil fuel consumption, the urgency to transition toward cleaner, more sustainable energy sources has reached unprecedented levels. Biomass, in its diverse forms

V. Varalakksmi · A. Arumugam (✉)
Advanced Biorefinery & Catalysis (ABC) Lab, Centre for Bioenergy, School of Chemical and Biotechnology, SASTRA Deemed University, Thirumalaisamudram, Thanjavur 613401, India
e-mail: aruchemxl@scbt.sastra.edu

S. Sudalai
Centre for Pollution Control and Environmental Engineering, School of Engineering and Technology, Pondicherry University, Puducherry 605014, India

S. Kumar et al. (eds.), *Clean Energy Transition-via-Biomass Resource Utilization*, Green Energy and Technology, https://doi.org/10.1007/978-981-97-6321-4_1

ranging from agricultural residues and forest byproducts to organic waste, presents an invaluable alternative with the potential to revolutionize our approach to energy production, waste management, and carbon sequestration [1].

Biomass, originating from a wide array of sources such as crops, forest residues, algae, and even organic waste, embodies the solar energy captured through photosynthesis. By converting this stored energy into various forms, such as heat, electricity, biofuels, and biogas, societies can significantly reduce their reliance on fossil fuels and consequently mitigate the release of greenhouse gases into the atmosphere [2]. Biomass resource utilization involves harnessing energy stored within organic materials through various processes such as combustion, fermentation, anaerobic digestion, etc. [53]. Unlike finite fossil fuels, biomass is a renewable resource that can be continually replenished through responsible land management and sustainable farming practices. Furthermore, its utilization emits significantly fewer greenhouse gases compared to conventional fossil fuels, making it a crucial component of strategies aimed at curbing global warming and meeting emission reduction targets outlined in international agreements like the Paris Agreement [3].

Beyond its contributions to climate change mitigation, the responsible use of biomass offers a range of other benefits. By leveraging agricultural residues and forestry waste that might otherwise go unused or contribute to environmental problems, societies can simultaneously reduce waste disposal challenges and generate valuable energy. This, in turn, bolsters energy security and supports local economies by creating jobs along the biomass supply chain [4].

Biomass resource utilization also aligns with circular economy principles, where waste is minimized, resources are maximally utilized, and environmental impacts are reduced. By diverting organic waste from landfills and incineration, biomass utilization minimizes methane emissions and concurrently generates clean energy [5]. Furthermore, innovative technologies enable the extraction of biogas, biofuels, and valuable biochemicals from biomass, paving the way for a sustainable transition in sectors ranging from transportation to chemical manufacturing [61].

The socio-economic benefits of biomass utilization are equally compelling. Localized biomass energy systems can stimulate rural economies by creating jobs in agriculture, forestry, and energy production. Moreover, biomass-derived energy can enhance energy security by diversifying energy portfolios, reducing dependence on imported fossil fuels, and enhancing energy self-sufficiency at various scales [22]. Moreover, biomass resource utilization has the potential to aid in land restoration and conservation efforts. In cases where deforestation or land degradation has occurred, strategically managed biomass production can aid in soil rehabilitation, prevent erosion, and promote reforestation. These efforts contribute to the preservation of biodiversity, enhance ecosystem services, and safeguard vital natural habitats.

As we navigate the complexities of climate change mitigation, the exploration of biomass resource utilization becomes more imperative than ever. However, it is essential to approach this strategy with caution and consider resource management to avoid potential negative impacts such as competition with food production or ecosystem disruption. By coupling scientific innovation, policy support, and community engagement, societies can tap into the immense potential of biomass while safeguarding the delicate balance of our planet's ecosystems [6].

In this chapter, we will delve deeper into the various facets of biomass resource utilization, examining its different applications, technological advancements, policy implications, and ways for successful integration into sustainable energy systems. Through these insights, we will illuminate the path forward for biomass as a promising avenue in the fight against climate change, offering a bridge to a more environmentally conscious and resilient future.

2 Biomass Utilization—Prospects and Challenges

Biomass utilization holds immense promise as a means to mitigate climate change by providing renewable energy, waste management solutions, and economic growth opportunities. The synergy between the prospects and challenges of biomass utilization underscores the need for a holistic and interdisciplinary approach [7]. By capitalizing on the benefits while navigating the complexities, societies can harness biomass's potential to contribute meaningfully to climate change mitigation, energy security, waste reduction, and economic growth (Fig. 1).

Fig. 1 Prospects of biomass utilization [42]

2.1 *Prospects*

2.1.1 Renewable Energy Source

Biomass is a renewable energy source as it can be produced and replenished relatively quickly compared to fossil fuels. Biomass energy can be harnessed through various processes, including combustion, gasification, and fermentation, to produce heat, electricity, and biofuels like biodiesel and bioethanol [5].

2.1.2 Reduction in Greenhouse Gas Emissions

Biomass can play a role in reducing greenhouse gas emissions when used for energy production, as it is considered carbon–neutral. The carbon dioxide released during biomass combustion is offset by the carbon dioxide absorbed by plants during their growth [47]. Biomass can replace fossil fuels in many applications, helping to mitigate climate change.

2.1.3 Waste Management

Biomass utilization can help manage organic waste materials, including agricultural residues, forestry waste, and food scraps, diverting them from landfills and reducing methane emissions. By converting these wastes into energy or biofuels, biomass can contribute to a more sustainable waste management system [58].

2.2 *Economic Benefits*

Biomass projects can stimulate local economies by creating jobs in biomass production, harvesting, transportation, and processing. Biomass can provide economic opportunities for rural communities, particularly in agriculture and forestry regions.

2.2.1 Energy Security and Storage

Biomass can contribute to energy security by diversifying energy sources and reducing dependence on imported fossil fuels. It can be used as a reliable source of baseload or intermittent power generation. Biomass can be used as a form of energy storage when converted into biofuels or bioenergy carriers like biogas or hydrogen [1]. This can help balance intermittent renewable energy sources such as wind and solar.

2.2.2 Technological Advancements

Ongoing research and development in biomass conversion technologies are leading to increased efficiency and cost-effectiveness. Advanced biomass processing techniques, such as torrefaction, pyrolysis, and anaerobic digestion, are being developed to enhance biomass utilization [32].

2.2.3 Carbon Sequestration

Biomass utilization contributes to carbon sequestration indirectly by promoting sustainable forestry and land management practices, such as reforestation. When biomass is used for energy or biofuel production, it can reduce net CO_2 emissions compared to fossil fuels, as long as the biomass source is managed sustainably [19]. Biomass can also be integrated with carbon capture technologies to further reduce emissions from biomass-based energy production.

2.3 *Challenges*

2.3.1 Competing Land Uses

The demand for biomass resources may compete with food production, raising concerns about food security [41]. Striking a balance between energy and food production is critical to prevent exacerbating existing food scarcity issues.

2.3.2 Environmental Impact

Unsustainable biomass extraction can lead to deforestation, habitat loss, and soil degradation [18]. Careful ecosystem management and land-use planning are necessary to avoid environmental harm and ensure the long-term sustainability of biomass resources. Biomass combustion can release pollutants such as particulate matter, nitrogen oxides (NOx), and carbon monoxide (CO), which can contribute to air quality issues. Careful emission control and monitoring are necessary to mitigate these effects.

2.3.3 Resource Availability and Variability

While biomass is renewable, its availability varies by region and season [50]. Balancing resource demand with supply requires sophisticated logistics and infrastructure to ensure a consistent and reliable energy supply. Different types of biomasses (e.g., wood, agricultural residues, algae) have varying compositions and

Fig. 2 Challenges of biomass utilization [15]

properties. This variability can affect the efficiency and performance of biomass conversion processes, requiring customized solutions for different feedstocks (Fig. 2).

2.3.4 Policy and Regulation

Establishing comprehensive regulatory frameworks is crucial to prevent overexploitation, ensure responsible sourcing, and maintain environmental integrity. Policies that incentivize sustainable practices and guide the responsible growth of the biomass industry are vital. Biomass utilization is subject to various regulations and policies that vary by region. These can impact the feasibility and profitability of biomass projects, so a clear and supportive regulatory environment is important [54].

2.3.5 Energy Conversion Efficiency and Density

Biomass conversion technologies, such as combustion, gasification, and fermentation, often have lower energy conversion efficiencies compared to fossil fuel technologies. This means that a significant portion of the energy in the biomass may be lost during conversion [12]. Biomass generally has a lower energy density compared to fossil fuels like coal and natural gas. As a result, a larger volume of biomass is required to generate the same amount of energy, which can pose challenges for transportation, storage, and handling [10].

3 Sustainable Factors of Biomass Utilization

In the context of the escalating climate crisis, the imperative to mitigate greenhouse gas emissions and transition to sustainable energy sources has never been more pressing. Among the array of options available, "Biomass Resource Utilization" stands out as a versatile and sustainable strategy that offers a multifaceted approach to address climate change while promoting socio-economic development (Fig. 3). The sustainable factors that underscore the potential of biomass utilization as a pivotal way to mitigate climate change are as follows.

3.1 *Renewability and Carbon Neutrality*

Central to the sustainability of biomass resource utilization is its renewable nature. Unlike fossil fuels, which are finite and contribute to the release of sequestered carbon into the atmosphere, biomass originates from organic matter that can be replenished through responsible land management and agricultural practices [46]. When biomass is combusted for energy, the carbon dioxide released is equivalent to the carbon dioxide absorbed during the plant's growth through photosynthesis. This closed-loop carbon cycle renders biomass energy generation nearly carbon–neutral, making it a crucial tool for reducing net emissions and achieving climate goals [15].

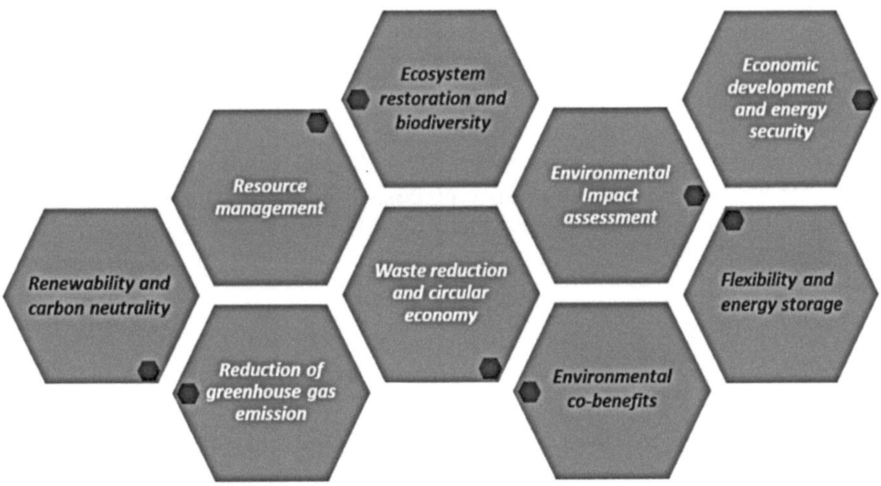

Fig. 3 Sustainable factors of biomass utilization [57])

3.2 Resource Management

Sustainable biomass utilization starts with responsible resource management. This involves ensuring that the biomass feedstocks, such as wood, crop residues, and organic waste, are harvested or collected in a manner that does not deplete the resource. Practices like selective logging in forestry and efficient waste collection and recycling systems contribute to sustainable resource management [20].

3.3 Reduction of Greenhouse Gas Emissions

The combustion of fossil fuels is a significant contributor to the accumulation of greenhouse gases in the atmosphere, resulting in global warming and climate disruptions. Biomass utilization offers a sustainable alternative by releasing significantly lower levels of carbon dioxide compared to fossil fuel combustion [35]. This reduction in emissions not only directly mitigates climate change, but also fosters cleaner air quality and improved public health, further enhancing the societal benefits of biomass energy.

3.4 Waste Reduction and Circular Economy

Biomass utilization aligns seamlessly with the principles of the circular economy by converting organic waste into valuable resources. Agricultural residues, food waste, and other organic byproducts can be diverted from landfills or incineration, preventing the release of methane and other harmful gases. Instead, these materials can be transformed into biogas, biofuels, and biochemicals through anaerobic digestion, fermentation, and other innovative processes. This waste-to-energy paradigm not only addresses waste management challenges, but also reduces the need for fossil fuel extraction [31].

3.5 Ecosystem Restoration and Biodiversity Preservation

Sustainable biomass utilization practices underscore the importance of responsible land management to avoid ecosystem degradation. By utilizing waste biomass and residues, the pressure on natural habitats and forests is reduced, which in turn safeguards biodiversity and promotes ecosystem health [27]. Moreover, strategic biomass production can aid in reforestation efforts, soil rehabilitation, and the prevention of erosion, contributing to the restoration of critical ecosystems and enhancing their resilience to climate impacts.

3.6 Environmental Impact Assessment

Before commencing biomass utilization projects, thorough environmental impact assessments (EIAs) should be conducted. These assessments evaluate potential adverse effects on air quality, water quality, soil health, and ecosystems [4]. They also identify mitigation measures to ensure that the project aligns with sustainability goals.

3.7 Economic Development and Energy Security

The integration of biomass resource utilization into energy systems has significant economic implications. Localized biomass energy production can stimulate rural economies by creating jobs in agriculture, forestry, and energy production [44]. It provides a decentralized energy solution that enhances energy security, reduces reliance on imported fossil fuels, and empowers communities to take control of their energy future.

3.8 Flexibility and Energy Storage

Biomass resource utilization offers a flexible energy source that can contribute to balancing the intermittency of renewable energy systems such as solar and wind. Biomass facilities can be ramped up or down relatively quickly to respond to changes in energy demand. Additionally, biomass-derived biofuels can serve as a form of energy storage, allowing for the stockpiling of energy during periods of low demand and its release when demand surges [8]. This flexibility enhances the stability of energy grids and supports the integration of variable renewable sources.

3.9 Environmental Co-benefits

Sustainable biomass utilization can offer various environmental co-benefits beyond climate change mitigation [25]. For instance, it can reduce the need for conventional fertilizers by recycling nutrient-rich organic waste back into the soil. This practice promotes soil health, minimizes water pollution, and conserves resources. In the broader context of combating climate change, the sustainable factors of biomass resource utilization extend far beyond carbon reduction. They touch on various dimensions of sustainability, including economic viability, social equity, ecosystem health, and technological innovation. By embracing these factors, societies can leverage biomass as a potent tool to mitigate climate change, drive sustainable development, and forge a path toward a more resilient and harmonious coexistence with our planet [22].

4 Policies for Sustainability

4.1 Renewable Energy Standards (RES) or Renewable Portfolio Standards (RPS)

These policies require utilities to obtain a certain percentage of their energy from renewable sources, including biomass. They create a reliable market for biomass-based energy, incentivizing its production and utilization. States and countries often set specific targets and deadlines for increasing the share of renewable energy in their energy mix.

4.2 Feed-in Tariffs (FiTs)

FiTs guarantee a fixed, premium price for renewable energy producers, making biomass energy projects financially attractive [24]. These incentives encourage private investments in biomass power plants, biofuel production facilities, and other biomass-based energy projects. FiTs provide predictability and stability in revenue for biomass energy producers.

4.3 Tax Incentives and Credits

Governments offer tax incentives, credits, or grants to biomass producers and users to promote sustainable practices and technology adoption [43]. Examples include investment tax credits, production tax credits, and accelerated depreciation for biomass-related investments. These financial incentives help offset the initial capital costs of biomass projects.

4.4 Carbon Pricing Mechanisms

Carbon pricing policies, such as carbon taxes and cap-and-trade systems, put a price on carbon emissions, making fossil fuels more expensive. Biomass, when sustainably managed and utilized, is considered carbon–neutral, as the carbon emitted during combustion is offset by the carbon absorbed during growth. This makes biomass a financially attractive option in carbon pricing markets.

4.5 *Environmental Regulations*

Governments impose environmental regulations on biomass utilization to minimize air and water pollution [56]. Emission limits, air quality standards, and wastewater treatment requirements ensure that biomass energy production and processing are environmentally responsible. Compliance with these regulations is crucial for the sustainability of biomass projects.

4.6 *Sustainable Forest Management Standards*

For woody biomass, sustainable forest management standards ensure that forests are harvested responsibly [16]. Selective logging, reforestation, and protection of sensitive ecosystems are essential components of these standards. Sustainable forestry practices maintain a healthy forest ecosystem and a sustainable supply of wood biomass.

4.7 *Land Use Policies*

To prevent conflicts with food production and habitat preservation, policies can limit the conversion of natural ecosystems or agricultural lands into biomass plantations. Sustainable land use practices help safeguard food security and biodiversity.

4.8 *Bioenergy Certification Systems*

Certification programs like the Roundtable on Sustainable Biomaterials (RSB) and Forest Stewardship Council (FSC) verify that biomass is sourced and processed sustainably [29]. Certified biomass products have a market advantage, as they demonstrate adherence to rigorous sustainability criteria.

4.9 Research and Development Funding

Governments allocate funds for research and development in biomass technology, sustainable practices, and innovation. These investments accelerate the adoption of advanced biomass utilization methods, improving efficiency and sustainability.

4.10 Waste-to-Energy Policies

Policies promoting the conversion of organic waste into energy, such as anaerobic digestion or gasification, reduce waste sent to landfills [34]. Organic waste-to-energy processes can provide a sustainable source of biomass feedstock and reduce methane emissions.

4.11 Community Engagement and Benefit Sharing

Policies encourage biomass projects to engage with local communities, create jobs, and share economic benefits. Benefit-sharing mechanisms, such as revenue sharing or community development funds, promote social sustainability and acceptance of biomass projects [17].

4.12 Monitoring and Reporting Requirements

Governments implement reporting and monitoring requirements for biomass utilization projects. These requirements ensure compliance with sustainability criteria, environmental standards, and emissions limits. Monitoring helps identify areas for improvement and assesses the ongoing sustainability of biomass projects.

5 Biomass Supply Chain

The biomass supply chain is a complex network that encompasses the entire journey of biomass, from its origination as organic matter to its conversion into energy, biofuels, or other valuable products. A well-designed and sustainable biomass supply chain is crucial for ensuring that biomass utilization effectively contributes to climate change mitigation while minimizing environmental and social impacts. This section delves into the intricate components of the biomass supply chain, focusing on

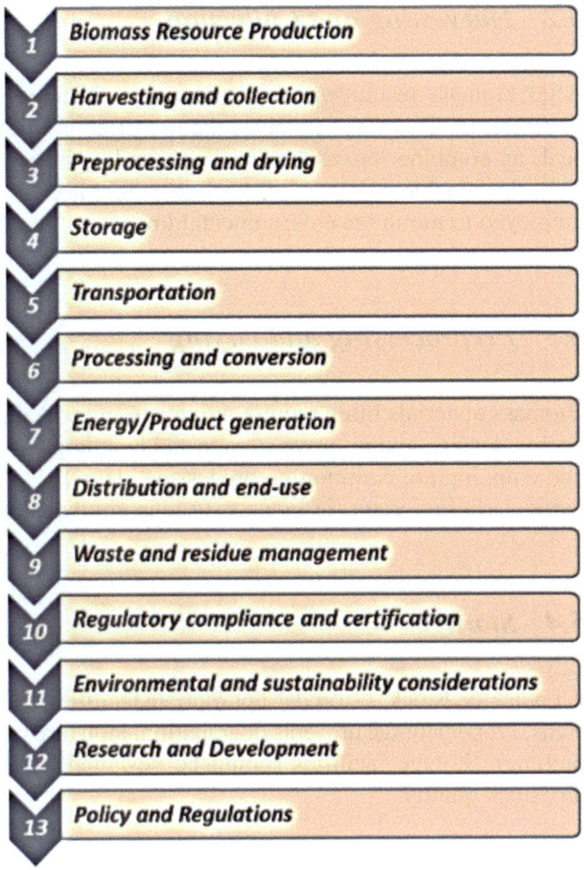

Fig. 4 Biomass supply chain [45]

its stages, challenges, and the sustainable practices necessary to achieve climate mitigation goals (Fig. 4).

5.1 Biomass Resource Production

The supply chain begins with the production of biomass resources, which can include crops (e.g., corn, sugarcane), forestry (e.g., wood, wood residues), agricultural residues (e.g., straw, corn stover), and organic waste materials (e.g., food waste, yard trimmings). Sustainable agricultural and forestry practices are crucial to ensure a continuous and environmentally friendly supply of biomass.

5.2 Harvesting and Collection

After biomass resources reach maturity or become available, they are harvested or collected. This stage involves the use of appropriate machinery and equipment, such as combine harvesters for crops or logging equipment for woody biomass [45]. Sustainable practices, like selective logging or crop residue management, are employed to minimize environmental impact.

5.3 Preprocessing and Drying

Biomass materials often require preprocessing to reduce moisture content, increase energy density, and improve storage and handling characteristics. Drying, grinding, and chipping are common preprocessing steps [26]. Reducing moisture content is particularly important for biomass used in combustion or gasification processes.

5.4 Storage

Biomass feedstock is stored in appropriate facilities, such as silos, bunkers, or covered areas. Proper storage prevents degradation, moisture uptake, and the growth of molds or fungi. Storage facilities should be designed to minimize losses and maintain feedstock quality.

5.5 Transportation

Biomass is transported from the production or collection sites to processing facilities or end-users. Transportation methods can include trucks, rail, ships, or even pipelines for certain types of biomasses. Efficient logistics and transportation planning are essential to minimize costs and environmental impacts [58].

5.6 Processing and Conversion

Biomass can undergo various processing and conversion methods depending on its intended use. This includes combustion for heat and power generation, gasification for syngas production, fermentation for biofuels, or chemical processes for bioproducts [38]. Advanced technologies like pyrolysis, torrefaction, or enzymatic hydrolysis may also be used to optimize biomass conversion.

5.7 Energy or Product Generation

Biomass is converted into energy (electricity, heat) or bioproducts (biofuels, chemicals) at processing facilities. These facilities can range from small-scale operations to large power plants and biofuel refineries. Combustion, gasification, anaerobic digestion, and biochemical processes are some of the common methods used for biomass utilization [55].

5.8 Distribution and End-Use

The energy or bioproducts generated from biomass are distributed to end-users through established distribution networks (e.g., electrical grids, and fuel distribution networks). End-users can include residential, commercial, industrial, and transportation sectors.

5.9 Waste and Residue Management

During the biomass conversion process, there may be byproducts or residues generated. These need to be managed properly, whether it's utilizing them for co-products, disposing of them responsibly, or returning them to the soil as agricultural or forestry residues [37].

5.10 Environmental and Sustainability Considerations

Throughout the supply chain, environmental sustainability practices, such as sustainable sourcing, carbon neutrality, and reduced emissions, are essential to ensure that biomass utilization has a positive impact on the environment and mitigates climate change.

5.11 Regulatory Compliance and Certification

Compliance with local, regional, and national regulations related to environmental standards, land use, and emissions is crucial. Some biomass supply chains may seek certification from organizations like the Forest Stewardship Council (FSC) or Roundtable on Sustainable Biomaterials (RSB) to demonstrate sustainable practices.

5.12 *Research and Development*

Ongoing research and development efforts are essential to improve the efficiency and sustainability of the biomass supply chain, as well as explore innovative technologies and feedstock sources.

5.13 *Policy and Regulation*

Government policies and regulations at various levels influence the biomass supply chain by addressing issues such as land use, environmental standards, incentives for renewable energy, and sustainability requirements [36].

6 Biomass Processing

6.1 *Thermochemical Conversion*

Biomass combustion is a widely used technique where organic materials are burned in the presence of oxygen to produce heat energy. This process is commonly used in biomass power plants and industrial boilers

$$\text{CxHyOz (biomass)} + O_2 \rightarrow CO_2 + H_2O + \text{heat}$$

During combustion, biomass releases carbon dioxide, water vapor, and heat energy. The heat is typically used to produce steam, which drives turbines to generate electricity or to provide heat for industrial processes [23].

Gasification is a process where biomass is heated in a controlled environment with limited oxygen (partial oxidation) or steam to produce a gas mixture called syngas (synthetic gas) [62]

$$C_6H_{10}O_5 + 2\,H_2O \rightarrow 6\,CO + 15\,H_2 + 2\,CO_2$$

Syngas consists of carbon monoxide (CO), hydrogen (H_2), carbon dioxide (CO_2), and traces of other gases. Syngas can be used for various applications, including electricity and heat generation, as well as for the production of biofuels and chemicals.

Pyrolysis involves heating biomass in the absence of oxygen, resulting in the decomposition of biomass into three primary products biochar (a solid carbon-rich material), bio-oil (a liquid), and syngas [60].

$$\text{CxHyOz (Biomass)} \rightarrow \text{Char (solid carbonaceous material)} + \text{Bio-oil (liquid)} + \text{Syngas (gaseous mixture)}$$

Biochar can be used as a soil amendment, bio-oil can be further processed into biofuels or chemicals, and syngas can be used for energy production.

6.2 *Biochemical Conversion*

Fermentation

Fermentation is a biological process where microorganisms, such as yeast or bacteria, convert sugars in biomass feedstock into biofuels like ethanol or biobutanol [52].

$$C_6H_{12}O_6 \rightarrow 2\,C_2H_5OH + 2\,CO_2$$

It is commonly used in the production of first-generation biofuels from crops like sugarcane and corn.

Enzymatic Hydrolysis

Enzymatic hydrolysis is a biochemical process that uses enzymes to break down complex carbohydrates in biomass, particularly cellulose and hemicellulose, into simpler sugars like glucose. These sugars can then be fermented into biofuels.

$$C_6H_{10}O_5 + n\,H_2O \rightarrow C_6H_{12}O_6 + n\,C_6H_{12}O_6$$

Enzymatic hydrolysis is a key step in the production of second-generation or advanced biofuels made from lignocellulosic biomass.

6.3 *Mechanical and Physical Processing*

Milling is a mechanical process that grinds biomass materials into smaller particles. This increases the surface area of the biomass, making it easier to handle and improving its accessibility for subsequent processing steps. Milling is commonly used in the production of wood pellets and cellulosic biofuels [59]. Size reduction techniques involve physically reducing the size of biomass materials through processes like chopping, shredding, or grinding. This reduces the overall bulk of the biomass and can make it more suitable for transportation and processing.

6.4 Chemical Processing

Acid hydrolysis is a chemical process where biomass is treated with acids to break down lignocellulosic materials into sugars [11].

$$C_6H_{10}O_5 + 6\,H_2SO_4 \rightarrow 6\,C_6H_{12}O_6 + 6\,H_2O$$

The resulting sugar-rich solution can be used for biofuel production or other biochemical processes.

Alkaline pretreatment involves the use of alkaline solutions (e.g., sodium hydroxide or ammonia) to remove lignin and hemicellulose from biomass feedstock.

$$\text{CxHyOz (lignocellulose)} + \text{NaOH} \rightarrow \text{Solubilization of lignin and hemicellulose}$$

This process enhances the accessibility of cellulose for enzymatic hydrolysis and improves overall biofuel yields.

6.5 Torrefaction

Torrefaction is a thermal processing technique where biomass is heated at relatively low temperatures (200–300 °C) in a low-oxygen environment. This removes moisture and volatile components from the biomass, resulting in a more energy-dense and stable material known as torrefied biomass. Torrefaction improves the grindability and storage stability of biomass [28].

6.6 Hydrothermal Carbonization

Hydrothermal carbonization involves heating biomass in the presence of water at elevated temperatures and pressures. This process converts biomass into hydrochar, a solid carbon-rich material with properties similar to coal. Hydrochar can be used as a solid fuel or soil conditioner [14].

6.7 Supercritical Methods

Supercritical fluids (typically carbon dioxide) are used to extract valuable compounds from biomass, such as essential oils, bioactive compounds, and specialty chemicals [39]. Biomass is exposed to high temperatures and pressures in the presence of supercritical water to produce hydrogen gas and other valuable products.

6.8 Anaerobic Digestion

Anaerobic digestion is a biological process that breaks down organic matter, such as biomass, in the absence of oxygen to produce biogas. The primary reaction in anaerobic digestion involves the conversion of organic compounds into biogas, primarily methane (CH_4) and carbon dioxide (CO_2) [10]. Here's the simplified reaction equation for anaerobic digestion.

$$C_6H_{12}O_6 \rightarrow 3\,CH_4 + 3\,CO_2$$

7 Product Application

7.1 Bioenergy Production

7.1.1 Biofuels

Biofuels are liquid or gaseous fuels produced from biomass through various conversion processes [2]. The two primary types of biofuels are bioethanol and biodiesel. Bioethanol is typically produced through a fermentation process using crops rich in sugars, such as corn, sugarcane, or wheat. The sugars are converted into ethanol by yeast or bacteria. Bioethanol is commonly blended with gasoline (E10, E15, E85) and used as a fuel for gasoline-powered vehicles. Ethanol is used as an industrial solvent and in the production of various chemicals. Biodiesel is produced through a chemical process called transesterification, where triglycerides from plant oils (e.g., soybean oil, canola oil) or animal fats are converted into biodiesel and glycerin. Biodiesel can be used as a renewable alternative to diesel fuel, either as a blend (e.g., B5, B20) or as 100% biodiesel (B100). Biodiesel can also be used for space heating and water heating in homes and businesses.

7.1.2 Biopower

Biopower refers to electricity generated from biomass [49]. There are two primary methods for producing biopower. Biomass is burned in a combustion chamber to produce heat, which is then used to generate steam. The steam drives turbines connected to generators, producing electricity. Biomass power plants are used to produce electricity for residential, commercial, and industrial purposes. Some biomass power plants also produce heat for district heating or industrial processes. Biomass is heated in a controlled environment with limited oxygen or steam to produce syngas (a mixture of carbon monoxide and hydrogen). The syngas is then used in a gas turbine or engine to generate electricity. Biomass gasification can be used in small-scale distributed power generation or larger power plants. Syngas can be used for combined heat and power (CHP) systems, providing both electricity and heat or cooling.

7.1.3 Bioheat

Bioheat refers to the use of biomass-derived heat for space heating, water heating, and industrial processes [30]. Biomass can be burned directly in stoves, boilers, or furnaces to generate heat. The heat is distributed through radiators, pipes, or other heating systems. Biomass heating systems can be used in homes for space heating and domestic hot water. Biomass heating is also applied in commercial buildings, greenhouses, and industrial facilities for process heating.

7.1.4 Biogas

Biogas is a gaseous biofuel produced through the anaerobic digestion of organic materials, such as agricultural residues, sewage sludge, and organic waste [7]. Microorganisms break down organic matter in the absence of oxygen, producing biogas primarily composed of methane (CH_4) and carbon dioxide (CO_2). Biogas can be used to generate electricity in gas engines or turbines. Biogas can be used for combined heat and power (CHP) applications, providing both electricity and heat. Compressed natural gas (CNG) derived from biogas can be used as a vehicle fuel.

7.1.5 Bio-oil

Pyrolysis is a process that converts biomass into biochar, bio-oil, and syngas. While bio-oil itself is not typically used as a standalone biofuel, it can serve as a feedstock for various applications, including biofuels and chemical production [33]. Biomass is heated in the absence of oxygen, resulting in the production of biochar, bio-oil, and syngas. Bio-oil can be further processed into transportation fuels, such as renewable diesel and jet fuel. Bio-oil can serve as a feedstock for the production of chemicals and materials.

7.2 Biochemicals Production

7.2.1 Bioplastics

- Polylactic Acid (PLA) is produced through the fermentation of plant starch, typically derived from corn. Used in the manufacturing of biodegradable plastics, packaging materials, disposable tableware, and textiles.

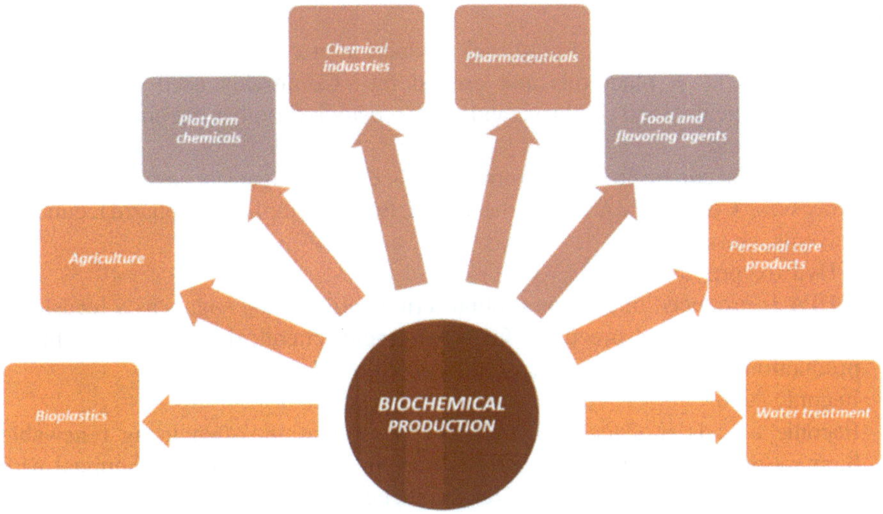

Fig. 5 Biochemical production utilizing biomass [40]

- Polyhydroxyalkanoates (PHA) bioplastics are synthesized by bacteria through the fermentation of sugars or plant oils. Find uses in packaging, agriculture (biodegradable mulch films), and medical devices.
- Polyethylene Furanoate (PEF) is made from plant-based feedstocks such as sugar or starch. A potential replacement for PET (polyethylene terephthalate) in beverage bottles, offering improved barrier properties and environmental benefits (Fig. 5) [3].

7.2.2 Biochemicals for Agriculture

- Biopesticides are often derived from naturally occurring substances or microorganisms, such as neem oil or Bacillus thuringiensis (Bt) bacteria [51]. Used to control pests and diseases in agriculture, reducing the need for chemical pesticides.
- Biostimulants are organic compounds that enhance plant growth and stress tolerance. Improve crop yield and quality, increase nutrient uptake, and enhance plant health.

7.2.3 Platform Chemicals

- Lactic Acid
  Produced through the fermentation of sugars, primarily glucose derived from biomass sources such as corn starch or sugarcane. Used in the production of biodegradable plastics, pharmaceuticals, food additives, and cosmetics.

- Glycerol
  Glycerol is a byproduct of biodiesel production or can be derived from triglycerides present in plant oils and animal fats. Commonly used in cosmetics, pharmaceuticals, food products, and as a feedstock for other chemicals [13].
- Succinic Acid
  Succinic acid can be produced from biomass-derived sugars through fermentation processes. Commonly used in the production of bioplastics, pharmaceuticals, solvents, and chemicals.
- 3-Hydroxypropionic Acid (3-HPA)
  3-HPA is produced through microbial fermentation using sugars as a feedstock. Used as a bio-based precursor for acrylic acid, chemical intermediates in the production of chemicals, in the production of bioplastics.
- Itaconic Acid
  Itaconic acid is produced via microbial fermentation, often using renewable biomass-derived sugars as a substrate. Itaconic acid is a platform chemical with diverse uses as polymers, plasticizers, and chemical derivatives.

7.2.4 Biochemicals for the Chemical Industry

- Bio-based Solvents
  Bio-based solvents like bioethanol and limonene are derived from biomass sources. Used in the chemical industry as replacements for traditional petroleum-based solvents.
- Bio-based Surfactants
  Surfactants, which are surface-active agents, can be produced from biomass-derived sources. Found in detergents, personal care products, and industrial applications for their emulsifying and cleaning properties [9].

7.2.5 Biochemicals for the Pharmaceutical Industry

- Artemisinin
  Extracted from the sweet wormwood plant (Artemisia annua). The key component in antimalarial drugs, helping combat drug-resistant malaria strains.
- Paclitaxel was originally isolated from yew trees, but can also be produced via microbial fermentation. Used in the treatment of various cancers, including breast and ovarian cancer.

7.2.6 Biochemicals for Food and Flavoring

- Vanillin is traditionally derived from vanilla beans, it can also be produced from lignin or ferulic acid. Commonly used as a flavoring agent in food and beverages.
- Citric Acid is produced through microbial fermentation of sugars. Widely used as an acidulant and preservative in the food and beverage industry [6].

7.2.7 Biochemicals for Personal Care Products

- Squalene is extracted from sugarcane, olives, or shark liver oil. Used in cosmetics and skin care products for its moisturizing and emollient properties.
- Hyaluronic Acid is produced via microbial fermentation. Used in skincare products, wound care, and pharmaceuticals for its hydrating and anti-aging properties [48].

7.2.8 Biochemicals for Water Treatment

- Chitosan is derived from chitin, a component of shrimp and crab shells. Used in water treatment processes to remove contaminants and pollutants due to its flocculating and adsorption properties [21].

8 Conclusion

In conclusion, the chapter on sustainable utilization of biomass resources has delved into the multifaceted world of biomass, exploring its significance in the context of sustainable development, energy production, and environmental conservation. Throughout this chapter, we have examined various aspects of biomass resources, ranging from their types and sources to the diverse array of applications and technologies that enable their responsible use. One of the central themes of this chapter has been the critical role that biomass plays in mitigating the environmental challenges of our time. As the world grapples with the consequences of climate change and the need to reduce greenhouse gas emissions, biomass resources emerge as a viable and renewable solution. By harnessing the energy stored in organic materials such as wood, crop residues, and algae, we can reduce our reliance on fossil fuels and contribute to a more sustainable energy landscape. Moreover, the carbon neutrality of biomass energy, when managed properly, can help offset emissions and promote a more balanced carbon cycle. The chapter has also highlighted the importance of sustainable biomass management. To ensure the longevity of biomass resources and prevent ecological degradation, we must adopt sustainable practices such as afforestation, reforestation, and responsible harvesting. Balancing the demands of biomass utilization with the need to conserve biodiversity and protect fragile ecosystems is a challenge that requires careful planning and regulatory oversight.

Furthermore, the integration of advanced technologies and processes, such as biorefineries, pyrolysis, and anaerobic digestion, has been explored as a means to maximize the value derived from biomass resources. These technologies not only enhance the efficiency of biomass conversion but also broaden the spectrum of valuable products that can be obtained, from biofuels and bioplastics to biochemicals and pharmaceuticals. Throughout the chapter, we have underscored the need for a holistic approach to biomass utilization—one that considers economic, social, and environmental dimensions. Sustainable biomass utilization should not only yield economic benefits but also empower local communities, create jobs, and improve livelihoods. It should be built upon principles of equity and social responsibility. In summary, the sustainable utilization of biomass resources represents a vital component of our transition toward a more environmentally responsible and renewable energy future. By recognizing the potential of biomass and embracing sustainable practices, we can reduce our ecological footprint, combat climate change, and promote a more resilient and equitable society. The challenges are numerous, but with continued research, innovation, and global cooperation, we can unlock the full potential of biomass resources while safeguarding the health of our planet for generations to come.

References

1. Akkala SR, Jeremias M, Kaviti AK, Sikarwar VS (2023) Reference module in earth systems and environmental sciences. In: Biomass conversion by gasification process. Elsevier. https://www.sciencedirect.com/science/article/pii/B9780323939409000177
2. Alam S, Tanveer S (2020) Chapter 5—Conversion of biomass into biofuel: a cutting-edge technology. In: Singh L, Yousuf A, Mahapatra DM (eds) Bioreactors. Elsevier, pp 55–74. https://www.sciencedirect.com/science/article/pii/B978012821264600005X
3. Ashter SA (2016) 1—Introduction. In: Ashter SA (ed) Introduction to Bioplastics Engineering. Plastics design library (Series). William Andrew Publishing, Oxford, pp 1–17. https://www.sciencedirect.com/science/article/pii/B9780323393966000014
4. Balaman SY (2019) Chapter 4—Sustainability issues in biomass-based production chains. In: Balaman SY (ed) Decision-making for biomass-based production chains. Academic Press, pp 77–112. https://www.sciencedirect.com/science/article/pii/B9780128142783000042
5. Ben-Iwo J, Manovic V, Longhurst PJ (2016) Biomass resources and biofuels potential for the production of transportation fuels in Nigeria. Renew Sustain Energy Rev 63:172–192. https://www.sciencedirect.com/science/article/pii/S1364032116301605
6. Berovic M, Legisa M (2007) Citric acid production. In: Raafat El-Gewely M (ed) Biotechnology annual review. Elsevier, pp 303–343. https://www.sciencedirect.com/science/article/pii/S1387265607130118
7. Bhatia SC (2014) 17—Biogas. In: Bhatia SC (ed) Advanced renewable energy systems. Woodhead Publishing India, pp 426–72. https://www.sciencedirect.com/science/article/pii/B9781782422693500176
8. Blanco H, Faaij A (2018) A review at the role of storage in energy systems with a focus on power to gas and long-term storage. Renew Sustain Energy Rev 81:1049–1086. https://www.sciencedirect.com/science/article/pii/S1364032117311310
9. Bois R et al (2020) Physicochemical, foaming and biological properties of lowly irritant anionic sugar-based surfactants. Colloids Surf A Physicochem Eng Aspects 607:125525. https://www.sciencedirect.com/science/article/pii/S0927775720311183
10. Carstensen FV et al (2013) Chapter 13—Economic analysis of bioenergy: an integrated multidisciplinary approach. In: Suib SL (ed) New and future developments in catalysis. Elsevier, Amsterdam, pp 297–323. https://www.sciencedirect.com/science/article/pii/B9780444538789000151
11. Chen H (2015) 3—Lignocellulose biorefinery feedstock engineering. In: Chen H (ed) Lignocellulose biorefinery engineering. Woodhead Publishing, pp 37–86. https://www.sciencedirect.com/science/article/pii/B978008100135600003X

12. Chen W-H et al (2021) Progress in biomass torrefaction: principles, applications and challenges. Prog Energy Combust Sci 82:100887. https://www.sciencedirect.com/science/article/pii/S0360128520300976
13. Chilakamarry CR, Mimi Sakinah AM, Zularisam AW, Pandey A (2021) Glycerol waste to value added products and its potential applications. Syst Microbiol Biomanufact 1(4):378–396. https://doi.org/10.1007/s43393-021-00036-w
14. Czerwińska K, Śliz M, Wilk M (2022) Hydrothermal carbonization process: fundamentals, main parameter characteristics and possible applications including an effective method of SARS-CoV-2 mitigation in sewage sludge. A review. Renew Sustain Energy Rev 154:111873. https://www.sciencedirect.com/science/article/pii/S1364032121011400
15. Diaz-Chavez R, Stichnothe H, Johnson K (2016) Chapter 4—Sustainability considerations for the future bioeconomy. In: Lamers P, Searcy E, Richard Hess J, Stichnothe H (eds) Developing the global bioeconomy. Academic Press, pp 69–90. https://www.sciencedirect.com/science/article/pii/B9780128051658000045
16. Douglas Brodie J, Tattersall (Tat) Smith C, Tappeiner JC, Boyle JR (2016) Economic considerations for sustainable forestry. In: Reference module in earth systems and environmental sciences. Elsevier. https://www.sciencedirect.com/science/article/pii/B9780124095489094847
17. Eswarlal VK, Vasudevan G, Dey PK, Vasudevan P (2014) Role of community acceptance in sustainable bioenergy projects in India. Energy Policy 73:333–343. https://www.sciencedirect.com/science/article/pii/S0301421514002353
18. Ferreira C et al (2022) Soil degradation in the European mediterranean region: processes, status and consequences. Sci Total Environ 805:150106. https://www.sciencedirect.com/science/article/pii/S0048969721051810
19. Green C, Byrne KA (2004) Biomass: impact on carbon cycle and greenhouse gas emissions. In: Cleveland CJ (ed) Encyclopedia of energy. Elsevier, New York, pp 223–236. https://www.sciencedirect.com/science/article/pii/B012176480X004186
20. Gupta J et al (2022) Agro-forestry waste management—a review. Chemosphere 287:132321. https://www.sciencedirect.com/science/article/pii/S0045653521027934
21. Issahaku I, Tetteh IK, Tetteh AY (2023) Chitosan and Chitosan derivatives: recent advancements in production and applications in environmental remediation. Environ Adv 11:100351. https://www.sciencedirect.com/science/article/pii/S266676572300011X
22. Johansson B (2013) Security aspects of future renewable energy systems–a short overview. Energy 61:598–605. https://www.sciencedirect.com/science/article/pii/S0360544213007743
23. Jouhara H et al (2018) Waste heat recovery technologies and applications. Therm Sci Eng Prog 6:268–289. https://www.sciencedirect.com/science/article/pii/S2451904918300015
24. Keyuraphan S, Thanarak P, Ketjoy N, Rakwichian W (2012) Subsidy schemes of renewable energy policy for electricity generation in Thailand. Proc Eng 32:440–448. https://www.sciencedirect.com/science/article/pii/S1877705812013409
25. Kılkış S et al (2022) Effective mitigation of climate change with sustainable development of energy, water and environment systems. Energy Convers Manage 269:116146. https://www.sciencedirect.com/science/article/pii/S0196890422009293
26. Kucuksayacigil F et al (2022) Optimal control to handle variations in moisture content and reactor in-feed rate. Energy 248:123650. https://www.sciencedirect.com/science/article/pii/S0360544222005539
27. Kumar R et al (2022) Long-term impacts of afforestation on biomass production, carbon stock, and climate resilience in a degraded semi-arid ravine ecosystem of India. Ecol Eng 177:106559. https://www.sciencedirect.com/science/article/pii/S0925857422000209
28. Limousy L, Jeguirim M, Labaki M (2017) Chapter 11—Energy applications of coffee processing by-products. In: Galanakis CM (ed) Handbook of coffee processing by-products. Academic Press, pp 323–367. https://www.sciencedirect.com/science/article/pii/B9780128112908000116
29. Mai-Moulin T, Hoefnagels R, Grundmann P, Junginger M (2021) Effective sustainability criteria for bioenergy: towards the implementation of the European renewable directive II.

Renew Sustain Energy Rev 138:110645. https://www.sciencedirect.com/science/article/pii/S1364032120309291
30. Malico I, Pereira RN, Gonçalves AC, Sousa A (2019) Current status and future perspectives for energy production from solid biomass in the European industry. Renew Sustain Energy Rev 112:960–977. https://www.sciencedirect.com/science/article/pii/S1364032119304150
31. Malinauskaite J et al (2017) Municipal solid waste management and waste-to-energy in the context of a circular economy and energy recycling in Europe. Energy 141:2013–2044. https://www.sciencedirect.com/science/article/pii/S0360544217319862
32. Mamvura TA, Danha G (2020) Biomass torrefaction as an emerging technology to aid in energy production. Heliyon 6(3):e03531. https://www.sciencedirect.com/science/article/pii/S2405844020303765
33. Matayeva A et al (2019) Chapter 12—Development of upgraded bio-oil via liquefaction and pyrolysis. Stud Surf Sci Catal 231–256. In: Albonetti S, Perathoner S, Quadrelli EA (eds) Horizons in Sustainable Industrial Chemistry and Catalysis. Elsevier. https://www.sciencedirect.com/science/article/pii/B9780444641274000124
34. Matsakas L et al (2017) Green conversion of municipal solid wastes into fuels and chemicals. Electron J Biotechnol 26:69–83. https://www.sciencedirect.com/science/article/pii/S0717345817300040
35. Melamed ML, Schmale J, von Schneidemesser E (2016) Sustainable policy—key considerations for air quality and climate change. Curr Opin Environ Sustain 23:85–91. https://www.sciencedirect.com/science/article/pii/S1877343516301087
36. Ghani MA, Muhammad Aslaam N, Vogiatzis C, Szmerekovsky J (2018) Biomass feedstock supply chain network design with biomass conversion incentives. Energy Policy 116:39–49. https://www.sciencedirect.com/science/article/pii/S030142151830051X
37. Muscat A, de Olde EM, de Boer IJM, Ripoll-Bosch R (2020) The battle for biomass: a systematic review of food-feed-fuel competition. Glob Food Secur 25:100330. https://www.sciencedirect.com/science/article/pii/S2211912418301366
38. Osman AI et al (2021) Conversion of biomass to biofuels and life cycle assessment: a review. Environ Chem Lett 19(6):4075–4118. https://doi.org/10.1007/s10311-021-01273-0
39. Pereira CG, Angela M, Meireles A (2010) Supercritical fluid extraction of bioactive compounds: fundamentals, applications and economic perspectives. Food Bioprocess Technol 3(3):340–372. https://doi.org/10.1007/s11947-009-0263-2
40. Pishvaee MS, Mohseni S, Bairamzadeh S (2021) Chapter 2—Biofuel supply chain structures and activities. In: Pishvaee MS, Mohseni S, Bairamzadeh S (eds) Biomass to biofuel supply chain design and planning under uncertainty. Academic Press, pp 21–36. https://www.sciencedirect.com/science/article/pii/B9780128206409000027
41. Popp J, Lakner Z, Harangi-Rákos M, Fári M (2014) The effect of bioenergy expansion: food, energy, and environment. Renew Sustain Energy Rev 32:559–578. https://www.sciencedirect.com/science/article/pii/S1364032114000677
42. Prasad RD, Raturi A (2021) Prospects of sustainable biomass-based power generation in a small island country. J Cleaner Prod 318:128519. https://www.sciencedirect.com/science/article/pii/S0959652621027281
43. Qadir SA, Al-Motairi H, Tahir F, Al-Fagih L (2021) Incentives and strategies for financing the renewable energy transition: a review. Energy Rep 7:3590–3606. https://www.sciencedirect.com/science/article/pii/S2352484721004066
44. Qamruzzaman Md, Karim S (2023) Does public-private investment augment renewable energy consumption in BIMSTEC nations? Evidence from symmetric and asymmetric assessment. Energy Strategy Rev 49:101169. https://www.sciencedirect.com/science/article/pii/S2211467X23001190
45. Rentizelas AA (2013) 2—Biomass supply chains. In: Rosendahl L (ed) Biomass combustion science technology and engineering. Woodhead Publishing Series in Energy. Woodhead Publishing, pp 9–35. https://www.sciencedirect.com/science/article/pii/B9780857091314500029

46. Sahoo G, Sharma A, Dash AC (2022) Biomass from trees for bioenergy and biofuels—a briefing paper. Mater Today Proc 65:461–467. https://www.sciencedirect.com/science/article/pii/S2214785322012846
47. Sanchez DL et al (2019) Chapter 4—Rightsizing expectations for bioenergy with carbon capture and storage toward ambitious climate goals. In: Pires JCM, da Cunha Gonçalves AL (eds) Bioenergy with carbon capture and storage. Academic Press, pp 63–84. https://www.sciencedirect.com/science/article/pii/B9780128162293000041
48. Shikina EV, Kovalevsky RA, Shirkovskaya AI, Toukach PhV (2022) Prospective bacterial and fungal sources of hyaluronic acid: a review. Comput Struct Biotechnol J 20:6214–6236. https://www.sciencedirect.com/science/article/pii/S2001037022005074
49. Simmons BA (2012) 31—Bioenergy from plants and plant residues. In: Altman A, Hasegawa PM (eds) Plant biotechnology and agriculture. Academic Press, San Diego, pp 495–505. https://www.sciencedirect.com/science/article/pii/B9780123814661000316
50. Singh J (1996) Economic and environmental aspects of bio-energy productivity. In: Chartier P et al (eds) Pergamon, Oxford, pp 1865–1868. https://www.sciencedirect.com/science/article/pii/B9780080428499500720
51. Singh S, Yadav R, Singh AN (2023) Chapter 27—Applications of waste-to-economy practices in the urban wastewater sector: implications for ecosystem, human health and environment. In: Singh P et al (eds). Elsevier, pp 625–646. https://www.sciencedirect.com/science/article/pii/B9780323904636000105
52. Sirohi R et al (2020) Green processing and biotechnological potential of grape pomace: current trends and opportunities for sustainable biorefinery. Bioresour Technol 314:123771. https://www.sciencedirect.com/science/article/pii/S0960852420310439
53. Srivastava RK et al (2021) Biomass utilization and production of biofuels from carbon neutral materials. Environ Pollut 276:116731. https://www.sciencedirect.com/science/article/pii/S0269749121003110
54. Sudha P, Ravindranath NH (1999) Land availability and biomass production potential in India. Biomass Bioenergy 16(3):207–221. https://www.sciencedirect.com/science/article/pii/S096195349800083X
55. Tursi A, Olivito F (2021) 1—Biomass conversion: general information, chemistry, and processes. In: Rahimpour MR, Kamali R, Makarem MA, Manshadi MKD (eds) Advances in bioenergy and microfluidic applications. Elsevier, pp 3–39. https://www.sciencedirect.com/science/article/pii/B9780128216019000017
56. Vallero DA (2008) 28—The elements of regulatory control. In: Vallero DA (ed) Fundamentals of air pollution (Fourth Edition). Academic Press, Burlington, pp 723–729. https://www.sciencedirect.com/science/article/pii/B9780123736154500297
57. Vijay V et al (2022) Sustainable utilization of biomass resources for decentralized energy generation and climate change mitigation: a regional case study in India. Environ Res 212:113257. https://www.sciencedirect.com/science/article/pii/S0013935122005849
58. Wang K, Tester JW (2023) Sustainable management of unavoidable biomass wastes. Green Energy Resour 1(1):100005. https://www.sciencedirect.com/science/article/pii/S2949720523000024
59. Williams O et al (2016) Influence of mill type on densified biomass comminution. Appl Energy 182:219–231. https://www.sciencedirect.com/science/article/pii/S0306261916312119
60. Yaashikaa PR, Senthil Kumar P, Varjani S, Saravanan A (2020) A critical review on the biochar production techniques, characterization, stability and applications for circular bioeconomy. Biotechnol Rep 28:e00570. https://www.sciencedirect.com/science/article/pii/S2215017X20300023
61. Yoshida H, Gable JJ, Park JK (2012) Evaluation of organic waste diversion alternatives for greenhouse gas reduction. Resour Conserv Recycl 60:1–9. https://www.sciencedirect.com/science/article/pii/S0921344911002448
62. Zhang Y et al (2019) Chapter 14—Gasification technologies and their energy potentials. In: Taherzadeh M, Bolton K, Wong J, Pandey A (eds) Sustainable resource recovery and zero waste approaches. Elsevier, pp 193–206. https://www.sciencedirect.com/science/article/pii/B9780444642004000141

Biomass Resources and Potential for Its Applications in Energy and Platform Chemicals with 3G Biorefinery Approaches

Md. Sourav Talukder, G. Gokul, Harveen Kaur, Swapnamoy Dutta, Km Smriti Jaiswal, Arun Prasath Ramaswamy, Boggavarapu Veera Venkata Kiran Krishna, Amit K. Jaiswal, V. Venkateswara Sarma, and Krishna Kumar Jaiswal

Abstract Most of the industrially produced energy materials and chemicals are derived from fossil fuel-based resources, which attribute adverse environmental impacts. Biomass resources have been identified as an alternative to shrink the reliance on fossil fuels. Also, sensible food/feed supplies, ecological alarms, climate discrepancy, and geopolitical concerns endorse to explore of 3G feedstocks in renewable energies and platform chemicals. This chapter provides an insight for the potential and sustainable transformation of abundant renewable resources to produce wide-ranging biofuels, platform chemicals, and high-value materials in 3G biorefinery. Algal feedstocks are the key component of the 3G biorefinery that strengthens the bio-renewables industry. It has perceived considerable success in research laboratories, leading to advanced technical progresses in microalgal biomass transformation.

Md. S. Talukder · G. Gokul · B. V. V. K. Krishna · K. K. Jaiswal (✉)
Bioprocess Engineering Laboratory, Department of Green Energy Technology, Pondicherry University, Puducherry 605014, India
e-mail: kkjindia@gmail.com; kkjindia.get@pondiuni.ac.in

H. Kaur
Department of Resource Management & Design Application, University of Delhi, New Delhi 110001, India

S. Dutta
Bredesen Center for Interdisciplinary Research and Graduate Education, University of Tennessee, Knoxville, TN 37996, USA

K. S. Jaiswal · A. P. Ramaswamy
Laboratory for Energy Materials and Sustainability, Department of Green Energy Technology, Pondicherry University, Puducherry 605014, India

A. K. Jaiswal
School of Food Science and Environmental Health, Technological University Dublin – City Campus, Central Quad, Grangegorman, Dublin 07 ADY7, Ireland

V. Venkateswara Sarma
Department of Biotechnology, Pondicherry University, Puducherry 605014, India

S. Kumar et al. (eds.), *Clean Energy Transition-via-Biomass Resource Utilization*, Green Energy and Technology, https://doi.org/10.1007/978-981-97-6321-4_2

Thus, this chapter signifies a conceivable picture of microalgal-based 3G biorefinery and future research directions to explore biomass resources in biofuels and platform chemicals in sustainability aspects via multidisciplinary technology routes.

Keywords Microalgae biomass · Sustainable transformation · Renewable biofuels · Platform chemicals · 3G biorefinery

1 Introduction

Biomass resources have been instigated enormously for energy purposes, mainly residues of organic biomass left over from forestry process wastes, agricultural wastes, and several energy crops. Although carbon dioxide has been released into the atmosphere during energy generation processes from biomass, it reabsorbs the atmospheric carbon dioxide emitted as it grows again, paying off its carbon debt and becoming carbon neutral [8]. Therefore, the overall bioenergy utilization processes have been considered carbon neutral in the long term. Also, for energy generation, still, biomass resources need to be utilized more efficiently than ever before. By growing more energy crops, the supply of operational bioresources can be potentially improved, enhancing the eventual utilization of bioenergy [57]. Thus, bioenergy-associated mitigation strategies will be positively impacted by the efficacious management of biomass resources. It is essential to understand that greenhouse gas emissions can be significantly reduced via such effective use of biomass and bioenergy. The carbon dioxide produced during the burning and curing gets absorbed and assimilated when the plant grows. Therefore, as a substitute for fossil energy, employing biomass for bioenergy generation is an excellent opportunity to be explored rapidly. The biomass resource estimate is a comprehensive statistical and spatial description of different varieties of biomass [80]. At the same time, the potential bioenergy estimates the promising energy generation capacity confined to biomass resources.

Along with offsetting the fossil fuels shortage, enhancing the involvement of bioenergy will be favorable to stabilize and control atmospheric greenhouse gas concentrations. In such a way, the framework convention goals on climate change will be approachable. Understanding bioenergy from the industrial link is the fundamental step through which estimation can be done on the potential of the bioenergy [58]. Simultaneously considering its impact on future development is also necessary. The bioenergy potential of biomass resources is subtracted by the number of resources utilized in other ways and lost in transportation, harvesting, etc., from the entire amounts of each feasible category of biomass that can be used for generating bioenergy. Afterward, the energy conversion coefficient is multiplied by the remaining energy amount and converted to a unit of energy. According to the data source and the technique applied to estimate the amount of bioresources. Mainly, two ways have been used to evaluate biomass and its energy potential [45]. Of those two methods, one method is based on statistical data evaluation for each type of biomass and other methods integrate remote sensing with geospatial information system (RS-GIS). Via processing and evaluating remote sensing data, the amount of green biomass can be effectively estimated. Determining the quantity and quality of vegetation in a specific zone through remote sensing and GIS is a new and suitable approach for biomass measurement [44].

Biomass sources are generally categorized into three different classifications, viz., municipal organic waste, energy crops (i.e., starch crops, oilseeds crops, etc.), and forestry wood waste (i.e., wood, forest products, shrubs, trees, and wood residues, etc., from forest clearing). Apart from combusting residual biomasses to generate energy, another route of energy generation is to produce biogas via landfilling biomass residuals [70]. One key aspect affecting energy production from biomass resources is the energy elements present in biomass, which can be specified by heating values. The higher and lower calorific values are used when analyzing the energy capacity of biomass. Renewable biomass resources have been shown to produce different high-value chemicals (e.g., sorbitol, lactic acid, succinic acid, levulinic acid, ethanol, etc.) in biorefineries [67]. In addition, different value-added chemicals and polymers have also been generated in commercialized production. Production of renewable platform chemicals is predicted to be augmented significantly in the upcoming times with boosted commercialization. Even though a single-product process is used to process materials and chemicals in the biorefinery, energy (biofuels) and bio-products are generated in an integrated biorefinery as it is considered a comparably advanced and suitable technique for the sustainable biomass valorization. In the diverse applicability of biomass resources, some biomass ash has shown excellent binding characteristics that can partially substitute the clinker content in cement, possibly lessening carbon dioxide emanations allied with traditional clinker build-ups [71]. To produce concrete, biomass ash can replace fine aggregates. Good mechanical resistance can be achieved in the mixture when an optimal amount of biomass ash has been added. In this chapter, we have deliberated on the prospects of producing various platform chemicals and high-value bio-products together with bioenergy in an integrated sustainable biorefinery. Hence, biofuels and platform production in sustainable biorefinery will substantially shrink greenhouse gas emissions, driving state-of-the-art advances in bioresources and the potential for their applications in energy and platform chemicals to expand the bio-economy.

2 Biomass Resources

Biomass resources comprise bio-renewable resources, such as the organic components of the earth of living plant or animal origin and their waste products. A wide variety of materials are found in biomass resources with distinctive chemical and physical characteristics. Dedicated energy crops or waste materials are considered major biomass resources [7]. Due to its abundance, residual biomass has been considered a promising substance that can generate sustainable energy. Specifically, there is no such exact standard biomass classification. Still, depending on their origin, source,

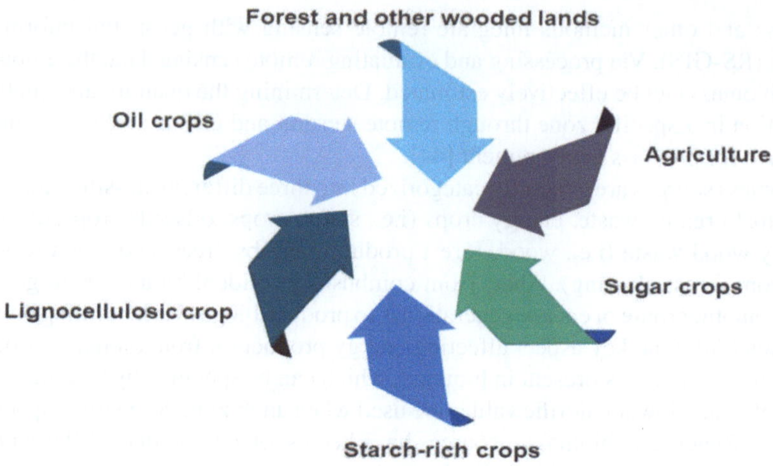

Fig. 1 Different types of biomass resources in clean energy transitions

or composition, a few categories are mentioned in the literature, e.g., energy crops, woody materials, agricultural residues, municipal wastes, and manure (Fig. 1). The enormous variety of biomass being used for bioenergy comes from plants resources along with the potential to mitigate climate variation [13].

2.1 *Agricultural Bioresources*

Agricultural sectors are the second largest global biomass source, accounting for almost 10% of overall biomass feedstock. Energy crops, waste from farm activities, and other crop-derived by-products (vegetative grasses, agroforestry, straw, and livestock by-products) are the three primary agricultural biomass sources that are/can be utilized for gaseous, solid, and liquid biofuel production. Energy crops and agricultural activities include several examples involving grains, sugar crops, oilseeds, and other crops [2].

2.1.1 Sugar-rich Crops

Among the sugar-rich crops, sugarcane is considered one of the crucial options necessary to process for generating bioenergy and biofuels. Additionally, compared to corn, sugarcane's production costs are more competitive and the energy efficiency/profitability of the entire bioethanol chains is almost ten times lesser, however the required climate is different. Also, sugar residue utilization for ethanol is efficacious for the biorefinery model, where food-grade crystallized sugar processing can also be effectively done [72]. No ethanol plant uses sugar beets directly, they use processed sugar juice and molasses acquired from agro-industries of sugar beets.

2.1.2 Starch-rich Crops

The cereal crops have been signified as the main source of feedstocks based on the starch of the grains for ethanol production, using corn as the main crop, which leads the world production of biofuels. Current technologies allow it possible to make the most of cereal crops to produce ethanol, and animal feeds in various ways, liable on the resources and facilities [2]. Ethanol from corn starch has been noted to be ~94% of entirely produced biofuels. The statistics of the bioenergy provides the proportion of corn used for ethanol.

2.1.3 Oilseeds Crops

Many oilseed crops have been grown worldwide and are considered essential due to their economic value. Expanding oilseed growing areas may be another approach used to meet the increased demand. Sunflower, rapeseed, cottonseed, soybean, palm, and peanut-like oilseed crops are getting cultivated to an enormous amount in recent times because of the extensive applications of their seeds. For bioenergy production purposes, the contribution of oilseeds is related to the production of biodiesel, which is mainly processed through hydrotreatment or via transesterification technique using such oil feedstocks. One of the most preferred feedstocks for processing biodiesel is the vegetable because of the optimum presence of triglycerides, composed of various fatty acids [46]. For biofuel production, firstly from the harvesting, oilseed crop residues are obtained, and then processed oilseed crops, are finally employed for fuel production. Through biological or thermochemical techniques, bioethanol, bio-butanol, biomethane, and bio-hydrogen-like different biofuels are produced using oilseed crop residues.

2.2 Lignocellulosic and Non-lignocellulosic Biomass Resources

Generally, dry plant biomass is considered lignocellulosic biomass, which is the most abundantly available feedstock for biofuel preparation. Hemicellulose and cellulose carbohydrate polymers are mainly present in lignocellulosic biomass feedstocks along with aromatic-rich lignin polymers. The chemical behavior of such polymers is distinctive, reflected by their individual characteristics [28]. Therefore, the lignocellulosic biomass process stands a very complex and challenging task. Recalcitrance is one condition that develops resistance to degradation or even separation. However, in several researches, it has been seen that via combinational usage of chemicals, heat treatment, enzymes, and microorganisms, it is possible to produce solid, high-value products by overcoming this recalcitrance. Mostly, five and six-carbon sugar monomers are present in these carbohydrate-rich polymers and such monomers of sugars (6-C and 5-C sugars) are covalently linked to lignin. Virgin biomass, energy crops, and waste (residual) biomass are the three main classifications of lignocellulosic biomass (such as trees, shrubs, and grasses). Residual biomasses (biomass wastes) are generated as a low-cost by-product from different agro-industrial sectors and forestry. Energy crops are being generated to serve as the feedstocks for second-generation biofuels [73]. Moreover, non-lignocellulosic biomass resources have comprised mainly saccharides, proteins, lipids, and inorganics, in addition to a fraction of lignin and cellulose. Due to these feedstocks' unique structures and compositions, manufactured carbon materials owns tremendous properties for the applications in catalysis, energy storage, sensors, etc. [69].

2.2.1 Cellulose

One of the highly abundant lignocellulosic compounds is cellulose, accounting for almost 40–60 wt. % consist of β-D-glucopyranose units linked with β-(1,4)-glucosidic bonds, with cellobiose as an elementary component repetitive. 500–1400 D-glucose units containing cellulose chains are ordered in a proper structural way to develop microfibrils packed jointly to create cellulose fibrils [31]. Cellulose is substantially resistant to enzymatic hydrolysis as the fibrils are entrenched in the lignocellulosic matrix. The glucose release is directly or positively correlated with the cellulose content of biomass [79]. In fact, the alteration of the degree of polymerization is always impacted by porosity and crystallinity-like structural parameters. It has been discovered that by utilizing γ-irradiation, cotton linters degree of polymerization can be reduced, which did not significantly impact the saccharification rate [37]. However, the degree of polymerization in cellulose has been negatively correlated with cellulose hydrolysis. As many strong hydrogen bonds are present in long cellulose chains, they are very challenging to hydrolyze. In contrast, weaker hydrogen bonding systems are present in shorter cellulose chains, which offer easier access for enzymes, as a result, hydrolysis can be done without much difficulty [60].

2.2.2 Hemicellulose and Acetyl Groups

Hemicelluloses are heterogeneous polysaccharides, usually present on average 20–35% in the lignocellulosic feedstock. Several monosaccharide subunits are contained in hemicelluloses, which merge to form xyloglucans, mannans, xylans, glucomannans, etc. Hemicelluloses work as a physical cover which limits the enzyme availability [30]. The pretreatment of hemicellulose with dilute acid blasting or nebulization is used for easy removal of them, which in turn is favorable in augmenting the enzyme availability in cellulose to improve the conversion of cellulose. The effect of hemicellulose on the lignocellulosic biomass challenge is still unclear, as hemicelluloses frequently remove some lignin. Few reports have indicated that hemicellulose scrapping was further effective than lignin scrapping in refining the enzymatic hydrolysis rate, while others indicated that lignin scrapping was much more critical. Acetyl groups can circumscribe the availability of cellulose by snooping with enzyme recognition. It could also hinder the conformation of the productivity between the cellulose and the cellulose catalytic sphere by adding the periphery of the cellulose chain or changing its hydrophobicity. Earlier reports regarding sludge stover have revealed that lessening the acetyl content improves the efficiency of the enzymes. Also, switchgrass, wheat straw, poplar wood, and bagasse have demonstrated the effectiveness of deacetylation in soft hemicellulose compared to soft cellulose. The impact of acetyl groups varies with the biomass material crystallinity, cellulose, and lignin content [82].

2.2.3 Lignin

After cellulose, lignin components are another vital component in lignocellulosic biomass, accounting for nearly 15–40% of the dry weight. Having aromatic and aliphatic structures within, lignin is regarded as a significant class of hydrocarbon polymers. Lignin has high molecular weight along with a 3-D amorphous hydrophobic and complex polymeric structure, and it is comprised of phenyl propane units, which are formed from methoxy- and hydroxyl-substituted phenylpropane units (sinapyl, coniferyl, and p-coumaryl alcohol). Lignin can physically limit the availability of polysaccharides, as it plays a physical protective role in blocking the interaction of enzymes to the cellulose. Previous studies have shown the negatively identified lignin with enzyme mildness in transgenic rice, poplar, miscanthus, and wheat straw. Lignin accumulation generally alters the lignin-carbohydrate matrix, intensifies the porosity, and diminishes the non-productive adsorption sites for the enzymes. Phenolic hydroxyl groups (lignin-deduced components) have been noticed to cause the reversible inhibition of celluloses. In general, lignin contributes explosively to lignocellulosic biomass defiance because of the chemical composition, beneficial structure, and limiting the availability of cellulose enzymes.

2.3 *Forest and Other Wooded Lands*

Globally, energy production using biomass, forest-related resources, and other wooded land have a significant contribution. As per the studies/data of the World Bioenergy Association, 88% of the biomass used for energy originated from forestry. It has been noticed that the unique compositions of the wood is well-preserved in firewood, black-liquor, charcoal, wood fuels, and even in other ethanol processed from the wood feedstock [25]. Direct woodfuel can be primary forest products (e.g., traditional coppices). However, from the technological perspective, natural firewood to biogas and liquid biofuel production must be improved to make the fuel yield quantity and quality more attractive for safe and suitable commercial usage. In recent studies, for producing ethanol-like liquid biofuels, novel lignocellulosic biomass pre-treatment/treatment processes are being investigated to improve fuel processing and yield in particular [5].

2.4 Dedicated Energy Crops

Owing to their high biomass yields, energy crops attracted the interest of several researchers around the globe. For most energy crops, their favorable harvesting ability at several times of the year is a great opportunity to explore their energy generation capability. Dedicated energy crops refer to crops grown specifically for the production of biomass that can be converted to energy. These crops are essential in developing renewable energy sources, as they offer low-carbon sustainable alternatives to fossil fuels (e.g., switch grass, miscanthus, and woody biomass). Switchgrass (a dedicated energy crop) is a perennial grass that is native to North America and can grow in a variety of soils and climates [11]. It is known for its high yield potential and low input requirements, making it an attractive crop for bioenergy production. The premier energy crop has been recognized as sugar cane (an excellent source of fermentable sucrose and lignocellulosic bagasse). Also, Miscanthus (a tall grass, native to Asia) has been shown to have high yield potential and low environmental impact [16]. Woody biomass, such as fast-growing trees, can also produce bioenergy. Dedicated bioenergy crops can benefit farmers and rural communities economically, reduce emission levels of greenhouse gases and ensure energy security. Although energy crops can contribute a considerable portion of future energy but the concerns regarding the impact of large-scale bioenergy crop harvesting on excess land use, biodiversity, and food security remain unclear.

2.5 Algal Bioresources

Due to having high carbon dioxide fixation efficiency and superior growth rate compared to most terrestrial plants, Algae is considered a crucial organism on the earth's surface. Unlike lignocellulosic biomass, Algal biomass is composed of different constituents such as protein, carbohydrate, lipid, and ash content. Elementally, algal or microalgal biomass has been comprised of elemental carbon (C), hydrogen (H), nitrogen (N), oxygen (O), and sulfur (S). Lignin is also found inside the cell of certain green/red algae [51]. According to recent reports, the major constituents of algae are O and C, which account for nearly 15–50% and 25–55%, respectively. Both H and N accounted for 5–10%. Many researchers have concentrated their research on biofuel production through algae biomass pyrolysis, due to their benefits, such as abundance, effortless storage, and better oil yield to tackle the current shortage of fossil fuels [65]. Also, the eutrophication of water bodies can be controlled or reduced as microalgae have an ability to assimilate phosphorous and nitrogen from the different water bodies.

2.6 Sewage Sludge

One of the major by-products obtained from the wastewater treatment plants is the sludge, which is in consideration of potential biomass. Its opulent generation around the world represents a serious environmental and economic liability. In wastewater treatment plants, the amount of sewage sludge generated is nearly 10–25 million tons per year, in which the moisture percentage is almost 80%. As pollution aggregate, sewage sludge contains various components such as organic, inorganic, and microbial loads [50]. Although H, N, O, and C are commonly present elements in sewage sludge obtained from any source, their proportions vary depending on the specific sources from which they are processed. The elemental composition variation can exhibit different benefits and drawbacks. Due to organic matter and valuable elements such as K, P, and N, many nutrients are readily available in sewage sludge. Such substances are useful for plants, so agricultural usage of sewage sludge as soil regenerators or fertilizers has grabbed much attention. In addition, due to having high decomposable organic content, sewage sludge feedstock has been utilized to generate renewable biofuels. For instance, the high ash components result from a high amount of C content in sewage sludge, 15–50% ash content is observed when C varies between 25 and 70%. However, not only C but also P, Ca, Si, Fe, K, Mg, etc., elements are responsible for high ash content in sewage sludge, which restricts it from several application purposes [21]. Furthermore, the sewage sludge treatment and management is a complex method as it contains accumulated heavy metals, pathogens, and hazardous organic compounds, which, if not adequately treated, can cause secondary pollution and harm the environment. Hence, exploring better treatment methods to remove toxic constituents will be a promising way to fully utilize the potential of sewage sludge and its components.

2.7 Manure

Animal and human manure are well known for being utilized as plant fertilizers, as these are comprised of P, N, K, H, and C like essential elements and vital organic matters, which are considered essential sources of nutrients. According to existing literature, in manure, on average, N, O, H, and C elements are present in the following proportions 2–5%, 5–7%, 30–35%, and 40–50%, respectively. Recently, animal husbandry is playing a central role in producing large amounts of manure [27]. Huge human manure is getting produced as the urban population is growing faster, but it can't be directly used in the agriculture field because of its adverse effects on soil, air, and water pollution. So, processing or better treatment methods of such feedstock are necessary to efficiently utilize its vital elements.

3 Microalgae as a Predominant Renewable Biomass Resource

Microalgae biomass growth has been carried out in different types of photobioreactors. The activation energy can be determined through the various analysis of the biomass obtained from the experimental setup. Also, biomass growth kinetics and thermal hydrolysis can be analyzed for the experimental cases with the carbohydrate concern. Stress convinced by phosphorus limitation in synthetic wastewater enhances intracellular carbohydrate production. This specifies the energy required initiating the thermal decline response increases as attention to carbohydrates increases [43]. However, in the hydrothermal pretreatment, the temperature in the reactor raises the conditions to extract carbohydrates from the microalgae biomass. Temperature decreases the rate of response to protein confinement. The elegant conditions for the confinement of carbohydrates in various biomasses during hydrothermal pretreatment have been studied. Numerous investigations have explored the growth characteristics and application of microalgal biomass in the current literature, and the implicit wide distribution of microalgae has not yet been explored well [34]. The standard model of microalgae growth calculates the periodic average yield of microalgae around the world based on climate and nature data and predicts the implied global microalgae biomass energy distribution. The report has tried to develop a reference for the future microalgae biomass energy development and the selection of microalgae culture storage points by estimating the regions with high microalgae biomass energy.

Considering the growth characteristics of microalgae and biotechnology, this study used microalgae biomass as a biomass energy indicator. The average periodic yield of microalgae has impacted each region of the world and the energy distribution of microalgae biomass in the areas with a high impact on microalgae biomass energy worldwide. At the same time, the practical impact of microalgae biomass energy development on the original energy structure has been further estimated in accordance with the situation of energy consumption [54]. Instantly, microalgae's most rapid growth characteristics and application technologies focus on the implicit global energy distribution of microalgae biomass. However, the researchers have deliberated on the specific regions with vast unused land with abundant solar energy to estimate the microalgal uniform productivity from unused land. Several reports have implicit the productivity and distribution of microalgae considering solar radiation and growth patterns of microalgae at ambient temperature. At the same time, the actual impact of microalgae biomass energy development has been further estimated in areas with high eventuality. Also, the noble applications with superior demand have been explored with ways to convert wet biomass into biofuels in case of energy uses.

Wastewater treatment is an essential step in reducing toxic pollutants, and currently, in civic areas, more wastewater treatment plants are being established. Wastewater treatment plants generally utilize step-by-step approaches, which include chemical, biological, or physical treatment techniques to eliminate pathogens' contaminating activity and organic waste. These procedures are generally categorized into several orders of wastewater treatment methods, such as tertiary, secondary, primary, and advanced. Different treatment processing approaches have been developed reliant on the features (such as wastewater source, characteristics, land availability, economic budget, and effluent quality) [9]. Traditional strategies for wastewater treatment involve a series of anaerobic and aerobic steps to convert wastewater compositions into safer (neutral) components for other reusing purposes or, if of no use, are cautiously discarded. These conventional techniques eliminate some proportion of phosphorous, nitrogen, carbon, and other nutrients, although high energy consumption is necessary during this process, which involves high costs. It is due to the varying requirement of mechanical and environmental conditions for different types of nutrient separation. An anaerobic system can be utilized for treating toxic and unusual substances for wastewater cleaning by adjusting between aerobic, anoxic, and anaerobic conditions. According to the report, about three of the gross global greenhouse gases have been emitted by traditional wastewater treatment plants [47]. And so, traditional treatments involve complex means to eliminate the significant environmental pollutants due to the emigration of greenhouse gases.

In addition, this exploration examines the point of view of integrating food processing waste as an organic carbon source. The food processing residues that have been investigated usually include milk and cookie manufacturing factories. Furthermore, the culture medium has been modified by combining BG-11 (standard algal culture media) and food processing wastes to be applied for optimal microalgae growth and biochemical enrichment. Compared to microalgae grown with BG-11, biscuit waste, and BG-11 have produced advanced biomass compositions with higher lipid, carbohydrate, and protein contents [55]. *Chlorella vulgaris* have shown an adequate capacity to absorb nutrients from the culture medium with a combination of food processing wastes and BG-11, enhancing its growth. However, in recent studies, thermochemical conversion models and approaches are being significantly focused on achieving higher productivity and making their conversion performance much more efficient. Apart from the experimental approaches of thermochemical conversion processes, several computational techniques related to machine learning methods have also been mentioned in recent reports that aim to study real-time performance monitoring, optimizing operational parameters, and process control. The status of exploration and development of processes have been reported in the literature for pyrolysis, combustion, torrefaction, gasification, carbonization, liquefaction, and biochemical operations [56]. Artificial neural networks have recently been broadly explored to learn correlations from extremely non-linear input data.

4 Carbon Metabolism in Microalgae

Microalgae cultivation has been fascinating to obtain high-density and inexpensive microalgae biomass, where the production cost is financially viable. Microalgae biomass and accumulation of intracellular components can be enriched through optimizing or altering the culture parameters (such as temperature, nitrogen concentration, the concentration of carbon sources, carbon dioxide, organic carbon, etc.). Microalgal cultivation has been commonly distinguished into three types based on the modes of using carbon sources photoautotrophic, heterotrophic, and mixotrophic. The alterations in culture modes can impact the growth and productivity of microalgae biomass. Photoautotrophs are light-based organisms and use carbon dioxide as the sole carbon for photosynthetic growth; yet, in some instances, they are restricted by light [26]. In highly dim light or complete darkness conditions, the heterotrophic culture of microalgae grows, which significantly depends on organic carbon.

Mixotrophic cultivation is the combination of autotrophic and heterotrophic culture modes, in which organic carbon uses during photosynthesis. It elucidates the complications associated with inadequacy of light and proliferates the biomass and functional biochemical components by inducing metabolism with carbon distribution in the cells. The low molecular weight organic carbons (such as glucose, acetate, glycerol, etc.) enhance microalgae growth during mixotrophic culture. However, the significance of microalgae cell on the uptake and metabolism of organic carbons depends on the different species (not all microalgae are appropriate for mixotrophic cultivation). During mixotrophic culture, microalgal cells are often affected by various vital conditions, such as nitrogen and organic carbon concentration. The utilization of organic/inorganic nutrients involve under light through photosynthesis and aerobic respiration. The advantages of the mixotrophic culture comprise the improvement in photo-inhibition, higher growth rates than autotrophic and heterotrophic cultivation, reduced nocturnal biomass loss, and less photooxidative damage during culture. The mixotrophic culture has been well-considered for higher growth rates and pigment production in microalgae.

However, autotrophic cultivation has limitations, such as the necessity of appropriate light, CO_2, organic carbon, and oxygen simultaneously, lowers energy conversion efficiency, and net CO_2 consumption. Moreover, an inexpensive glycerol by-product from the biodiesel industry could be used as a potential substrate of microalgae cultivation to augment biomass generation. Extremely high CO_2 concentrations can inhibit the growth of microalgae. Still, according to studies on the physiological mechanism of microalgae, elevated CO_2 enhances photosynthesis and increases protein and lipid production. It is well known that biomass and product yields depend on culture parameters, which improves via the regulation of carbon metabolism [81]. The efficiency of light harvesting, CO_2 fixation, and energy conversion still limit the growth rate of microalgae. Therefore, heterotrophic culture has been accepted as proficient and viable approach for certain microalgae to produce different biofuels and bio-products using carbon substances (e.g., glucose) as a sole source.

5 Production of Biofuels and Platform Chemicals

The transition from conventional petrochemical manufacturing to carbon–neutral bio-based chemical manufacturing faces a key challenge to develop a sustainable global supply. Productivity and yield are the primary concerns to meet artificial conditions of elementary chemicals, environmental sustainability, and feedstock supplementation. Design-figure-test-learn (DFTL) natural engineering cycles have motivated the exploration of developing colorful chemicals through the turmoil of factory biomass to lead and establish global foundries [1]. The application of industrial biomass as an essential feedstock for reactionary coffers has been promoted, reflecting the social infiltration of the conception of the bio-economy. Therefore, carbons recycling are being developed to reduce CO_2 emigration. Even so, the competition of arable land with the food product and the scarcity of fresh water are essential limitations for effectively exercising industrial biomass. Environmental issues, such as destroying ecosystems from excessive deforestation, are significant concerns about using industrial biomass as feedstocks for non-food stores [1]. Therefore, in addition to the churning of biomass in factories, bio-production via photosynthetic microorganisms needs to be advanced to meet artificial requirements and minimize environmental impacts to institute bio-economic society.

Feedstocks are usually fractionated in biorefinery to be reused in various essential intermediaries [23]. These intermediaries function as a platform to further process and convert into intended commercial product (e.g., energies and platform chemicals). Intermediaries are inevitably not a distinct emulsion but often a mixture of different compounds of variable purity. Accordingly, biorefinery has been categorized on the basis of crucial features; such as feedstock(s), platform(s), and process(s), and product(s) [29]. Herewith, it is an overview of the most applicable platforms with chemicals in biorefinery, such as pyrolysis oil platform, biomass balance approach, CO_2 platform, syngas platform, sugar platform, products of sugars fermentation, products of sugars chemical transformation, bio-oil platform, lignin platform, and organic solutions platform. Various reports and research papers have extensively evaluated the potential for producing polymers and chemicals from biomass. In 2004, the US Department of Energy have circulated a statement listing twelve essential chemical compounds that it considered to be potential components for the future [18]. The list has been revised and informed in 2010 with the addition of new prospects (Table 1).

The BREW project has explored the potential of marketplace of bulk chemicals manufactured from the renewable sources by means of biotechnological approaches (Table 2). The bio-renewable platform chemicals (such as ethanol, glycerin, and lactic acid) constitute majority of renewable chemicals retailed in 2018. There is an extensive marketplace of platform chemicals. The most substantial growth has been observed for the biopolymers (e.g., polylactic acid, polyhydroxyalkanoate, bioethylene, etc.) used to make bio-based plastic products.

Table 1 Bio-based potential platform chemicals (12 chemical building blocks) recognized by the Department of Energy (DOE)

S. no	In the year 2004	Revised in the year 2010
1	1,4 Succinic, fumaric and malic acids	Ethanol
2	2,5 Furan dicarboxylic acid	Furfural
3	3 Hydroxy propionic acid	Hydroxymethylfurfural
4	Aspartic acid	2,5-Furandicarboxylic acid
5	Glucaric acid	Glycerol
6	Glutamic acid	Isoprene
7	Itaconic acid	Succinic acid
8	Levulinic acid	3-Hydroxypropionic acid/Aldehyde
9	3-Hydroxybutyrolactone	Levulinic acid
10	Glycerol	Lactic acid
11	Sorbitol	Sorbitol
12	Xylitol/Arabinitol	Xylitol

Table 2 Bio-based chemicals evaluated for market penetration and reference materials

S. no	Bio-based chemical	Reference platform chemicals
1	Acetic acid (CH_3COOH)	CH_3COOH
2	Adipic acid [$(CH_2)_4(COOH)_2$]	$(CH_2)_4(COOH)_2$
3	n-butanol (C_4H_9OH)	C_4H_9OH
4	Ethylene ($H_2C{=}CH_2$)	$H_2C{=}CH_2$
5	Mono ethylene glycol [$(CH_2OH)_2$]	$(CH_2OH)_2$
6	Ethyl lactate ($CH_3CHCO_2CH_2CH_3$)	Ethyl acetate ($CH_3CO_2CH_2CH_3$)
7	2,5-Furandicarboxylic acid (FDCA)	Terephthalic acid [$C_6H_4(CO_2H)_2$]
8	Polyhydroxyalkanoates (PHA)	High density polyethylene (HDPE)
9	Polylactic acid (PLA)	Polyethylene terephthalate (PET) and polystyrene (PS)
10	Succinic acid [$(CH_2)_2(CO_2H)_2$]	Maleic anhydride [$C_2H_2(CO)_2O$]

6 Strategies to Enhance the Accumulation of Biomass and Bio-products

Advances in microbial physiology approaches to accumulate microalgal biomass are being decoupled from genetic tools and the fermentation models have proliferated valued products to drop commercial prices. The culture techniques design to sustain energy (light) and nutrient equilibriums to create an appropriate cell growth environment [78]. Strategies to interrupt the equilibrium anticipate forming the stress environment to synthesize the preferred product.

6.1 Key-engineering Schemes

Microalgae have been considered as favorable cell factories for metabolic engineering schemes. Recently, the whole genome sequence proclamation in various microalgal strains assists in the regulations of metabolic pathways as well as genetic engineering strategies. Mostly, the metabolic engineering targets to guide the carbon and energy flux for directed pathways and products. However, the synergistic impacts for the two fluxes are in sometimes concern [36].

6.1.1 Carbon Allocation Strategies

Microalgae drive the partitioning of carbon (5–10%) into various fatty acids and terpenoids necessary for constant proliferations. The Calvin-Benson-Bassham cycle and the TCA cycle are associated with the CO_2 and glucose as a carbon source in the carbon flux. It is also associated with several metabolic pathways, such as the PP pathway. Therefore, the two cycles indicate noteworthy roles in microalgal metabolisms [66]. Engineered lipid and carotenoid biosynthesis pathways are active arenas of their nutritional and pharmaceutical roles. The paybacks of microalgal biodiesel also fascinate various researchers.

6.1.2 Calvin-Benson-Bassham Cycle Engineering Strategies

The Calvin-Benson-Bassham cycle regulates the redox process. Post-translational reversible-redox alterations signify essential role to regulate cellular metabolisms via cysteine residues transformations into diverse forms. It has been notorious early for enzymes regulation in the Calvin-Benson-Bassham cycle. Furthermore, usually enzymatic reactions are regulated by thioredoxins (TRX). TRX activates almost 15 enzymes (e.g., RPE, RPI, FBA, TPI, PGK, etc.) including RCA and CP12 in Calvin-Benson-Bassham cycle [6]. TRX reduces the photosynthetic electron transfer chain via ferredoxin/thioredoxin reductase oxidation. TRX plays a role in the connection among chloroplasts and mitochondria via NAD(H)/NADP(H). The gapA and gapB genes encoding photosynthetic glyceraldehyde-3-phosphate dehydrogenase get influenced due to the TRX at several points. And so, Calvin-Benson-Bassham cycle efficiently regulates using CP12 through the NAD(H)/NADP(H) relationship.

6.1.3 Tricarboxylic Acid Cycle Engineering Strategies

The TCA cycle is activated as two architectures that depend on whether the terminal electron acceptor is available in the reaction. The TCA cycle can be divided into the reductive branch from oxaloacetate to succinyl-CoA and the oxidative unit from citrate to 2-oxoglutarate. When NAD^+ is available for electrons, the TCA cycle oxidizes the substrate to CO_2 and produces NADH. Otherwise, the TCA cycle acts to produce intermediates for anabolism, such as citric acid and l-glutamic acid [42]. However, it is analyzed that the intermediate performance in the TCA cycle is lower than the theoretical maximum yields. The carbon turnover rate in the TCA cycle is limited. The TCA engineering cycle tends to allocate carbon in specific products at the expense of energy generation. Therefore, the intermediate biosynthesis pathway that requires a low energy level to alleviate the undesirable effect of energy is conferred. Replacing the NAD+ electron acceptor to lower the Gibbs free energy of the reaction could induce the reaction to occur more rapidly [3]. The glycosylate shunt in the TCA cycle can produce succinate without energy input and build a bridge between the reductive and oxidative branches. In microalgae, up-regulated glyoxylate shunt can increase carbon turnover in the TCA and CBB cycles, and thus the efficiency of carbon fixation can be improved.

6.1.4 Increasing Energy Capture and Delivery

Solar energy absorption and electron supply are limited during light reactions. Increasing the amount of energy flow by engineering the LHC is done to improve the efficiency of light reactions. The state of the antenna pigments is essential in a reaction to light since the pigments absorb and transfer protons to PSII and PSI [24]. During microalgal culture, photo-inhibition is amplified as cell density increases. Antenna pigments trap photons and block the photons to microalgae deep within the culture system. In the function of trapping photons, the quantity and variety of pigments are the key factors in increasing the light-harvesting efficiency. Although high light potentially generates an energy sink and strengthens energy dissipation, increased pigment compositions can alleviate light and produce high-value products for commercial production. Reducing antenna size by genetic modification approaches promises to improve photosynthetic efficiency at high cell density in terms of avoiding photo-inhibition [14]. The approaches have been shown to enhance biomass with an 11.5–111.5% increase in mass cultures of microalgae.

6.2 *Optimizing Growth Conditions*

Microalgae growth conditions are key factors in increasing biomass and product yields. However, the conditions differ significantly for the various microalgae strains. Therefore, we deliberate these conditions in the aspect of methods to improve the environment, and then the methods can be applied to various microalgae.

6.2.1 Light

Light is the critical factor to increase biomass accumulation and product biosynthesis. In mass culture, enhancing light intensity is the traditional way to provide enough protons to increase biomass, while the effect is unsatisfactory due to photo-inhibition as cell number increases. Furthermore, considerable heat will be generated if the light comes into close contact with the microalgae. Alternatively, an innovative light source, possessing a narrow spectral output to overlap the photosynthetic absorption spectrum of the microalgae used, is needed to increase light energy conversion and eliminate heat. Therefore, examining the optimal light wavelength with LED light is the other way to increase biomass with high photosynthetic efficiency, which is also an active field in recent years [12]. However, although blue light application is effective in increasing cell size and red light is favorable for high division rate, single light application is still challenged by low cell activity, mainly due to the fact that other wavelengths of light are also essential for cell metabolism. Therefore, mixing red and white light in a 1:1 ratio increased and maintained cell activity. The desired mix induces cells to be active with the highest photosynthetic efficiency. The industrial yield of microalgae by autotrophic culture is limited by cell concentration and light supply. Open ponds receiving natural light are common for mass culture, while the biomass is around 0.5 g/L. Therefore, various PBR designs have been proposed to increase light utilization. The tubular bioreactor increases the average biomass to 5 g/L than the open raceway pond. A vertical column bioreactor has been established for *Fistulifera* sp. and a panel bioreactor for *Chlorella vulgaris* with a specific growth rate of 0.29–0.57 g/L/day and 0.131 g/L/day, respectively [66].

6.2.2 Nutrients

Under suitable environmental conditions, sufficient nutrients extend the steady-state growth of microalgae with a high growth rate. However, the initial nutrient concentrations are limited by the microalgae used. Consequently, in most cases, the concentrations are insufficient for the microalgae to grow steadily for a long time. The additional operation promises to prevent the decrease in cell activity. The stable concentration of nutrients in the cultivation process is essential in maintaining the growth rate at a high level and achieving high biomass [22]. Nutrient addition is usually done during cultivation. To avoid the dilution rate effect, fed-batch culture is feasible to enhance biomass. Nutrient limitation is the conventional stress-based strategy to increase product yield since nutrients such as nitrogen, sulfur, and phosphorus are essential for the biosynthesis of proteins, lipids, and other central intermediates. Their deficiency in media can increase intracellular nutrient availability of carbon by the degradation of macromolecular substances, leading to growth retardation or even cell death. Nitrogen starvation is a widely used strategy to increase starch and lipid accumulation. Once responding to nitrogen starvation, microalgae redirect protein into carbon skeletons to the central carbon metabolism intermediate of pyruvate, which could be used to synthesize starch and lipids as storage forms [15].

Since microalgae require less energy to guide per carbon molecule towards starch than lipids, they would prefer to guide carbon molecules towards starch. Therefore, nitrogen deprivation effectively induces carbon into lipids, while energy limits conversion efficiency.

6.2.3 Temperature

Temperature changes also influence nutrient absorption, mainly by regulating cell growth enzymes. Typically, the cell growth rate increases as the temperature rises up to a threshold value, i.e., optimal for cell growth, and then decreases when the temperature exceeds this value. The traditional temperature for the development of most microalgae is considered to be between 14 and 35 °C. The optimum temperature for cell growth is highly dependent on the species used. A comparative study involving the effect of temperature on the growth of microalgae suggested that maximum biomass concentrations can be obtained at 14 °C for *Nannochloropsis oculata*, 26 °C for *Dixioniella grisea*, 14 °C for *Rhodomonas salina*, and 14 °C for *Isochrysis galbana* after 14 days of culture [59]. Although the species are the same in the cultivation process, other environmental conditions still influence the optimum temperature. For example, earlier research has reported that the optimal temperature for the growth of *C. vulgaris* was between 25 and 30 °C [32]. Therefore, it was essential to optimize the temperature conditions in microalgae cultivation.

6.2.4 pH

pH has been considered a critical environmental factor that controls cell metabolism and biomass accumulation. All microalgae strains seem to have a small optimal pH range for a high cell growth rate. Growth of *C. vulgaris* at acidic (3.0–6.2), neutral (7.5–8.0), and alkaline (8.3–9.0) pH suggested that neutral pH resulted in the maximum rate of cell growth [40]. The similar neutral pH range was also reported to be suitable for *Nannochloropsis salina*, *Chlamydomonas applanata*, and *Dunaliella bardawil*. Furthermore, a series of experiments have revealed that the alkaline pH was in the optimal range for the growth of *Chlorella ellipsoidea*, *Dunaliella salina*, and *Nannochloropsis salina*. Although acidic pH is not suitable for most microalgae strains, pH still has a strong effect on promoting *Euglena mutabilis* cell growth. Environmental pH is known to influence nutrient uptake and photosynthetic efficiency; its optimization is beneficial to increase the concentration of biomass. In addition, other environmental factors such as salts, pH, and reactive oxygen species (ROS) also affect the biosynthesis of the product [74]. Salts in the environment influence the osmotic pressure of cells.

6.3 *Maximizing Product*

Microalgae products can be classified into primary metabolites such as H_2 and lutein and secondary metabolites such as astaxanthin and β-carotene. High temperature, light intensity, appropriate light wavelength, and nutrient limitation are conditions frequently used to promote product biosynthesis [52]. Most conditions result in high carbon availability, which is the key factor in increasing product biosynthesis. However, these conditions are often different from the conditions for biomass accumulation.

7 Biochemical Composition of Microalgae and Effects on Biofuel Production

The efficient generation of biomass from microalgae and cyanobacteria compared to terrestrial plants using solar energy photosynthesis are promising to produce biochemical components (e.g., carbohydrate, proteins, and lipids), platform chemicals, and valuable natural compounds such as carotenoids, phycocyanin, and polyunsaturated fatty acids [76]. Biofuels and bio-based chemicals have been tested using molecular refinement tools to develop mass cultivation of many microalgae (including *Chlamydomonas* sp. and *Chlorella* sp.) and cyanobacteria (including *Synechocystis* sp. and *Synechococcus* sp.). The biofuels and platform chemicals productivity using microalgae or cyanobacteria is inadequate for profitable marketplace with limited response to socio-industrial demands.

Multi-omics pools the omics analyses, similar to genomics, metabolomics, transcriptomics, lipidomics, and proteomics. It has been applied to relate preparatory molecular mechanism of cell physiology, which contributes to identify new pathways and/or rate-limiting responses in the biosynthesis of targeted product. Metabolomics determine the regression step, making a critical contribution to formulating lucid genetic engineering and determining optimum culture environments. By looking at the microalgal or cyanobacteria photosynthetic product, this work highlights recent advancements in metabolomics. The operational methods to connote the essential metabolic processes and dynamics can be engineered to increase the product yields. Algae produce lipids, proteins, and carbohydrates using light, CO_2, and nutrients during photosynthesis. The relative amounts of these metabolic products are closely related to environmental and nutrient conditions, including the intensity of sunlight, CO_2 levels, pH, temperature, available nutrients, and the presence (or absence) of other organisms. Carbon, hydrogen, and oxygen are non-mineral nutrients necessary for algal growth. Macronutrients include nitrogen, phosphorous, sulfur, potassium, and magnesium. Micronutrients such as iron and manganese are also required in small amounts (2.5–30 ppm), while other trace elements such as cobalt, zinc, boron, copper, and molybdenum are essential elements (4.5–25 ppm) [19].

Inclusively, environmental conditions (especially light and temperature) and the availability of non-mineral nutrients, macronutrients, and micronutrients significantly impact the biochemical composition of microalgae. Other factors such as pH and toxic metals are also important factors influencing algae growth and metabolism. All these factors can affect photosynthesis, thus altering carbon fixation and allocation in different macromolecules. In turn, the macromolecular composition of the cell determines its utility in biofuel products. The temperature may be one of the most critical environmental factors affecting algae growth rate, cell size, biochemical composition, and nutrient conditions. In the United States, algae grow in a wide range of temperatures (15–40 °C), depending on the strain, region, and season. Below optimum growth temperatures, the growth rate increases with increasing temperature but decreases significantly above the species- or strain-specific optimum. Light intensity affects algae growth and biochemical compositions through its impact on photosynthesis and during the photoautotrophic growth phase, and organisms use light energy to convert carbon dioxide into organic compounds, especially sugars. One of the essential factors in algae culture is pH, as it determines the solubility and void of CO_2 and essential nutrients, which can significantly affect algae metabolism. Due to algae uptake of inorganic carbon, the pH can increase considerably in algae cultures. Maximum algae growth occurs around a neutral pH, although the optimum pH is the original culture pH to which an alga is acclimated to grow [20]. Exposing algae to altered salinity levels than their natural (or acclimated) can change growth rate and alter the biochemical compositions (e.g., higher salinity increases the lipid content of algae). Considerable variation in the biochemical composition of algae can be observed under conditions of nutrient limitation, depending on which nutrient is limited and to what extent. In general, the growth rate of algae is proportional to the uptake rate of the most limiting nutrient under optimal conditions of temperature and pH. The Michaelis–Menten equation generally describes it [61].

8 Application of Microalgae Biomass in Different Bioenergy and Biofuels

Photosynthetic microalgae can proliferate and generate enormous amounts of biomass compared to traditional agriculture. Microalgae have become a hot topic of interest to produce different renewable energy. There are some project schemes to apply microalgae research on hydrogen and methane generations. Since then, microalgae have been used to develop the techniques of wastewater remediation and carbon sequestration. Microalgae are a feedstock for 3G biofuels and do not compete with terrestrial foods/crops [68]. Microalgae culture has been carried out in open/closed/undeveloped lands. Edible oils (e.g., oiled obtained from palm, soybean, corn, wheat, moringa, vegetable, etc.), are the favorable feedstock of first-generation (1G) biofuels. First-generation biofuels have been reported to have detrimental consequences (such as food security, water availability, global food markets,

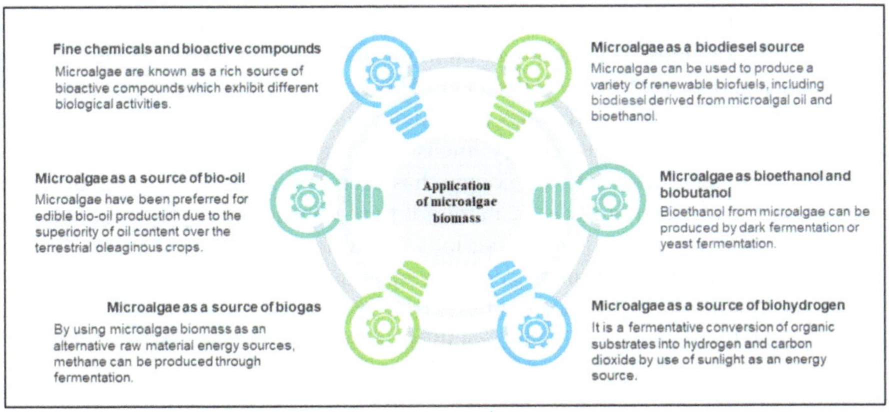

Fig. 2 Diverse applications of renewable and sustainable biomass of microalgae

and deforestation). Therefore, 1G biofuels have sparked much debate. In addition, second-generation (2G) biofuels have been manufactured using non-edible oils (e.g., *Jatropha curcas*, *Simarouba glauca*, *Pongamia pinnata*, etc.), forest residues. The lignocellulosic biomasses require large tracts of land cover. Recently, technologies those are viable to process waste products at a single point to generate biofuels still need to be improved in producing second-generation biofuels. Therefore, a genetically modified microalga signifies prospects to produce new generation of fourth-generation (4G) biofuels [63]. Figure 2 illustrates the diverse applications of renewable and sustainable biomass of microalgae.

Microalgae-derived fuel has benefits, including biodegradability, environmental friendliness, and superior energy scenarios. The microalgal biofuels can be blended with petro-diesel, jet fuels, and gasoline to be used for the fuels of transportation. Microalgae produce a significant extent of oil in comparison to any other renewable feedstock. To cultivate microalgae, light source, CO_2 and nutrients are preconditions. Nutrient supplements can be obtained using wastewater streams. Also, CO_2 can be provided from the source of the fossil fuels combustions. Moreover, by means of this system to transform CO_2 into platform chemicals and fuels devoid of contaminating the environment with economic benefits is a rewarding approach to minimize greenhouse gas emanations.

8.1 Microalgae as a Biodiesel Source

Biodiesel has been manufactured via transesterification reaction, in which the lipid molecules react with alcohol to produce methyl esters and glycerol, usually assisted by a catalyst. The conventional methods of transesterification require the separation and refining steps of microalgae biomass, including biomass drying, extraction of oils, and disruption of cells [77]. The refining steps greatly confine the marketability of 3G biodiesel since dependence on cell size and strength of the cell wall of certain microalgae, energy-intensive disruption of cell, and oil extraction methods. An adverse effect of wet biomass has been observed for oil extraction due to the existence of water, and necessitating of time-consuming drying process. The form of the catalyst determines the various transesterification mechanisms in the conventional transesterification process. Several studies have evaluated the suitable operational conditions to produce biodiesel from microalgae by enzymatic catalysis, acid catalysis, homogeneous alkaline catalysis, or heterogeneous catalysis [41].

8.2 Microalgae as a Bio-alcohol Source

One of the important biofuels, i.e., bioethanol has been categorized into bioethanol from sugar/starch, bioethanol from lignocellulosic, and bioethanol from microalgae. Due to the abundance and ease, sugarcane has been generally used as a feedstock for bioethanol. Lignocellulosic starches and biomass are inexpensive; though, it is hydrolyzed for the fermentable sugars production. Microalgae can be fermented to form essential sugars from polysaccharides [33]. Different microalgae carbohydrates (e.g., cellulose, starch, agar, laminarin, mannitol, etc.) are being used to produce bio-alcohol, such as ethanol, propanol, and butanol. The fermentation yield can be determined according to the preferences of microalgae. In bio-alcohol manufacture, the direct use of harvested wet biomass is possible; however, dry biomass is supposed to work significantly better.

8.3 Microalgae as a Biohydrogen Source

Hydrogen has become a leading sustainable bioenergy source due to the involvement of innovative studies, processing, and technical developments for sustainable generation. Biohydrogen generation is being carried out in a photobiological reaction involving water splitting under sunlight. The generation of biohydrogen, ethanol, and the fraction of CO_2 depends on the starch formed via oxidative reaction under different conditions microbial communities. Several ways have been reported to produce hydrogen, including electrolysis of water, steam methane reforming, gasification of coals, thermal processes, oxidation of oil/natural gases, and indirect and direct biophotolysis [4]. Photo-fermentation and dark fermentation processes generate hydrogen from organic compounds. Photo-fermentation faces challenges to produce hydrogen inefficiently due to the release of O_2 which simultaneously shuts mechanism of hydrogen production due the inhibition of hydrogenase enzyme. During the dark fermentation of fructose and xylose, the obligate anaerobes release hydrogen and oxygen. *Chlamydomonas* sp. is used extensively to investigate the hydrogen production ability; however, it generates very small amount of hydrogen and oxygen [35].

8.4 Microalgae as a Biomethane/Biogas Source

Pre-treated or untreated biomass of microalgae and/or residues can be applied as a feedstock of methane-producing substrate for anaerobic digestion, which generates methane, carbon dioxide, and digestate (fertilizer). Typically, the microalgae biomass needs to be pre-treated (e.g., chemical, mechanical, biological, or thermal methods) prior to start the methane generation. The biological pretreatment of microalgae biomass involves the enzymatic action (fungi or bacteria as pretreatment enzyme source) on cell walls [10]. Although enzymes significantly increase methane productivity, their cost precludes their use in real-world situations. However, crude enzymes are also capable to be utilized for the pretreatment of the microalgae biomass. The crude enzymes extracted from fungi demonstrate high efficacy in microalgal cell wall digestion to release soluble sugars. Fungal broth containing laccase from *Taramites versicolor* effectively increases biogas production from microalgae by ~74% [64]. Although the use of crude enzymes supports the alleviation of constraints of the high costs. An additional step is required to produce and extract the enzymes. To address this problem, the incubation of microalgal biomass along with enzyme-producing organism is required at the manufacturing facility. However, due to the high costs of biomass harvesting, cultivating, and refining steps, biomethane produced from the anaerobic digestion requires a fair comparison for commercially available fuels.

8.5 Microalgae as a Bio-oil Source

Dark, viscous, and energy-dense fuels are being popularized to produce crude bio-oil from microalgae biomass. Bio-oils from microalgae biomass have been produced via hydrothermal liquefaction or pyrolysis process. Bio-oil yields of various microalgae have been reported to be 5–25% greater in comparison to the lipid contents. Microalgal wet biomass processing attracts as it reduces the energy requisite for drying process. In pyrolysis (slow, medium, or fast), high temperatures ($\leq$700 °C) heating of organic compounds carried out under oxygen-restricted environments. The chemical components of bio-oil are comprised of aldehydes, acids, and cresols. The bio-oils derived from the lignocellulosic bioresources transformation are rich in oxygen elemental compositions but complex and vicious [75]. Therefore, hydrogenation and cracking steps are compulsory to improve the bio-oils. On the other hand, the bio-oils made from microalgae biomass pyrolysis consist of linear hydrocarbons and nitrogenous species. It is produced due to the pyrolysis of proteins and lipids. In the hydrothermal liquefaction (a thermochemical process), hot compressed

Table 3 Different microalgal strains in the production of high-value products

Strains	Products	Applications
Arthrospira (*Spirulina*) sp.	Phycocyanin, protein, vitamin B_{12}	Health, cosmetics, antioxidant capsules
Chlorella sp.	Carbohydrates	Animal nutrition, health drinks, food supplements, feed surrogates
Dunaliella salina	β–carotene	Health, food supplements, feed
Haematococcus pluvialis	Astaxanthin	Health, pharmaceuticals, feed additives
Phaedactylum tricomutum	Fatty acids	Nutrition, biofuels
Nanochloropsis oculata *Nanochloropsis* sp.	Eicosapentaenoic acid	Nutrition, feed for larvae and juvenile marine fish
Porphyridium cruentum	Polysaccharides	Pharmaceuticals, cosmetics
Phaeodactylum tricornutum *Nannochloropsis* sp. *Nitzchia* sp.	Eicosapentaenoic acid	Food supplements, nutrition

or subcritical water transforms moist microalgae into the liquid biofuels, which is usually performed at 300–350 °C and 5–20 MPa, with a mass fraction of microalgae (5–50%) in sludge feeds [48].

8.6 *Microalgae as a Source of Bioactive and Fine Chemical Compounds*

Microalgae are of great industrialized significance in the arena of non-combustible resources. Microalgal biomass has been used as biochemical and bio-product assets (e.g., proteins, lipids, carbohydrates, cellulose, starch, polyunsaturated fatty acids, vitamins, foods, cosmetics, pesticides, organic fertilizers, livestock feeds, animal feeds, pigments, natural dyes, nutraceuticals, etc.), and even in the generation of combustion-based electricity as a few of the industrially important products. Table 3 illustrates the use of different microalgae strain to produce valuable products.

9 3G Biorefinery and High-value Microalgae Biomass for Platform Chemicals

3G biorefineries have been introduced to produce platform chemicals, bioenergy, and value-added products (e.g., amino acids, vitamins, bio-pigments, antioxidants, etc.), which are believed to lead toward a better circular economy in the future. The main goal of 3G biorefineries is to use atmospheric carbon dioxide, light, and inorganic nutrients from wastewater resources for renewable energy. The bio-electricity generated from photovoltaic cells and wind mills have been considered as sustainable source along with to produce various valuable products. The microbial cell factories are engaged in this entire production process. 3G biorefinery significantly reduces the processing cost of feedstocks and poses far fewer stresses to the security of food and water supplies than biorefineries of the first two generations (1G and 2G). For 3G biorefineries, microalgae can be efficiently used as a base feedstock [39]. Mainly, unicellular microalgae biomass contains different inorganic and organic compounds. It is also essential to distinguish the composition of proteins, carbohydrates, and lipids in various microalgae species, which, if used correctly, will result in the favorable production of substances from the biorefinery. Moreover, parameters such as the type of culture system used, light availability, nutrient availability (e.g., nitrogen and phosphorous), temperature, CO_2 and O_2 levels, which can impact the necessary microalgae growth to achieve a better product. Recent studies have also mentioned various pre-treatment processes, such as physical, chemical, physicochemical, and biological processes, which can be implemented to obtain improved feedstock conversion and keep intact the cell biomass composition and sugar/lipid recovery [62].

Several companies are already working on commercial or pilot scale established on 3G biorefinery using microalgae as feedstock (Table 4). Earlier, Fermentalg has implemented efficient approaches using *Schizochytrium* sp. to produce different commercial products [17]. In another, LanzaTech and INEOS have applied acetogens to generate ethanol. It is also clearly indicated in the literature that by using the 3G biorefinery approach and lactic acid bacteria, lactic acid-like value-added compounds can be produced from residual mediums containing macroalgal xylose and hydrolyzed galactose. A technical–economic evaluation of cellulosic waste from red macroalgae using 3G biorefinery has been conducted to produce bioethanol and lactic acid [53]. In the study, the established 3G biorefinery approaches to becoming a genuine economy due to the involvement of on-site seed train to on-site manufacture of cellulase enzyme, yeast, and LAB to hydrolyze and ferment for inexpensive biorefinery processes to emphasize the marketable/economic viability in industrial-scale. *Ulva lactuca* has been used in the developed 3G biorefinery integration that involved the co-fermentation of reducing sugars to produce bioethanol and lactic acid, which have a minimal environmental impact with profitability [49]. However, bioethanol production can be higher using the 3G biorefinery with an alternative to reduce/replace gasoline. It can solve the adverse impacts (~70%) generated by GHGs emanation from combustible gases. Cost is another factor restricting the advancement of 3G biorefinery. Therefore, considering its advantages and disadvantages, commercial and laboratory scale approaches can optimize it for future applications.

Table 4 Commercial or pilot plant-based companies working on 3G biorefinery processes using microalgae as a feedstock

Company	Location	Products	Technology	Applications
LanzaTech	USA	Ethanol	Acetogens gas fermentation	Renewable energy
INEOS	USA	Ethanol	Gas fermentation	Renewable energy
Fitoplancton Marino	Spain	Proteins	Microalgae	Food industry
Fermentalg	France	Proteins and fatty acids	Microalgae	Food industry
Oakbio	USA	*n*-butanol and bioplastics	Oakbio's proprietary microbes	Energy and packaging
Pond Technologies	Canada	Biofuels	Microalgae	Renewable energy
Cellana	USA	Biofuels	Algae	Energy
Algenol	USA	Ethanol	Microalgae	Renewable energy

10 Conclusion

The biorefinery has yet to get an encouragement to extend research trends towards sustainable bio-based products. Sustainable biorefinery essentially produces renewable biofuels and high-value bio-products in integrated approaches. The prominent step of 3G biorefinery towards sustainable products of the microalgal biomass lies in preliminary fractionations using minimal adjustable parameters. Advances in fractionation of microalgal biomass for diverse commercial products in microalgae-based 3G biorefinery are in early stages. The integration of biorefinery potentially offers biofuels and platform chemicals at reasonable costs. It is a good time for 3G biorefinery to develop accessible mechanisms to produce large-scale renewable biofuels, foods, and platform chemicals to shrink reliance on petro-products to subsidize the establishing sustainable society in global considerations.

Acknowledgements We sincerely appreciate the facilities and resources of the Pondicherry University (A Central University), Puducherry, India.

References

1. Amer B, Baidoo EE (2021) Omics-driven biotechnology for industrial applications. Front Bioeng Biotechnol 9:613307

2. Antar M, Lyu D, Nazari M, Shah A, Zhou X, Smith DL (2021) Biomass for a sustainable bioeconomy: an overview of world biomass production and utilization. Renew Sustain Energy Rev 139:110691
3. Asada K (2019) Production and action of active oxygen species in photosynthetic tissues. In: Causes of photooxidative stress and amelioration of defense systems in plants. CRC press, pp 77–104
4. Avargani VM, Zendehboudi S, Saady NMC, Dusseault MB (2022) A comprehensive review on hydrogen production and utilization in North America: prospects and challenges. Energy Convers Manage 269:115927
5. Awasthi MK, Singh E, Binod P, Sindhu R, Sarsaiya S, Kumar A, Zhang Z et al (2022) Biotechnological strategies for bio-transforming biosolid into resources toward circular bio-economy: a review. Renew Sustain Energy Rev 156:111987
6. Balparda M, Bouzid M, del Pilar Martinez M, Zheng K, Schwarzländer M, Maurino VG (2023) Regulation of plant carbon assimilation metabolism by post-translational modifications. Plant J
7. Barot S (2022) Biomass and bioenergy: resources, conversion and application. Renew Energy Sustain Growth Assess 243–262
8. Barrett S (2009) The coming global climate–technology revolution. J Econ Perspect 23(2):53–75
9. Behera M, Nayak J, Banerjee S, Chakrabortty S, Tripathy SK (2021) A review on the treatment of textile industry waste effluents towards the development of efficient mitigation strategy: an integrated system design approach. J Environ Chem Eng 9(4):105277
10. Bhushan S, Jayakrishnan U, Shree B, Bhatt P, Eshkabilov S, Simsek H (2023) Biological pretreatment for algal biomass feedstock for biofuel production. J Environ Chem Eng 11(3):109870
11. Blanco-Canqui H (2016) Growing dedicated energy crops on marginal lands and ecosystem services. Soil Sci Soc Am J 80(4):845–858
12. Borella L, Ortolan D, Barbera E, Trivellin N, Sforza E (2021) A multiwavelength model to improve microalgal productivity and energetic conversion in a red-blue light emitting diodes (LEDs) continuous photobioreactor. Energy Convers Manage 243:114330
13. Calvin K, Cowie A, Berndes G, Arneth A, Cherubini F, Portugal-Pereira J, Smith P et al (2021) Bioenergy for climate change mitigation: Scale and sustainability. GCB Bioenergy 13(9):1346–1371
14. Carvalho AP, Silva SO, Baptista JM, Malcata FX (2011) Light requirements in microalgal photobioreactors: an overview of biophotonic aspects. Appl Microbiol Biotechnol 89:1275–1288
15. Chen R, Yamaoka Y, Feng Y, Chi Z, Xue S, Kong F (2023) Co-expression of lipid transporters simultaneously enhances oil and starch accumulation in the green microalga Chlamydomonas reinhardtii under nitrogen starvation. Metabolites 13(1):115
16. Chung JH, Kim DS (2012) Miscanthus as a potential bioenergy crop in East Asia. J Crop Sci Biotechnol 15:65–77
17. Conceição GR, de Jesus Santos JR, da Silva CS, Chinalia FA (2022) Microalgae as agents for upcycling industrial wastewater: energy patents and scientific literature overview. Res Soc Dev 11(15):e230111537097–e230111537097
18. de Jong E, Higson A, Walsh P, Wellisch M (2012) Bio-based chemicals value added products from biorefineries. IEA Bioenergy, Task42 Biorefinery 34:1–33
19. Dey S, Mazumder I, Roy C, Ali OA, Kumar S, Pattnaik R (2023) Biofuel production from algal biomass: technologies and opportunities addressing the global energy crisis. In: Green approach to alternative fuel for a sustainable future. Elsevier, pp 45–58
20. Fernandes F, Silkina A, Gayo-Peláez JI, Kapoore RV, de La Broise D, Llewellyn CA (2022) Microalgae cultivation on nutrient rich digestate: the importance of strain and digestate tailoring under pH control. Appl Sci 12(11):5429
21. Folgueras MB, Diaz RM, Xiberta J, Alonso M (2007) Effect of inorganic matter on trace element behavior during combustion of coal—sewage sludge blends. Energy Fuels 21(2):744–755

22. Gao F, Li C, Yang ZH, Zeng GM, Feng LJ, Liu JZ, Cai HW et al (2016) Continuous microalgae cultivation in aquaculture wastewater by a membrane photobioreactor for biomass production and nutrients removal. Ecol Eng 92:55–61
23. Goswami L, Kushwaha A, Singh A, Saha P, Choi Y, Maharana M, Kim BS et al (2022) Nano-biochar as a sustainable catalyst for anaerobic digestion: a synergetic closed-loop approach. Catalysts 12(2):186
24. Guidi L, Tattini M, Landi M (2017) How does chloroplast protect chlorophyll against excessive light. Chlorophyll 21:21–36
25. Guo M, Song W, Buhain J (2015) Bioenergy and biofuels: history, status, and perspective. Renew Sustain Energy Rev 42:712–725
26. Gupta V, Ratha SK, Sood A, Chaudhary V, Prasanna R (2013) New insights into the biodiversity and applications of cyanobacteria (blue-green algae)—prospects and challenges. Algal Res 2(2):79–97
27. Gómez-Brandón M, Juárez MFD, Domínguez J, Insam H (2013) Animal manures: recycling and management technologies. In: Biomass now-cultivation and utilization. InTech, Rijeka, Croatia, pp 237–272
28. Hart LF, Hertzog JE, Rauscher PM, Rawe BW, Tranquilli MM, Rowan SJ (2021) Material properties and applications of mechanically interlocked polymers. Nat Rev Mater 6(6):508–530
29. Hingsamer M, Jungmeier G (2019) Biorefineries. In: The role of bioenergy in the bioeconomy. Academic Press, pp 179–222
30. Houfani AA, Anders N, Spiess AC, Baldrian P, Benallaoua S (2020) Insights from enzymatic degradation of cellulose and hemicellulose to fermentable sugars–a review. Biomass Bioenergy 134:105481
31. Huang X, Ren J, Ran JY, Qin CL, Yang ZQ, Cao JP (2022) Recent advances in pyrolysis of cellulose to value-added chemicals. Fuel Process Technol 229:107175
32. Jayaseelan A, Panchamoorthy GK, Nithianantharaj V (2022) An eco-friendly and economical approach for removal of Remazol Blue, malachite green and rhodamine B dyes from wastewater using bio-char derived from Chlorella vulgaris biomass. Curr Anal Chem 18(3):370–382
33. Jeyakumar N, Hoang AT, Nižetić S, Balasubramanian D, Kamaraj S, Pandian PL, Nguyen XP et al (2022) Experimental investigation on simultaneous production of bioethanol and biodiesel from macro-algae. Fuel 329:125362
34. Kandasamy S, Zhang B, He Z, Bhuvanendran N, EL-Seesy AI, Wang Q, Dar MA et al (2022) Microalgae as a multipotential role in commercial applications: current scenario and future perspectives. Fuel 308:122053
35. King SJ, Jerkovic A, Brown LJ, Petroll K, Willows RD (2022) Synthetic biology for improved hydrogen production in Chlamydomonas reinhardtii. Microb Biotechnol 15(7):1946–1965
36. Ko YS, Kim JW, Lee JA, Han T, Kim GB, Park JE, Lee SY (2020) Tools and strategies of systems metabolic engineering for the development of microbial cell factories for chemical production. Chem Soc Rev 49(14):4615–4636
37. Kovalev GV, Bugaenko LT (2003) On the crosslinking of cellulose under exposure to radiation. High Energy Chem 37:209–215
38. Kruyeniski J, Ferreira PJ, Carvalho MDGVS, Vallejos ME, Felissia FE, Area MC (2019) Physical and chemical characteristics of pretreated slash pine sawdust influence its enzymatic hydrolysis. Ind Crops Prod 130:528–536
39. Kulyal L, Jalal P (2022) Bioenergy, a finer alternative for India: scope, barriers, socio-economic benefits and identified solution. In: Bioresource technology reports, 100947
40. Kumar R, Hegde AS, Sharma K, Parmar P, Srivatsan V (2022) Microalgae as a sustainable source of edible proteins and bioactive peptides-current trends and future prospects. Food Res Int 111338
41. Lee JWY, Chia WY, Ong WJ, Cheah WY, Lim SS, Chew KW (2022) Advances in catalytic transesterification routes for biodiesel production using microalgae. Sustain Energy Technol Assess 52:102336
42. Lillefosse HH, Clausen MR, Yde CC, Ditlev DB, Zhang X, Du ZY, Liaset B et al (2014) Urinary loss of tricarboxylic acid cycle intermediates as revealed by metabolomics studies: an

underlying mechanism to reduce lipid accretion by whey protein ingestion? J Proteome Res 13(5):2560–2570
43. Liu R, Li S, Tu Y, Hao X, Qiu F (2022) Recovery of value-added products by mining microalgae. J Environ Manage 307:114512
44. Liu C, Liu Y, Chen Y, Zhao C, Qiu J, Wu D, Tang K et al (2023) A state-of-the-practice review of three-dimensional laser scanning technology for tunnel distress monitoring. J Perform Constr Facil 37(2):03123001
45. Long H, Li X, Wang H, Jia J (2013) Biomass resources and their bioenergy potential estimation: a review. Renew Sustain Energy Rev 26:344–352
46. Maheshwari P, Haider MB, Yusuf M, Klemeš JJ, Bokhari A, Beg M, Jaiswal AK et al (2022) A review on latest trends in cleaner biodiesel production: role of feedstock, production methods, and catalysts. J Clean Prod 355:131588
47. Maity JP, Bundschuh J, Chen CY, Bhattacharya P (2014) Microalgae for third generation biofuel production, mitigation of greenhouse gas emissions and wastewater treatment: present and future perspectives–a mini review. Energy 78:104–113
48. Marinič D, Grilc M, Hočevar B, Delrue F, Likozar B (2023) Liquefaction, cracking and hydrogenation of microalgae biomass resources to CO_2 negative advanced biofuels: mechanisms, reaction microkinetics and modelling. Renew Energy 203:382–393
49. Mehariya S, Karthikeyan OP, Bhatia SK (eds) (2022) Algal biorefineries and the circular bioeconomy: algal products and processes. CRC Press
50. Mian MM, Liu G, Fu B (2019) Conversion of sewage sludge into environmental catalyst and microbial fuel cell electrode material: a review. Sci Total Environ 666:525–539
51. Del Mondo A, Sansone C, Brunet C (2022) Insights into the biosynthesis pathway of phenolic compounds in microalgae. Comput Struct Biotechnol J 20:1901–1913
52. Mulgund A (2022) Increasing lipid accumulation in microalgae through environmental manipulation, metabolic and genetic engineering: a review in the energy NEXUS framework. Energy Nexus 100054
53. Ocreto JB, Chen WH, Rollon AP, Ong HC, Pétrissans A, Pétrissans M, De Luna MDG (2022) Ionic liquid dissolution utilized for biomass conversion into biofuels, value-added chemicals and advanced materials: a comprehensive review. Chem Eng J 136733
54. Peter AP, Koyande AK, Chew KW, Ho SH, Chen WH, Chang JS, Show PL et al (2022a) Continuous cultivation of microalgae in photobioreactors as a source of renewable energy: current status and future challenges. Renew Sustain Energy Rev 154:111852
55. Peter AP, Tan X, Lim JY, Chew KW, Koyande AK, Show PL (2022b) Environmental analysis of Chlorella vulgaris cultivation in large scale closed system under waste nutrient source. Chem Eng J 433:134254
56. Ramos A, Monteiro E, Rouboa A (2022) Biomass pre-treatment techniques for the production of biofuels using thermal conversion methods–a review. Energy Convers Manage 270:116271
57. Rani GM, Pathania D, Umapathi R, Rustagi S, Huh YS, Gupta VK, Chaudhary V et al (2023) Agro-waste to sustainable energy: a green strategy of converting agricultural waste to nano-enabled energy applications. Sci Total Environ 875:162667
58. Ranjbari M, Esfandabadi ZS, Quatraro F, Vatanparast H, Lam SS, Aghbashlo M, Tabatabaei M (2022) Biomass and organic waste potentials towards implementing circular bioeconomy platforms: a systematic bibliometric analysis. Fuel 318:123585
59. Ren X, Liu Y, Fan C, Hong H, Wu W, Zhang W, Wang Y (2022) Production, processing, and protection of microalgal n-3 PUFA-rich oil. Foods 11(9):1215
60. Sang S, Zhuang X, Chen H, Qin Y, Cao J, Fan F, Lan T (2022) Effect of supramolecular structural changes during the crystalline transformation of cellulose on its enzymatic hydrolysis. Ind Crops Prod 180:114687
61. Sardari D (2023) Mathematical basis for diatom growth modeling. In: The mathematical biology of diatoms, pp 121–139
62. Sharma S, Show PL, Aminabhavi TM, Sevda S, Garlapati VK (2023) Valorization of environmental-burden waste towards microalgal metabolites production. Environ Res 227:115320

63. Shukla S, Shukla R, Singh N, Hemansi, Saini JK (2022) Next generation biofuels from macroalgae: prospects and challenges. Microb Biotechnol Renew Sustain Energy 55–75
64. Siddiki SYA, Mofijur M, Kumar PS, Ahmed SF, Inayat A, Kusumo F, Mahlia TMI et al (2022) Microalgae biomass as a sustainable source for biofuel, biochemical and biobased value-added products: an integrated biorefinery concept. Fuel 307:121782
65. Sidhu C, Kirstein IV, Meunier CL, Rick J, Fofonova V, Wiltshire KH, Teeling H et al (2023) Dissolved storage glycans shaped the community composition of abundant bacterioplankton clades during a North Sea spring phytoplankton bloom. Microbiome 11(1):1–18
66. Sun H, Zhao W, Mao X, Li Y, Wu T, Chen F (2018) High-value biomass from microalgae production platforms: strategies and progress based on carbon metabolism and energy conversion. Biotechnol Biofuels 11:1–23
67. Takkellapati S, Li T, Gonzalez MA (2018) An overview of biorefinery-derived platform chemicals from a cellulose and hemicellulose biorefinery. Clean Technol Environ Policy 20:1615–1630
68. Thanigaivel S, Priya AK, Dutta K, Rajendran S, Vasseghian Y (2022) Engineering strategies and opportunities of next generation biofuel from microalgae: a perspective review on the potential bioenergy feedstock. Fuel 312:122827
69. Tiwari SK, Bystrzejewski M, De Adhikari A, Huczko A, Wang N (2022) Methods for the conversion of biomass waste into value-added carbon nanomaterials: recent progress and applications. Prog Energy Combust Sci 92:101023
70. Tonini D, Hamelin L, Astrup TF (2016) Environmental implications of the use of agro-industrial residues for biorefineries: application of a deterministic model for indirect land-use changes. GCB Bioenergy 8(4):690–706
71. Tosti L, van Zomeren A, Pels JR, Dijkstra JJ, Comans RN (2019) Assessment of biomass ash applications in soil and cement mortars. Chemosphere 223:425–437
72. Usmani Z, Sharma M, Awasthi AK, Lukk T, Tuohy MG, Gong L, Gupta VK et al (2021) Lignocellulosic biorefineries: the current state of challenges and strategies for efficient commercialization. Renew Sustain Energy Rev 148:111258
73. Velvizhi G, Goswami C, Shetti NP, Ahmad E, Pant KK, Aminabhavi TM (2022) Valorisation of lignocellulosic biomass to value-added products: paving the pathway towards low-carbon footprint. Fuel 313:122678
74. Wen Y, Yan J, Yang B, Zhuang Z, Yu Y (2022) Reactive oxygen species on transition metal-based catalysts for sustainable environmental applications. J Mater Chem A 10(37):19184–19210
75. Wu Y, Wang H, Li H, Han X, Zhang M, Sun Y, Xu X et al (2022) Applications of catalysts in thermochemical conversion of biomass (pyrolysis, hydrothermal liquefaction and gasification): a critical review. Renew Energy
76. Xie Y, Khoo KS, Chew KW, Devadas VV, Phang SJ, Lim HR, Show PL et al (2022) Advancement of renewable energy technologies via artificial and microalgae photosynthesis. Biores Technol 363:127830
77. Yasin NHM, Aziz NNC, Azmai MBA, Hanapi MFM (2023) Transesterification method of microalgae biomass to produce fatty acid methyl esters. J Chem Technol Biotechnol
78. Yin Z, Zhu L, Li S, Hu T, Chu R, Mo F, Li B et al (2020) A comprehensive review on cultivation and harvesting of microalgae for biodiesel production: environmental pollution control and future directions. Biores Technol 301:122804
79. Yoo CG, Meng X, Pu Y, Ragauskas AJ (2020) The critical role of lignin in lignocellulosic biomass conversion and recent pretreatment strategies: a comprehensive review. Biores Technol 301:122784
80. Yu S, Zheng Y, Hu X, Shu K (2022) Spatial impacts of biomass resource endowment on provincial green development efficiency. Renew Energy 189:651–662
81. Zhang S, Liu Z (2021) Advances in the biological fixation of carbon dioxide by microalgae. J Chem Technol Biotechnol 96(6):1475–1495
82. Zhu L, O'Dwyer JP, Chang VS, Granda CB, Holtzapple MT (2010) Multiple linear regression model for predicting biomass digestibility from structural features. Biores Technol 101(13):4971–4979

Agro-forestry Biomass as a Potential Bioresource for Climate Change Mitigation

Anurag Jaswal, Piyush Pratap Singh, Sharad Lande, and Tarak Mondal

1 Introduction

1.1 Background

Energy demand and the evolutionary trajectory of human civilization are fundamentally intertwined. The discovery and exploitation of emerging energy technologies have been instrumental in driving exceptional progress, both socio-culturally and economically, across historical epochs. This extraordinary growth in energy demand is primarily attributable to an unprecedented upsurge in industrialization and urbanization, which, in turn, has led to significant amelioration in quality of life and an explosive demographic expansion, imposing a tremendous strain on global energy resources. With the population expected to grow further at an ever-increasing rate, the energy demand will also continue to do so, with estimates suggesting a rise of 40–50% over the current levels by 2040 [1, 2]. For the past 100 years or so, the majority of the energy required for powering our homes, businesses and industries and propelling our vehicles has come from burning fossil fuels. While the share of renewable sources such as solar and wind energy in the global energy mix has increased significantly, fossil fuels continue to be in the pole position, supplying nearly 82% of the global demand [3]. However, the burning of fossil fuels for energy production is associated with large-scale environmental damage in the form of air, water and land pollution as well as global warming. Consequently, there have been growing calls for shifting

A. Jaswal · P. P. Singh · T. Mondal (✉)
Catalysis and Sustainable Energy Research Laboratory, Department of Chemical Engineering, Indian Institute of Technology Ropar, Rupnagar, Punjab 140001, India
e-mail: tarakmondal@iitrpr.ac.in

P. P. Singh · S. Lande (✉)
Reliance Research and Development Center, Reliance Industries Limited, Ghansoli, Navi Mumbai, Maharashtra 400701, India
e-mail: sharad.lande@ril.com

S. Kumar et al. (eds.), *Clean Energy Transition-via-Biomass Resource Utilization*, Green Energy and Technology, https://doi.org/10.1007/978-981-97-6321-4_3

the current paradigm of energy production from the utilization of fossil fuels towards the utilization of renewable sources.

In recent years, the identification of renewable alternatives to fossil fuels has increasingly become the focus of not just academic and industrial research but also policy makers. While a variety of potential candidates such as solar power, wind power and hydropower have been considered, lignocellulosic biomass is one alternative that has shown early promise. The utilization of biomass as an alternative feedstock to fossil fuels for energy production has numerous advantages. To begin with, the replenishment times of biomass are orders of magnitude smaller than those for fossil fuels, which makes it a renewable and sustainable material [4]. Additionally, biomass is available in huge abundance and is well distributed across different geographies and countries [5] as opposed to fossil fuels whose reserves are finite and distributed unevenly across the globe [6]. Most importantly, it is considered carbon–neutral since the materials that comprise biomass consume CO_2 during their growth periods [7]. Yet another major upside of using biomass is that it provides an excellent strategy for the management of waste materials such as crop residues, an issue of critical importance for developing economies [8]. To sum up, setting up a biomass-based energy system provides twin benefits of waste management and green energy production while also providing an avenue for job creation. Among various feedstocks, lignocellulosic biomass materials such as agricultural crop residues, forestry wastes and municipal solid wastes are most suited as feedstocks for energy production owing to their (i) inedible nature, (ii) availability in great abundance and (iii) lower production costs and faster growth periods [9]. It is an extremely complex natural polymer that is the building block of cell walls and is made up of three major components—cellulose, hemicellulose and lignin (Fig. 1). The contents of these fractions, and their structures vary greatly depending upon the type of biomass [10].

1.2 Role of Agro-forestry Wastes in Climate Change

Rapid population growth over the past century has given rise to a tremendous demand for basic necessities and raw materials. Consequently, there has been an equally massive upsurge in various sectors such as agriculture and forestry to meet the needs of the growing population. Every year, millions of tonnes of waste residues are generated as a result of these activities, which pose a major problem, especially for burgeoning economies. While some amount of the produced agricultural residues end up being utilized either in domestic and/or industrial sectors for various applications, a considerable amount remains unutilized. These surplus crop residues are eventually burned on farm by the farmers, a phenomenon common in Asian countries such as India, to prepare the land for the next cycle of sowing [11]. Similarly, in the case of forestry residues, they are either left behind to decompose naturally or are used in power generation in industries and as firewood for domestic consumption in rural areas [12]. The residues left in the forests can catch fire, resulting in forest fires that would disturb the local environment and result in the emission

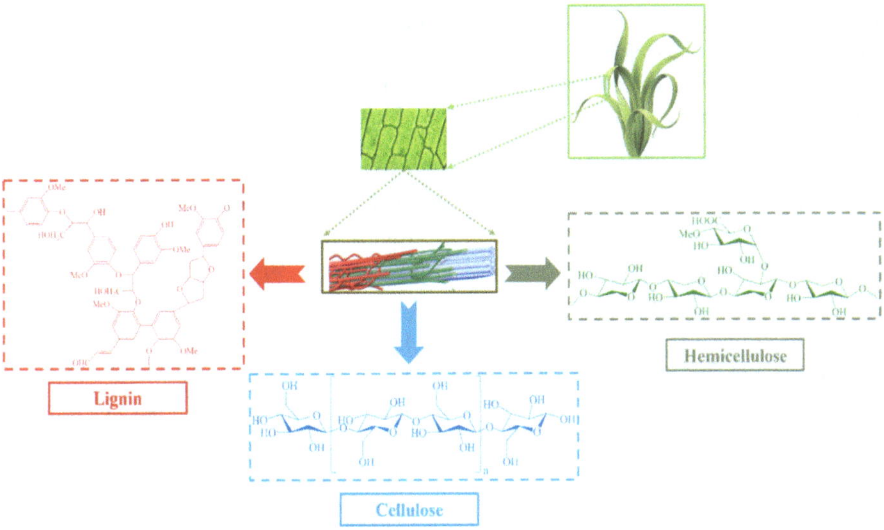

Fig. 1 Chemical composition of lignocellulosic biomass

of huge amounts of greenhouse gases and particulate matter [12]. Both agricultural and forestry residues represent carbon and energy-rich resources that are not only available in huge abundance at low costs but also do not compete with food. Most importantly, they consume CO_2 during their growth periods, making them carbon–neutral since any CO_2 released during their eventual utilization through any strategy would be recycled. Consequently, the utilization of agro-forestry residues as a feedstock represents a very attractive opportunity from a sustainability standpoint for mitigating climate change resulting from the current industrial paradigm of the production of chemicals, fuels and energy.

2 Agro-forestry Residues—Feedstocks and Their Composition

2.1 Agricultural Residues

Agricultural residue is a term that refers to materials generated during the course of harvesting and processing of cereals, vegetables and fruits as well as wastes from farms, poultry and slaughterhouses [13]. Among these, the residues generated from cereal crops such as rice and wheat are by far the largest contributors [14]. For the purpose of classification, crop-based residues can be categorized as either primary or secondary residues. Primary residues are wastes left behind in the fields after the harvesting of major crops such as rice and wheat and include stalks, straws, stovers and leaves [15]. Meanwhile, secondary residues are those generated during subsequent post-harvest processing of crops and include cobs, husk, sugarcane bagasse and molasses [15]. Table 1 shows the composition of major agricultural wastes. Figure 2 displays some of the most common agro-residues.

Table 1 Proximate and ultimate analysis of major agricultural residues [2, 16, 17]

Residue	Proximate analysis (%)				Ultimate analysis (%)					Heating value (MJ/kg)
	Moisture	Volatile matter	Ash	Fixed carbon	C	H	N	S	O	
Rice straw	5.2	74.47	6.26	14.09	44.28	6.83	1.28	0	47.9	17.4
Rice husk	7.36	69.75	20.07	2.82	45.18	5.08	1.62	0.07	48.05	18.55
Wheat straw	7.66	78.33	7	7	48.24	5.64	0.56	–	45.55	15.29
Corn cob	10.2	80	5.7	4.2	44.2	5.9	0.4	0.08	44.2	15.5
Corn stover	8.3	73	11.1	7.9	45.7	6.3	0.4	0.5	45.7	18
Sorghum stalk	8.7	66	8.8	16.5	44.4	6.2	0.5	0.9	44.4	16.9
Mustard stalk	9.7	70	7.9	12.3	43.8	5.9	0.3	0.3	43.8	15.9
Cotton stalk	8.9	71	3.5	16.6	46.8	6.4	0.3	0.2	46.8	17.3
Sugarcane bagasse	10	76	4.4	9.6	43.2	6.2	0.4	0.8	43.2	15.5

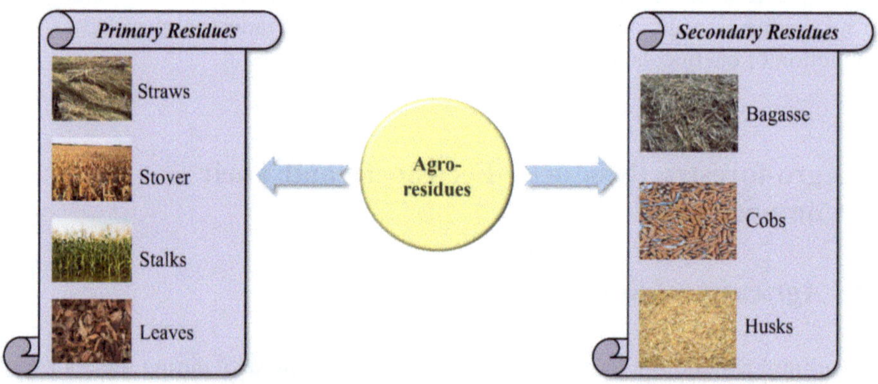

Fig. 2 Types of common agricultural residue feedstock

2.2 *Forestry Residues*

The residues from the forestry sector can generally be classified as either forestry residues or residues from wood processing and cutting industries. Forestry residues primarily refer to wastes such as branches, foliage, dead, damaged and/or small-diameter trees, tree tops and stumps that are generated during clear cutting, commercial harvesting of timber, pruning and thinning operations. Additionally, it may also include trees that have been damaged by natural phenomena such as insect outbreaks and fires [18, 19]. The complete removal of these materials is often not possible due to the difficulties arising from their remote geographic location. In addition, their removal can harm the soil and fauna of the local area, as a result of which they are left to naturally decompose in the forests [20]. The other category refers to the huge amounts of wastes such as sawdust, shaving, logs, wood chips and barks generated during processing operations [18, 20]. Figure 3 displays some of the most common forest residues (Table 2).

3 Strategies for Agro-forestry Residue Utilization

Over the years, several strategies have been developed for converting biomass into energy and/or fuels which can be broadly classified into two categories: thermochemical and biochemical [23]. Thermochemical routes involve the conversion of biomass through thermal degradation at moderate to high temperatures. Currently, the most commonly used thermochemical strategies are combustion, gasification, pyrolysis and hydrothermal liquefaction. On the other hand, biochemical routes utilize enzymes or microorganisms to convert biomass to valuable products such as biogas and bioethanol under much milder conditions in comparison to those used in thermochemical routes [24] (Fig. 4).

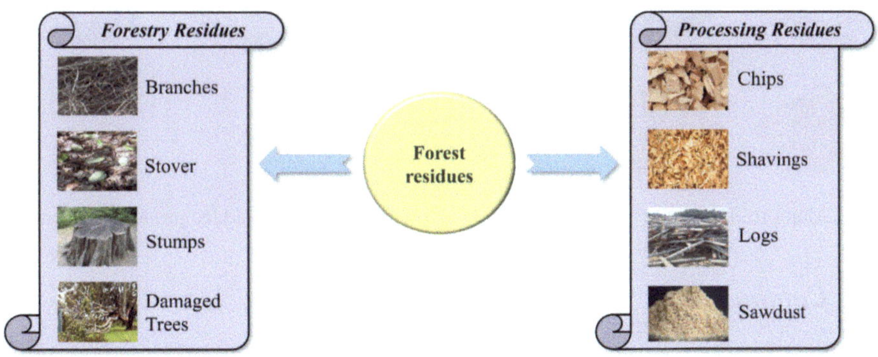

Fig. 3 Types of forest residue feedstocks

Table 2 Proximate and ultimate analysis of some common forestry residues [21, 22]

Residue	Proximate analysis (%)				Ultimate analysis (%)					Heating value (MJ/kg)
	Moisture	Volatile matter	Ash	Fixed carbon	C	H	N	S	O	
Wood bark	8.8	66.6	1.6	31.8	53.1	6.1	0.2	–	40.6	20.5
Wood chips	7.3	82.29	10.15	0.26	46.4	5.9	0.085	–	47.62	20.34
Softwood	8.8	70	1.7	28.1	52.1	6.1	0.2	–	41.6	20
Red oak wood	3.8	81.9	1.7	12.6	48.7	6.8	0.07	–	44	18.7
Pine chips	–	72.4[a]	5.95[a]	21.65[a]	49.66	5.67	0.51	0.076	38.1	19.79
Pine pruning waste	–	82.1[a]	2.67[a]	15.13[a]	50.55	6.12	0.45	–	40.2	19.19[b]
Willow	10.1	74.2	1.4	14.3	49.8	6.1	0.6	0.06	43.4	–
Christmas tree	37.8	46.1	3.2	12.9	54.5	5.9	0.5	0.42	38.7	–

[a]Calculated on dry basis. [b]LHV reported instead of HHV

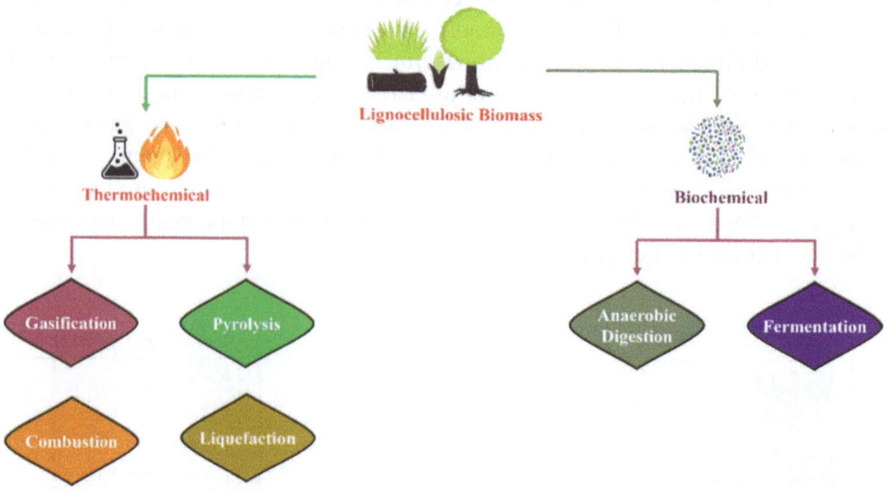

Fig. 4 Major routes for the conversion of biomass feedstocks to value-added products and energy

3.1 Combustion

Biomass combustion has for long been the predominant method for generating energy. For several millennia, mankind burned biomass for keeping warm, cooking food and seeking protection from predators. Even to this day, biomass combustion continues to be the largest contributor to global renewable energy production [25]. Biomass combustion has increasingly emerged as an attractive technology for heat and power generation for domestic as well as industrial sectors owing to the fact that it is a low-cost technology with high reliability whose power output can be adjusted (dispatchable source) and that can be operated over a wide range of scales ranging from a few kilowatts to hundreds of megawatts [26]. During combustion, biomass rapidly reacts with oxygen in either an open or controlled environment, resulting in the liberation of huge amounts of heat due to the exothermic nature of the process [27]. It is a highly complex process that involves a series of physical and chemical processes like drying, devolatilization, volatile burning, gasification and char combustion [28]. Generally, woody biomass materials such as briquettes, chips, fillings, logs, pellets, shavings and bark are preferred as fuels for combustion due to their lower ash content and nitrogen contents [29]. Herbaceous materials such as miscanthus and switch grass are also utilized as feedstocks for combustion; however, their utilization leads to increased emission of NO_x and SO_x gases and particulates in addition to greater ash formation, deposits and corrosion [29]. For combustion, biomass feedstocks with moisture contents lower than 50% are preferred [27] and an excess amount of air is used to achieve complete combustion to CO_2 and water [28]. Incomplete combustion, meanwhile, can lead to the emission of pollutants such as NO_x, SO_x, CO, tar and soot [25]. Of late, biomass co-firing with coal has also gained significant traction due to its potential to reduce the emission of greenhouse gases like CO_2 as well as SO_x and NO_x gases [30] while also providing a great opportunity for waste management through the utilization of energy. In the context of process technologies, for heat/power production at the domestic level, combustion is carried out using wood stoves, boilers, log boilers, pellet stoves and burners. Meanwhile, at the industrial level, combustion is carried out in fixed bed, fluidized bed and entrained flow reactors [26, 29].

3.2 Gasification

Gasification is one of the most popular and widely used thermochemical process technologies for generating energy from biomass owing to its high conversion efficiency [31]. It is a process where biomass is partially oxidized to a mixture of gases—H_2, CO, CO_2 and CH_4—commonly referred to as synthesis gas along with ash, char and tar at very high temperatures in the presence of a gasifying agent such as oxygen, air and steam [31]. Like combustion, gasification is a complex multi-step process during which biomass goes through four different steps: drying, pyrolysis, combustion and reduction [32]. The gaseous products can be further converted to either chemicals or liquid biofuels through routes such as the Fischer–Tropsch process or into heat and energy in power generation through combustion [33, 34]. The composition of the syngas produced is dependent on the gasifying agents used [33]. Air is the most commonly used gasifying agent due to its low cost and ease of availability [34] but it yields syngas of lower calorific value with just 8–14 vol% of H_2 and large amounts of nitrogen [31]. Using pure oxygen as the gasification agent yields syngas with the highest calorific value; however, the high cost of the air purification unit required to obtain pure oxygen limits its use [32]. Steam gasification, meanwhile, produces a high calorific value syngas with H_2 concentrations up to 60 vol% [31], as a result of which it has been increasingly preferred over other agents. Besides gasifying agents, low moisture content and high hemicellulose + cellulose contents of the biomass feedstock are also known to favour the production of high-quality syngas [34]. Depending upon the type of biomass fuel and its properties as well as operating conditions such as temperature and pressure, gasification can be carried out in fixed bed and/or moving bed, fluidized bed and entrained flow reactors with fixed bed reactors being preferred for commercial applications [32, 35].

3.3 Pyrolysis

Pyrolysis is a route that can not only be considered as an initial stage for other thermochemical technologies such as gasification and combustion but also as a separate process in its own right for the production of liquid biofuels as well as value-added chemicals [36]. Alongside gasification, pyrolysis is currently one of the most extensively investigated thermochemical routes and involves the thermal decomposition of biomass in the absence of oxygen from temperatures as low as 300 °C to as high as 800–900 °C [27, 28, 37]. At such high temperatures, the feedstock goes through an array of chemical reactions such as dehydration, decarbonylation, decarboxylation and depolymerization, to name a few, and ultimately gets converted to three products—biochar (solid), bio-oil (liquid) and non-condensable gases [38]. The product yields depend upon the structure and complexity of the biomass feedstock along with process conditions such as temperature, heating rate and residence time [39, 40]. Depending upon the heating rates and residence time, pyrolysis can be classified as slow, fast and flash [26, 37, 40]. Slow pyrolysis is often carried out at low temperatures at low heating rates and long residence times ranging from hours to days and yields biochar as the major product [25, 26]. Fast pyrolysis, meanwhile, uses higher temperatures as well as heating rates and shorter residence times in comparison to slow pyrolysis and, therefore, yielding bio-oil as the primary product [25, 26]. Flash pyrolysis uses extremely high temperatures (up to 1000 °C) and heating rates and very low residence times (<0.5 s) and gives non-condensable gases as the main product [25]. Among the three products, bio-oil has received the most attention and has been touted as a renewable drop-in fuel of the future [38]. It is a dark brown liquid with a smoky odour that contains hundreds of oxygenates belonging to different families such as acids, aldehydes, furans and phenols. Additionally, it possesses a considerable amount of water (up to 60 wt.%). Due to its complex chemical composition, high acidic and water content and low heating value and stability render bio-oil unusable directly for application and necessitate the requirement of extensive and expensive downstream upgradation [38, 41]. Upgradation techniques such as steam reforming [8, 42] and catalytic cracking [40, 43] have been developed for upgrading bio-oil into valuable resources such as H_2 and value-added chemicals and/or transportation fuels, respectively. The non-condensable gases—consisting mainly of CO, CO_2, CH_4 and H_2—can be used for power generation and production of value-added chemicals [38], while biochar can be used as an adsorbent for wastewater and also for soil amendment [40]. Lastly, in the context of reactor configurations, biomass pyrolysis has been investigated in a variety of reactor types such as fixed bed, fluidized bed, entrained flow, microwave, ablative, and auger reactors [26, 28, 40].

3.4 Hydrothermal Liquefaction

Hydrothermal liquefaction (HTL) is an emerging thermochemical route in which biomass undergoes decomposition in the presence of a solvent at temperatures ranging from 200 to 500–700 °C and pressures ranging from 5 to 35 MPa [44–46]. Similar to pyrolysis, this route also yields solid, liquid as well as gaseous products depending on the operating temperature. Solid products (hydrochar) are favoured at lower temperatures (180–250 °C), liquids (bio-crude/bio-oil) at moderate (250–370 °C) and gaseous products (syngas) at high temperatures (370–750 °C) [28, 45]. Thus, depending on the primary product, hydrothermal biomass processing technology can be classified as hydrothermal carbonization (HTC), HTL or hydrothermal gasification (HTG) [28, 45]. Among these, HTL has been touted as a promising technology and an alternative to pyrolysis for the production of high-quality oil. The most important advantage that HTL has over other commercially used technologies is that it can process almost any type of feedstock such as lignocellulosic biomass, algae and sewage sludge directly in wet form [46, 47]. Combustion, pyrolysis and gasification, on the other hand, use biomass feedstocks with moisture levels generally below 20% [47]. Consequently, HTL consumes a lesser amount of energy and represents a more efficient strategy for biomass conversion. Bio-crude, the main product obtained from HTL, is a dark brown, hydrocarbon-rich liquid that somewhat resembles crude oil in physical appearance and bio-oil in chemical make-up. Its carbon content and heating value are higher than that of pyrolysis bio-oil while its oxygen and water contents are lower than those of pyrolysis oil [25, 45, 46]. Owing to the presence of hydrocarbons, bio-crude has been proposed as a green alternative to crude oil, and can potentially be distilled to produce jet fuel, gasoline and diesel [46]. However, bio-crude may also contain a significant amount of nitrogen and sulphur in addition to oxygen, which coupled with complex chemical composition, necessitate the requirement of downstream upgradation and purification before it can be utilized for any purpose. Upgradation generally involves the removal of oxygen, nitrogen and sulphur through processes such as catalytic cracking, hydrodeoxygenation, hydrodenitrogenation and hydrodesulphurization [25, 44–46]. As with other thermochemical technologies, HTL is a complex process where biomass conversion to hydrochar, bio-crude and syngas occurs through steps such as decomposition, depolymerization, pyrolysis, reforming, gasification and repolymerization [44–46]. The yield and composition of bio-crude obtained with HTL are highly dependent on process parameters such as temperature, pressure, particle size, residence time and solvent used [44].

3.5 Anaerobic Digestion

Anaerobic digestion is an established and cost-effective biochemical approach that is capable of a variety of organic wastes such as lignocellulosic biomass, food wastes and wastewater [48]. In this process, organic wastes are decomposed by microorganisms in an oxygen-free environment, yielding biogas which mainly contains methane, carbon dioxide and some small amounts of nitrogen, oxygen, water vapour and hydrogen [49, 50]. In addition, it may also contain impurities such as ammonia and hydrogen sulphide [49, 50]. Apart from biogas, this process also yields a solid product called the digestate which refers to the decomposed or digested feed material. The digestate is rich in nutrients as it contains almost all of the nutrients present in the feed such as nitrogen, potassium and phosphorus and, therefore, can be used as a fertilizer [51]. The conversion of organic wastes takes place through four consecutive steps—hydrolysis, acidogenesis, acetogenesis and methanogenesis—each of which is catalyzed by a different set of microorganisms [50]. The amount of methane in biogas yield is used to evaluate the productivity of a particular feed material and is known to vary with the composition of biomass feed and its maturity, seasonal variation in feed, geographical location, inputs such as fertilizer and water [52]. Additionally, the ratio of carbon to nitrogen content of the feed also influences the yield of biogas, with the recommended values being between 20 and 30. Values lower than 20 result in increased ammonia production which inhibits the digestion process, while values greater than 30 lead to greater amounts of carbon in the digestate [27]. To enhance the digestion of biomass feedstock and biogas yield, the feedstock is exposed to pre-treatment strategies which could either be physical, chemical, biological or thermochemical [53]. Depending upon the concentration of the solids, current anaerobic digestion technology can be defined as wet (liquid), semi-dry (semi-solid) and dry (solid state) [48]. Up until now, anaerobic digestion was mostly carried out using a single type of feedstock, e.g., animal manure in small-scale biogas facilities. In recent years, the digestion of two or more feedstocks of different types in a single reactor has been noted to generate greater amounts of biogas [52]. In operational aspects, the yield of biogas is dependent on several variables such as temperature, pH, carbon-to-nitrogen ratio and hydraulic retention time [53].

3.6 Fermentation

Like anaerobic digestion, fermentation is a well-established biochemical conversion technology that has been used for the production of bioethanol, a potential biofuel that has been touted as an alternative to petroleum-derived transportation fuels [54].

Generally, any type of biomass feedstock that contains sugar can be used for the production of ethanol through fermentation [55]. The conversion of lignocellulosic biomass to bioethanol occurs in three steps: (i) pre-treatment, (ii) hydrolysis and (iii) fermentation [25, 54]. The recalcitrant nature of lignocellulosic biomass makes it immune to any chemical or microbial attack [7]. Consequently, pre-treatment is needed to open up its structure by fractionation and separation of the hemicellulose, cellulose and lignin fractions [55, 56]. The primary aim of the pre-treatment step is to delignify the biomass and gain access to the sugar fractions. Owing to the structural and chemical complexity of the biomass feedstock, this step is the most important and also expensive in the entire process. The pre-treatment strategies for this step can be categorized as physical, chemical, physicochemical and biological [55, 56]. In the next step, the polysaccharides in the pre-treated biomass are converted to their monomers in the presence of acids or enzymes which act as catalysts [54]. Acid hydrolysis can be carried out without prior pre-treatment of biomass feedstock using strong mineral acids such as hydrochloric acid and sulphuric acid. However, it leads to the corrosion of reaction equipment, and the generation of large amounts of wastes and inhibitors while requiring a neutralization step [54, 56]. Enzymatic hydrolysis is generally preferred as it can be carried out under mild conditions and high recovery of sugars [57]. In the last step called fermentation, the monomeric five- and six-carbon sugars are converted to a number of products like ethanol, acids and gases in the presence of microorganisms such as bacteria and yeast [56, 57]. As of now, there are four technologies for fermentation separate hydrolysis and fermentation, simultaneous hydrolysis and fermentation, and consolidated bioprocessing [54, 56, 57]. The fermentation process can be carried out in batch, fed-batch and continuous modes, with the fed-batch mode being preferred in industries. The process is affected by a number of parameters such as temperature, pH, tolerance of microorganisms to inhibitors, their growth rate and productivity [54, 56, 57]. Figure 5 displays the major biomass conversion technologies and their products.

4 Challenges, Possible Solutions and Future Perspectives

4.1 Availability of Biomass Feedstocks

The availability of a certain biomass feedstock is one of the most important factors to consider when evaluating its potential as a source of energy. This is because for future biorefineries set up for converting biomass to energy and/or valuable products to be economically viable and competitive, feedstock would need to be readily available at a low cost within a small radius [58]. Every year, millions of tonnes of agricultural crop and forestry residues are generated globally, and their production is only expected to increase with increasing agricultural and forestry activities. The major challenge here is to accurately map the utilization levels of these residues so that an estimate can be made for the surplus amounts that would be available for

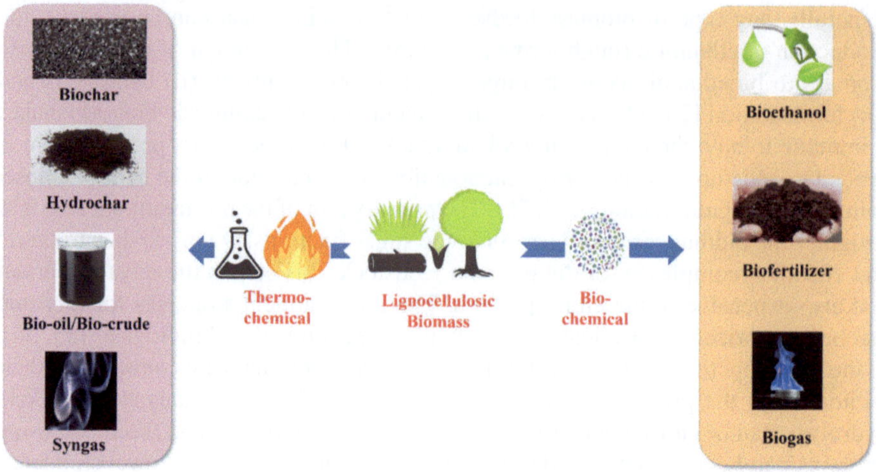

Fig. 5 Commonly used biomass conversion technologies and their products

producing chemicals, energy and/or fuels. Here, however, lies another problem which is the lack of any reliable data on the availability of various types of agro-forestry residues country-wise and globally. Furthermore, even in the case where availability has been reported, there is always some inconsistency between the data reported by various authors and/or agencies. Consequently, the first step in this direction should be to develop methods and/or tools for accurately assessing biomass potential across different geographies and countries.

4.2 *Biomass Supply Chain*

While there may be a lack of reliable data on biomass availability, the utilization of biomass in industries has been quite small in comparison to the estimates that have been suggested over the years for global biomass production. Consequently, this has led to a growing consensus that the logistics of maintaining a robust supply chain, rather than the availability, is a greater challenge for biorefineries [59]. In general, the supply chain would encompass collection, storage and transportation. Among these operations, collection is one that poses the most problems due to its decentralized nature, risks of feedstock being contaminated, and wide fluctuations in the quantity and quality of biomass [60]. Storage is yet another area that is likely to pose major problems. Biomass feedstocks, especially agricultural residues, tend to exhibit a great seasonal variation in quality and quantity. This in turn demands setting up storage facilities with large inventories capable of storing for long periods of time in order to ensure year-round operations of biorefinery with minimum disturbances [61]. Additionally, the low density of these materials and the high moisture content pose a major challenge during the collection and storage stages as well as transportation by reducing the loading capacity per batch [59]. The high cost of the biomass supply chain is one of the biggest barriers to the commercialization of biomass-based processes currently [62]. Therefore, it is imperative to bring down the costs of the logistical operations in order to boost the competitiveness and profitability of these processes. Constant supply of feedstock could be ensured by signing long-term binding contracts with farmers and landowners [59]. Collection could be simplified by tasking local entities or co-operatives with the collection and delivery of a fixed amount of biomass for a sustained time period [2]. Collected biomass can be densified through briquetting, pelletizing, etc. to reduce transportation costs and storage space. Exploring cheaper storage options, wherever possible, and drying the feedstocks could also help significantly down the costs of logistics while ensuring long-term storage without any losses [62]. Finally, the supply chain can be made more robust by selecting a biomass feedstock that is available year-round; however, it must also be stated here that this is rarely the case for any type of biomass [59].

4.3 Biomass Variability

In the past years, a tremendous amount of work has been done in improving harvesting, collection, storage and transportation of biomass feedstocks. However, there is one particular area that has not received much attention—variation in the quality of the biomass feedstock. Biomass materials are known to exhibit a wide variety of physical, chemical and structural heterogeneity, which greatly complicates their valorization [63]. The variation observed in biomass materials arises from a number of factors that include the type of biomass, storage and handling conditions, environmental factors and conditions during pre-processing operations [64]. Traditionally, any issues stemming from the quality of the selected biomass have been mitigated by reducing the costs of the logistical operations [65]. Among these properties, moisture is by far the most problematic one that causes a host of problems related to handling, grinding and feeding the biomass material. Ash content is yet another property that is known to influence the overall process by reducing conversion in most of the available conversion routes and affecting the composition and quality of products [63, 64]. The contents of cellulose, hemicellulose and lignin are also known to influence the performance of a particular biomass in the selected conversion route. Overall, these micro- and macro-scale attributes of biomass feedstocks influence strongly their reactivity in a myriad of ways, which in turn causes a huge variation in the composition, quality and/or yield of the products. As a consequence, it is imperative to minimize the effects of variation in feedstocks so that conversion processes can be carried out economically to function with minimal disturbances. In this direction, it has been proposed that physical and chemical pre-treatment strategies could be employed to improve feedstock properties as well as the quality of products [66]. The US Department of Energy (US DoE) has also proposed a concept called regionally distributed feedstock pre-processing depots [63, 67]. These depots would function as biomass collection for a single or multiple types of feed materials. They would also house pre-processing facilities where the collected biomass feedstocks would be physically or chemically treated to achieve the quality and composition required for a particular conversion technology [67]. Lastly, another strategy for mitigating the effects of variation in feed quality is that of blending, where high- and low-quality feedstocks can be mixed and/or blended in such a way that the effects of the undesirable components in the combined feed are reduced to the bare minimum [63].

4.4 Challenges Related to Conversion Technologies

Despite the extraordinary progress in the development of routes for biomass conversion in recent years, there exists plenty of room for improvement from both technological and economic points of view. While operation at ambient temperature and pressure conditions presents biochemical routes with a distinct advantage, most of these routes often require additional pre-processing [25, 68]. This pre-processing involves treatment with chemicals, during which some of the fractions of the biomass feedstock are separated out due to being unsuitable for valorization and disposed of as wastes [68]. Additionally, the pre-treatment steps consume a lot of energy which offsets the energy saving provided by mild operation conditions [68]. Meanwhile, thermochemical routes are much popular for biomass conversion from a commercial perspective due to their simple operation, and their ability to convert all biomass fractions [68] into a wide variety of products such as chemicals, and gaseous and liquid fuels in very short amounts of time. However, some of these routes such as pyrolysis and HTL are highly endothermic in nature, thereby hindering their commercialization [2]. Besides the issue of high energy requirement, another major issue is that of expensive product upgradation. Both pyrolysis and HTL produce bio-oil that requires extensive downstream upgradation at high temperatures and pressures in the presence of a catalyst. The syngas produced by gasification also needs downstream purification to separate out the nitrogen when air is used as a gasifying agent. Additionally, gasification also suffers from problems with tar formation, low H_2 yields as well as impurities such as the presence of NH_3 and H_2S in the produced syngas [32]. The problem of heterogeneity in biomass composition is also a major hurdle in the implementation of thermochemical technologies at commercial scales. In this direction, carrying out pyrolysis or HTL of a blended feedstock containing biomass and an additive that is also a waste or by-product such as plastics might prove to be an effective technique for producing high-quality bio-oil [69]. Physical and chemical pre-processing of biomass feedstocks could also be used for improving feedstock quality and producing upgraded bio-oil [41]. Using steam instead of the cheaper air as the gasification agent could yield higher yields of both H_2 as well as syngas [32]. Incorporating catalysts in gasification could also help produce high-quality syngas while lowering tar formation [32].

5 Industrial Perspectives on Current Conversion Technologies

The industrial outlook for today's biomass conversion technologies, encompassing both biochemical and thermochemical approaches, showcases a robust landscape of innovation, progress and potential. Thermochemical routes such as pyrolysis and gasification have witnessed remarkable advancement in recent years. They have benefitted from innovations in reactor design, catalyst development and heat recovery systems that have led to increased conversion efficiencies, as well as higher yields and consistent quality of target products. These advancements have led to the commercial deployments of these two routes, examples of which include BTG's Rotating Cone Reactor (RCR) technology for fast pyrolysis [70], Ensyn's Rapid Thermal Processing technology [70, 71], and biomass gasification combined heat and power (CHP) technologies of Nexterra and H3-CHP Clean Energy Systems [72, 73]. Similarly, biochemical routes such as anaerobic digestion and fermentation have greatly benefitted from genetic engineering, metabolic pathway optimization and novel bioreactor designs that have driven the creation of high-performance microorganisms. This in turn has led to more efficient and cost-effective conversion of substrate to biofuels, and biochemicals with higher yields and shorter processing times. Commercial technologies such as Aanergia's Organics Extrusion Press (OREX) [74] and bioethanol technologies developed by Abengoa and POET are a shining example of the feasibility of industrials-scale implementation of anaerobic digestion and fermentation [75, 76].

From an economic standpoint, the deployment of biomass conversion technologies is influenced by various factors such as feedstock availability, regulatory frameworks and market demand for bio-based products. While initial capital investments for infrastructure and equipment can be substantial, advancements in research and development can drive down costs and enhance overall process efficiency. In this direction, the integration of biomass conversion technologies within industrial infrastructure will prove crucial for cost reduction and scalability. Additionally, the economic feasibility of these technologies is also intimately linked to government incentives and policies around the world. Governments have implemented a spectrum of measures such as research grants, tax incentives and subsidies to accelerate the adoption of biomass-based technologies. Programs such as the United States' Renewable Fuel Standard, China's National Strategic Emerging Industries Development Plan and India's National Biofuel Policy are some examples of incentives that are aimed at driving the demand for biomass-based products and providing market stability. As governments increasingly align their policies with environmental goals, biomass-based technologies are poised to benefit greatly. However, the realization of their full potential for a greener industrial landscape hinges on cross-disciplinary collaboration between researchers—academic and industrial, economists and policy makers.

6 Conclusion and Future Potential

Widespread environmental damage due to the overutilization of fossil fuels and fears of their reserves running out in the near future has forced countries around the world to search for alternatives. Biomass has emerged as a promising alternative and will play a critical role as a source of renewable energy in the transition from a fossil-based economy to a sustainable, green economy. Lignocellulosic biomass materials such as agro-forestry residues present an exciting opportunity with twin benefits of waste management and feedstock for producing renewable chemicals, fuels and/or energy. Lately, tremendous progress has been made in developing strategies for converting lignocellulosic biomass. Despite this recent headway, the utilization of biomass remains rather low due to several challenges. As a result, major technological breakthroughs will be required if biomass-based industries are to become truly viable. In this regard, it would be absolutely crucial to develop policies for systematizing biomass collection, handling, storage and transport to have a robust supply chain. Emphasis needs to be placed on developing efficient and cost-effective pre-processing technologies in order to solve the problem of feedstock variation and ensuring that the quality of feed material meets the demands of the processes. Work also needs to be done on improving the currently available technologies for biomass conversion, in particular on enhancing their flexibility towards processing different types of feedstocks, conversion efficiencies and product quality. Improved product yields and quality would drastically bring down the huge investment needed for separation and further upgradation and in turn, boost the competitiveness of the processes. Finally, efforts also need to be made to assess environmental impact of the processes. While the technologies may still be in a nascent stage, the pace at which researchers around the world are working, a major breakthrough that would large-scale biomass utilization a reality in is not-so-distant future is not far off.

References

1. Joshi G, Pandey JK, Rana S, Rawat DS (2017) Challenges and opportunities for the application of biofuel. Renew Sustain Energy Rev 79:850–866. https://doi.org/10.1016/j.rser.2017.05.185
2. Tagade A, Kirti N, Sawarkar AN (2021) Pyrolysis of agricultural crop residues: an overview of researches by Indian scientific community. Bioresour Technol Rep 15:100761. https://doi.org/10.1016/j.biteb.2021.100761
3. BP, Statistical review of world energy 2022 (2022). https://www.bp.com/content/dam/bp/business-sites/en/global/corporate/pdfs/energy-economics/statistical-review/bp-stats-review-2022-full-report.pdf
4. Fiorentino G, Ripa M, Ulgiati S (2017) Chemicals from biomass: technological versus environmental feasibility. A review. Biofuels Bioprod Biorefin 11:195–214. https://doi.org/10.1002/BBB
5. Belyakov N (2019) Chapter Nineteen—Bioenergy. https://doi.org/10.1016/B978-0-12-817012-0.00033-5

6. Serrano-Ruiz JC, West RM, Dumesic JA (2010) Catalytic conversion of renewable biomass resources to fuels and chemicals. Annu Rev Chem Biomol Eng 1:79–100. https://doi.org/10.1146/annurev-chembioeng-073009-100935
7. Jaswal A, Singh PP, Mondal T (2022) Furfural-a versatile, biomass-derived platform chemical for the production of renewable chemicals. Green Chem 24:510–551. https://doi.org/10.1039/d1gc03278j
8. Singh PP, Jaswal A, Nirmalkar N, Mondal T (2023) Synergistic effect of transition metals substitution on the catalytic activity of LaNi0.5M0.5O3 (M = Co, Cu, and Fe) perovskite catalyst for steam reforming of simulated bio-oil for green hydrogen production. Renew Energy 207:575–587. https://doi.org/10.1016/j.renene.2023.03.057
9. National Science Foundation (2008) Breaking the chemical and engineering barriers to lignocellulosic biofuels, Washington DC, USA
10. Wang F, Ouyang D, Zhou Z, Page SJ, Liu D, Zhao X (2021) Lignocellulosic biomass as sustainable feedstock and materials for power generation and energy storage. J Energy Chem 57:247–280. https://doi.org/10.1016/j.jechem.2020.08.060
11. Dutta A, Patra A, Hazra KK, Nath CP, Kumar N, Rakshit A (2022) A state of the art review in crop residue burning in India: previous knowledge, present circumstances and future strategies. Environ Challenges 8:100581. https://doi.org/10.1016/j.envc.2022.100581
12. Bisht AS, Thakur NS (2019) Small scale biomass gasification plants for electricity generation in India: resources, installation, technical aspects, sustainability criteria & policy. Renew Energy Focus 28:112–126. https://doi.org/10.1016/j.ref.2018.12.004
13. Bhuvaneshwari S, Hettiarachchi H, Meegoda JN (2019) Crop residue burning in India: policy challenges and potential solutions. Int J Environ Res Public Health 16. https://doi.org/10.3390/ijerph16050832
14. Tripathi N, Hills CD, Singh RS, Atkinson CJ (2019) Biomass waste utilisation in low-carbon products: harnessing a major potential resource. NPJ Clim Atmos Sci 2. https://doi.org/10.1038/s41612-019-0093-5
15. Pattanaik L, Pattnaik F, Saxena DK, Naik SN (2019) Biofuels from agricultural wastes. Elsevier Inc. https://doi.org/10.1016/B978-0-12-815162-4.00005-7
16. Raj T, Kapoor M, Gaur R, Christopher J, Lamba B, Tuli DK, Kumar R (2015) Physical and chemical characterization of various Indian agriculture residues for biofuels production. Energy Fuels 29:3111–3118. https://doi.org/10.1021/ef5027373
17. Das N, Jena PK, Padhi D, Kumar Mohanty M, Sahoo G (2023) A comprehensive review of characterization, pretreatment and its applications on different lignocellulosic biomass for bioethanol production. Biomass Convers Biorefin 13:1503–1527. https://doi.org/10.1007/s13399-021-01294-3
18. Braghiroli FL, Passarini L (2020) Valorization of biomass residues from forest operations and wood manufacturing presents a wide range of sustainable and innovative possibilities. Curr For Rep 6:172–183. https://doi.org/10.1007/s40725-020-00112-9
19. Kizha AR, Han H-S (2015) Forest residues recovered from whole-tree timber harvesting operations. Eur J For Eng 1:46–55
20. Arâmburu A, Lunkes N, de Cademartori PHG, Gatto DA, Missio AL, Delucis RA (2022) Forestry wastes: technical concepts, economic circularity, and sustainability approaches. In: Jacob-Lopes E, Zepka LQ, Deprá MC (eds) Handbook of waste biorefinery, pp 369–415. https://doi.org/10.1007/978-3-031-06562-0_14
21. Vassilev SV, Baxter D, Andersen LK, Vassileva CG (2010) An overview of the chemical composition of biomass. Fuel 89:913–933. https://doi.org/10.1016/j.fuel.2009.10.022
22. Dhyani V, Bhaskar T (2018) A comprehensive review on the pyrolysis of lignocellulosic biomass. Renew Energy 129:695–716. https://doi.org/10.1016/j.renene.2017.04.035
23. Lee SY, Sankaran R, Chew KW, Tan CH, Krishnamoorthy R, Chu D-T, Show P-L (2019) Waste to bioenergy: a review on the recent conversion technologies. BMC Energy 1:1–22. https://doi.org/10.1186/s42500-019-0004-7
24. Tursi A (2019) A review on biomass: importance, chemistry, classification, and conversion. Biofuel Res J 6:962–979. https://doi.org/10.18331/BRJ2019.6.2.3

25. Osman AI, Mehta N, Elgarahy AM, Al-Hinai A, Al-Muhtaseb AH, Rooney DW (2021) Conversion of biomass to biofuels and life cycle assessment: a review. Springer International Publishing. https://doi.org/10.1007/s10311-021-01273-0
26. Solarte-Toro JC, González-Aguirre JA, Poveda Giraldo JA, Cardona Alzate CA (2021) Thermochemical processing of woody biomass: a review focused on energy-driven applications and catalytic upgrading. Renew Sustain Energy Rev 136. https://doi.org/10.1016/j.rser.2020.110376
27. Awasthi MK, Sar T, Gowd SC, Rajendran K, Kumar V, Sarsaiya S, Li Y, Sindhu R, Binod P, Zhang Z, Pandey A, Taherzadeh MJ (2023) A comprehensive review on thermochemical, and biochemical conversion methods of lignocellulosic biomass into valuable end product. Fuel 342:127790. https://doi.org/10.1016/j.fuel.2023.127790
28. Elgarahy AM, Hammad A, El-Sherif DM, Abouzid M, Gaballah MS, Elwakeel KZ (2021) Thermochemical conversion strategies of biomass to biofuels, techno-economic and bibliometric analysis: a conceptual review. J Environ Chem Eng 9:106503. https://doi.org/10.1016/j.jece.2021.106503
29. Nussbaumer T (2003) Combustion and co-combustion of biomass: fundamentals, technologies, and primary measures for emission reduction. Energy Fuels 17:1510–1521. https://doi.org/10.1021/ef030031q
30. Xu Y, Yang K, Zhou J, Zhao G (2020) Coal-biomass co-firing power generation technology: current status, challenges and policy implications. Sustainability 12. https://doi.org/10.3390/su12093692
31. Narnaware SL, Panwar NL (2022) Biomass gasification for climate change mitigation and policy framework in India: a review. Bioresour Technol Rep 17:100892. https://doi.org/10.1016/j.biteb.2021.100892
32. Tezer Ö, Karabağ N, Öngen A, Çolpan CÖ, Ayol A (2022) Biomass gasification for sustainable energy production: a review. Int J Hydrogen Energy 47:15419–15433. https://doi.org/10.1016/j.ijhydene.2022.02.158
33. Nunes LJR (2022) Biomass gasification as an industrial process with effective proof-of-concept: a comprehensive review on technologies, processes and future developments. Results Eng. 14:100408. https://doi.org/10.1016/j.rineng.2022.100408
34. Mishra S, Upadhyay RK (2021) Review on biomass gasification: gasifiers, gasifying mediums, and operational parameters. Mater Sci Energy Technol 4:329–340. https://doi.org/10.1016/j.mset.2021.08.009
35. Akbarian A, Andooz A, Kowsari E, Ramakrishna S, Asgari S, Cheshmeh ZA (2022) Challenges and opportunities of lignocellulosic biomass gasification in the path of circular bioeconomy. Bioresour Technol 362:127774. https://doi.org/10.1016/j.biortech.2022.127774
36. Wang G, Dai Y, Yang H, Xiong Q, Wang K, Zhou J, Li Y, Wang S (2020) A review of recent advances in biomass pyrolysis. Energy Fuels 34:15557–15578. https://doi.org/10.1021/acs.energyfuels.0c03107
37. Zhang X, Brown RC (2019) Introduction to thermochemical processing of biomass into fuels, chemicals, and power. https://doi.org/10.1002/9781119417637.ch1
38. Kumar R, Strezov V, Weldekidan H, He J, Singh S, Kan T, Dastjerdi B (2020) Lignocellulose biomass pyrolysis for bio-oil production: a review of biomass pre-treatment methods for production of drop-in fuels. Renew Sustain Energy Rev 123. https://doi.org/10.1016/j.rser.2020.109763
39. Azizi K, Keshavarz Moraveji M, Abedini Najafabadi H (2018) A review on bio-fuel production from microalgal biomass by using pyrolysis method. Renew Sustain Energy Rev 82:3046–3059. https://doi.org/10.1016/j.rser.2017.10.033
40. Amenaghawon AN, Anyalewechi CL, Okieimen CO, Kusuma HS (2021) Biomass pyrolysis technologies for value-added products: a state-of-the-art review. Springer Netherlands. https://doi.org/10.1007/s10668-021-01276-5
41. Kumar R, Strezov V (2021) Thermochemical production of bio-oil: a review of downstream processing technologies for bio-oil upgrading, production of hydrogen and high value-added products. Renew Sustain Energy Rev 135:110152. https://doi.org/10.1016/j.rser.2020.110152

42. Singh PP, Nirmalkar N, Mondal T (2022) Catalytic steam reforming of simulated bio-oil for green hydrogen production using highly active LaNixCo1−xO3 perovskite catalysts. Sustain Energy Fuels 6:1063–1074. https://doi.org/10.1039/d1se01786a
43. Bridgwater AV (2012) Review of fast pyrolysis of biomass and product upgrading. Biomass Bioenerg 38:68–94. https://doi.org/10.1016/j.biombioe.2011.01.048
44. Mishra RK, Kumar V, Kumar P, Mohanty K (2022) Hydrothermal liquefaction of biomass for bio-crude production: a review on feedstocks, chemical compositions, operating parameters, reaction kinetics, techno-economic study, and life cycle assessment. Fuel 316:123377. https://doi.org/10.1016/j.fuel.2022.123377
45. Yang C, Wang S, Yang J, Xu D, Li Y, Li J, Zhang Y (2020) Hydrothermal liquefaction and gasification of biomass and model compounds: a review. Green Chem 22:8210–8232. https://doi.org/10.1039/d0gc02802a
46. Basar IA, Liu H, Carrere H, Trably E, Eskicioglu C (2021) A review on key design and operational parameters to optimize and develop hydrothermal liquefaction of biomass for biorefinery applications. Green Chem 23:1404–1446. https://doi.org/10.1039/d0gc04092d
47. Siwal SS, Sheoran K, Saini AK, Vo DVN, Wang Q, Thakur VK (2022) Advanced thermochemical conversion technologies used for energy generation: advancement and prospects. Fuel 321:124107. https://doi.org/10.1016/j.fuel.2022.124107
48. Sawatdeenarunat C, Surendra KC, Takara D, Oechsner H, Khanal SK (2015) Anaerobic digestion of lignocellulosic biomass: challenges and opportunities. Bioresour Technol 178:178–186. https://doi.org/10.1016/j.biortech.2014.09.103
49. Bajpai P (2020) Biomass conversion processes. https://doi.org/10.1016/b978-0-12-818400-4.00005-0
50. Wainaina S, Awasthi MK, Sarsaiya S, Chen H, Singh E, Kumar A, Ravindran B, Awasthi SK, Liu T, Duan Y, Kumar S, Zhang Z, Taherzadeh MJ (2020) Resource recovery and circular economy from organic solid waste using aerobic and anaerobic digestion technologies. Bioresour Technol 301:122778. https://doi.org/10.1016/j.biortech.2020.122778
51. Ma H, Guo Y, Qin Y, Li YY (2018) Nutrient recovery technologies integrated with energy recovery by waste biomass anaerobic digestion. Bioresour Technol 269:520–531. https://doi.org/10.1016/j.biortech.2018.08.114
52. Kunatsa T, Xia X (2022) A review on anaerobic digestion with focus on the role of biomass co-digestion, modelling and optimisation on biogas production and enhancement. Bioresour Technol 344:126311. https://doi.org/10.1016/j.biortech.2021.126311
53. Bandgar PS, Jain S, Panwar NL (2022) A comprehensive review on optimization of anaerobic digestion technologies for lignocellulosic biomass available in India. Biomass Bioenerg 161:106479. https://doi.org/10.1016/j.biombioe.2022.106479
54. Devi A, Bajar S, Kour H, Kothari R, Pant D, Singh A (2022) Lignocellulosic biomass valorization for bioethanol production: a circular bioeconomy approach. Bioenergy Res 15:1820–1841. https://doi.org/10.1007/s12155-022-10401-9
55. Lin Y, Tanaka S (2006) Ethanol fermentation from biomass resources: current state and prospects. Appl Microbiol Biotechnol 69:627–642. https://doi.org/10.1007/s00253-005-0229-x
56. Sharma B, Larroche C, Dussap CG (2020) Comprehensive assessment of 2G bioethanol production. Bioresour Technol 313:123630. https://doi.org/10.1016/j.biortech.2020.123630
57. Balat M (2011) Production of bioethanol from lignocellulosic materials via the biochemical pathway: a review. Energy Convers Manag 52:858–875. https://doi.org/10.1016/j.enconman.2010.08.013
58. Thompson DN, Campbell T, Bals B, Runge T, Teymouri F, Ovard LP (2013) Chemical preconversion: application of low-severity pretreatment chemistries for commoditization of lignocellulosic feedstock. Biofuels 4:323–340. https://doi.org/10.4155/bfs.13.15
59. Valdivia M, Galan JL, Laffarga J, Ramos JL (2016) Biofuels 2020: biorefineries based on lignocellulosic materials. Microb Biotechnol 9:585–594. https://doi.org/10.1111/1751-7915.12387

60. Hassan SS, Williams GA, Jaiswal AK (2019) Lignocellulosic biorefineries in Europe: current state and prospects. Trends Biotechnol 37:231–234. https://doi.org/10.1016/j.tibtech.2018.07.002
61. Liu Z, Wang S, Ouyang Y (2017) Reliable biomass supply chain design under feedstock seasonality and probabilistic facility disruptions. Energies 10. https://doi.org/10.3390/en10111895
62. Rentizelas AA, Tolis AJ, Tatsiopoulos IP (2009) Logistics issues of biomass: the storage problem and the multi-biomass supply chain. Renew Sustain Energy Rev 13:887–894. https://doi.org/10.1016/j.rser.2008.01.003
63. Li C, Aston JE, Lacey JA, Thompson VS, Thompson DN (2016) Impact of feedstock quality and variation on biochemical and thermochemical conversion. Renew Sustain Energy Rev 65:525–536. https://doi.org/10.1016/j.rser.2016.06.063
64. Williams CL, Westover TL, Emerson RM, Tumuluru JS, Li C (2016) Sources of biomass feedstock variability and the potential impact on biofuels production. Bioenergy Res 9:1–14. https://doi.org/10.1007/s12155-015-9694-y
65. Kenney KL, Smith WA, Gresham GL, Westover TL (2013) Understanding biomass feedstock variability. Biofuels 4:111–127. https://doi.org/10.4155/bfs.12.83
66. Wagle A, Angove MJ, Mahara A, Wagle A, Mainali B, Martins M, Goldbeck R, Paudel SR (2022) Multi-stage pre-treatment of lignocellulosic biomass for multi-product biorefinery: a review. Sustain Energy Technol Assess 49
67. Lamers P, Roni MS, Tumuluru JS, Jacobson JJ, Cafferty KG, Hansen JK, Kenney K, Teymouri F, Bals B (2015) Techno-economic analysis of decentralized biomass processing depots. Bioresour Technol 194:205–213. https://doi.org/10.1016/j.biortech.2015.07.009
68. García-Velásquez CA, Cardona CA (2019) Comparison of the biochemical and thermochemical routes for bioenergy production: a techno-economic (TEA), energetic and environmental assessment. Energy 172:232–242. https://doi.org/10.1016/j.energy.2019.01.073
69. Dewangan A, Pradhan D, Singh RK (2016) Co-pyrolysis of sugarcane bagasse and low-density polyethylene: influence of plastic on pyrolysis product yield. Fuel 185:508–516. https://doi.org/10.1016/j.fuel.2016.08.011
70. Czernik S, Bridgwater AV (2004) Overview of applications of biomass fast pyrolysis oil. Energy Fuels 18:590–598. https://doi.org/10.1021/ef034067u
71. Ensyn (n.d.) Ensyn RTP technology overview. http://www.ensyn.com/technology.html#:~:text=Ensyn'sRTPprocessisthe,ingredients%2Cchemicalsandheatingfuels
72. H3-CHP Clean Energy Systems (n.d.) Biomass gasification technology. https://h3chp.com/gasification/
73. Nexterra (n.d.) Nexterra biomass gasification CHP technology. http://www.nexterra.ca/files/gasification-technology.php
74. Anaergia (n.d.) OREX. https://www.anaergia.com/solutions/municipal-solid-waste/organics-extraction/
75. Abengoa (n.d.) Energy recovery from waste and biomass. https://www.abengoa.com/web/en/negocio/energia/residuos/
76. POET (n.d.) Biothanol. https://poet.com/bioethanol

Vegetable Oil Seeds as Biomass Resource for Potential Applications in Biofuel and Biomaterials

Gandhi Bhukya and Shiva Shanker Kaki

Abstract The ever-increasing awareness about the environmental and global warming problems has also led us to increase the use of alternative and sustainable sources for energy and material-related applications. Apart from energy applications, the use of sustainable resources has also gained attention in other potential applications worldwide, especially in industries involved in the production of biobased fuels and biomaterials. Among the sources used for biomaterials and related products, the use of vegetable oil seeds as a biomass resource is of great prominence due to the availability of a wide range of sources and varieties of plant-based seeds. The present chapter focuses on the utilization of castor seed oil and its products (an agricultural-based feedstock) as a biomass source for a wide range of oleochemical products for potential applications ranging from biolubricants to biomaterials.

Keywords Sustainable resources · Biomass · Vegetable oil seeds · Castor · Biolubricants · Biomaterials

1 Introduction

The environmental concerns have raised the interest to focus on biobased products that have ecological compatibility, lower toxicity and renewability compared to fossil petrochemical-derived products [1]. This has now become the most prioritized field of material science, and the development of any material especially for energy applications is now screened for its production criterion to be based on renewable resources of which biomass is of prominence [2]. It is well known that any renewable organic material that is derived from plants and animals is called biomass. The use of biomass-derived fuel for energy and transport applications is gaining more attention due to its sustainable nature and also because of the increasing environmental concerns and harmful emissions due to the use of non-renewable fossil fuels [3].

G. Bhukya · S. S. Kaki (✉)
Department of Oils, Lipid Science & Technology, CSIR-Indian Institute of Chemical Technology, Hyderabad, Telangana 500007, India
e-mail: sskaki@iict.res.in

S. Kumar et al. (eds.), *Clean Energy Transition-via-Biomass Resource Utilization*, Green Energy and Technology, https://doi.org/10.1007/978-981-97-6321-4_4

The use of sustainable feedstock is the need of the hour as it plays an important role in environmental, economic and social studies with sustainability being a growing field because of its interdisciplinary nature. Among the biomass reserves of nature, oil seeds are one of the important sources of biomass which are essentially natural products derived from plants that not only serve as the main source of energy but also provide a wide range of platform chemicals or oleochemicals [4]. The seeds from plants and trees are considered more sustainable and renewable bio-feedstocks compared to animal-based fats. The fruits yield seeds which are composed of a seed coat as an outer layer and kernel are present inside which is the main source for fixed oil. It is important to differentiate between fixed oil and essential oil as an essential oil is generally extracted by steam distillation and some other techniques from various plant parts like leaves, flowers, roots and other aerial parts except from seeds. Essential oils are volatile and possess aroma which is characteristic of the plant from which it is extracted and are used in cosmetics, food flavors, fragrances, perfumes and pesticides [5, 6]. On the other side, seed oils from seeds are extracted by solvent or by mechanical crushing and the oil obtained is generally non-volatile. After oil extraction from the seed, the deoiled cake is also utilized for a number of applications as a source of protein, ash, minerals, compost and also as a fertilizer. The seed coat as such and the seed ash are also exploited for the preparation of heterogeneous catalysts [7].

2 Seed Oils

The oil which is extracted from the kernel of the seeds is the main source of fatty acids, phytosterols, phospholipids, waxes, vitamins, etc. all of which have both edible and industrial applications. The main building blocks of vegetable oils are fatty acids, and chemically vegetable oils are composed of a mixture of triglycerides as a major component. These triglycerides have three molecules of fatty acids esterified to the three hydroxy functionalities of glycerol forming a triacylglycerol structure. Fatty acids in vegetable oils have different chain lengths and unsaturation and often differ in composition between different plant species. Figure 1 shows the representative structure of triglyceride and some commonly found selected fatty acids in vegetable oils.

The seed oils are classified under the biomass from field and plantation, whereas the oil cake is classified as biomass from industrial processing. Especially forest-based seed oils are considered as a renewable biomass source for nutritional, fuel and industrial applications [8]. The increasing usage of vegetable oils which are natural agricultural feedstocks for energy uses and as oleochemicals would cause a decrease in the emission of greenhouse gases that will help in the reduction of environmental pollution [9]. One of the growing applications of vegetable oil in the clean energy sector is the use of biodiesel as an alternative fuel to fossil fuels [10]. The biodiesel or the alkyl ester of fatty acids is produced by the transesterification of vegetable or animal oils and fats which results in the formation of biodiesel and

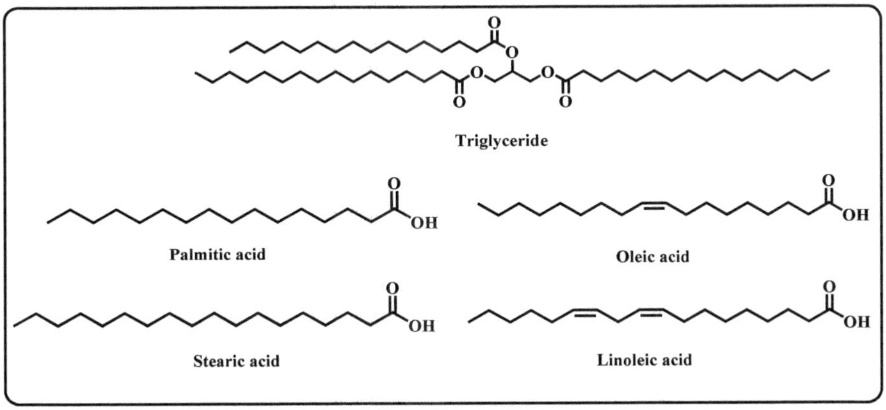

Fig. 1 Structures of triglyceride and some common fatty acids

glycerol as shown in Fig. 2. Chemically, biodiesel is composed of alkyl esters of long-chain fatty acids esterified to alcohols like methanol or ethanol in most cases.

When compared to diesel fuel, biodiesel has favorable characteristics like a lower level of toxic emissions, being renewable in nature, biodegradable and there is no need to make changes in the existing diesel engines [11]. These features can contribute in some magnitude to mitigate the climate change caused by fossil fuel sources. However, as of now improvements have to be made in terms of the cost of biodiesel by scouting the process of production with more feedstocks that are economical [12]. Another biofuel derived from triglycerides of vegetable oil seeds and animal fats is hydro-treated vegetable oil which is also a renewable diesel fuel but is different from biodiesel [13]. Though biodiesel and hydro-treated vegetable oil diesel are derived from vegetable oil feedstock, the process of making them is different as biodiesel is produced by transesterification of oil resulting in glycerol as a by-product which is not produced in hydro-treated vegetable oil diesel. Hydro-treated vegetable oil fuel is produced by the catalytic hydrocracking of vegetable oil where hydrogen is used to convert the triglyceride into glycerides (mono and diglycerides) and fatty acids as the intermediates. These intermediates when subjected to a pressure of 3 MPa and

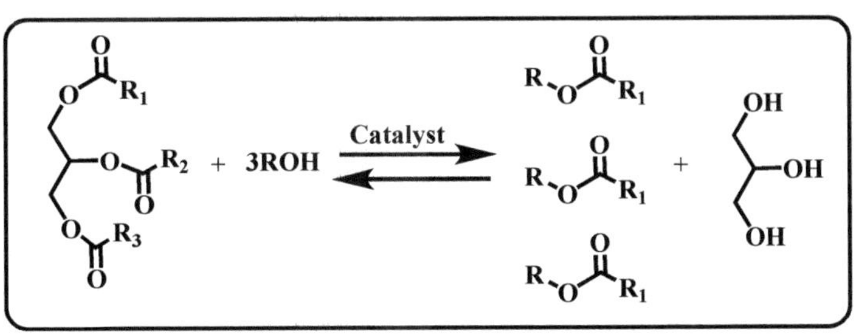

Fig. 2 Transesterification of triglycerides with alcohol

temperature range of 300–360 °C undergo hydrodeoxygenation, decarboxylation and decarbonylation resulting in alkanes as the main fuel products along with water, carbon dioxide and carbon monoxide as side-products [14]. Hydro-treated vegetable oil fuel is gaining interest as a sustainable fuel as it is primarily derived from seed oil-based biomass and has better low-temperature properties and higher oxidative stability compared to biodiesel. However, lubricity is the only parameter that is better for biodiesel compared to hydro-treated vegetable oil and studies suggest that the lubricity can be modified by employing suitable additives [15].

Globally oleochemicals are viewed as renewable, natural, organic, green, safe and biodegradable substrates by researchers and consumers which adhere to the principles of Green Chemistry especially # 7. However, the food versus fuel requirements must be addressed as the over-usage of edible seed oils should not compromise the food security options as the seed oils are sources for a number of nutrients in addition to providing energy. In view of these aspects, tree-borne oil seeds and other seeds that cannot be used for edible purposes are being considered for sustainable fuel and specialty chemical applications. One such renewable feedstock is castor oil which is a non-edible oil that is commercially available as a renewable resource for the chemical industry. Due to their distinctive properties and extensive range of applications, castor oil and its derivatives are among the most popular and important vegetable oils. This chapter provides a brief review of the production of oleochemical products with biofuel and biomaterial applications from castor oil and its products as biobased feedstock with a focus on the reports from the year 2000 onwards. There are many reports and reviews available on vegetable oils as a source of different fatty acids, squalene, phytosterols, carotenoids and other minor constituents that have a wide range of applications in many industries like pharmaceutical, cosmetic and biodiesel sectors [16]. Hence, the present chapter is divided into two parts where the first part focuses on castor oil and its characteristics with studies of castor-based products for energy applications like biolubricant and the second part describes the biomaterial applications focusing on oleochemical derivatives from castor oil including some recent studies on the biological activity studies of a few novel small molecules derived from castor fatty acids.

3 Castor Oil

Castor (*Ricinus communis* L.) is one of the non-edible oil seed crops that has unique characteristics that are important for the chemical industry. The plant usually grows up to a height of nearly 12 m and is smaller in height in temperate zones depending on the climatic conditions. The leaves of the castor plant are green with a peculiar star shape growing between 10–40 cm, and the flowers are reddish or yellowish green and found as clusters at the tip of stem branches. The seeds are found in the pod or capsule of the stem which appears red, purple or green with a soft thorny outer covering (Fig. 3).

Fig. 3 Castor plantation field

The castor plant can be grown on lands with low water availability, and it can withstand diverse environmental conditions which makes castor an attractive candidate with immense potential. Due to the constant pressure for renewable fuels and biomass resource utilization, castor has become one of the potential sources of biofuel and plays a major role in the agricultural economy by earning foreign exchange through the export of castor seeds and oil. India is a top global producer of castor oil followed by China and Brazil, and it is estimated that 85% percent of the global castor seed output is produced by India. From the recent reports, it is evident that, in the world market, there has been a steady increase in the demand for castor oil and its products [17]. The mature and dried castor seeds after harvesting are subjected to extraction processes like mechanical extraction and other extraction methods for obtaining the oil. In terms of environmental factors, extraction by mechanical method like hydraulic pressing is more suitable as it involves less investment, reduced cost and is safer compared to other methods of extraction of oil. However, for recovering the maximum amount of oil from the seeds, it is suggested to use a combination of the hydraulic press method and solvent extraction processes [18]. The oil extracted is crude and is reported to contain free fatty acids, partial glycerides, phospholipids and other minor components which have to be removed by refining involving different steps that are well documented [19]. The deoiled meal or cake is known to contain a toxic alkaloid ricin that has to be detoxified in order to be used in other applications. The refined oil that has been processed can now be employed for industrial uses such as polymers, lubricants, cosmetics and pharmaceuticals.

Castor oil is known to be used therapeutically as medicinal oil for centuries due to its laxative effects [20, 21]. Castor oil is reported to have nauseant properties which can be due to its fatty acid composition and hence is classified as non-edible oil. It is reported that the seed contains 35–55% oil and plays a vital role in the Indian vegetable oil economy. Eco-friendliness, low costs, biodegradability, non-competition with food and the renewable nature of castor oil can be considered as the main reasons for this huge demand. It is now estimated that castor oil has more than 700 industrial applications [22].

Extraction of castor oil from castor seeds in a laboratory is usually carried out by solvent extraction using Soxhlet extraction with a suitable solvent. The oil content of castor varies from 35–55% depending on the geographical region. It is observed that the average oil content in castor seeds of Indian origin is about 45–50%. The solvent extraction process for castor oil on a lab scale is shown in Fig. 4.

The fatty acid composition is determined by gas chromatography after the oil is converted to its fatty acid methyl esters. The fatty acid composition of castor oil is well studied and is composed of fatty acids like palmitic acid (C16:0), stearic acid (C18:0), oleic acid (C18:1), linoleic acid (C18:2) and ricinoleic acids (C18:1; OH), and the fatty acid distribution is shown in Table 1.

It can be observed that ricinoleic acid is the major fatty acid in castor seed oil compared to other fatty acids. The chemistry of castor oil is primarily focused on ricinoleic acid because of its high concentration in the oil and the presence of three functional groups in the acid (Fig. 5). Because ricinoleic acid has an ester linkage, a double bond and a hydroxy group, it is a good renewable resource for numerous chemical reactions, modifications and transformations that can be carried out for the generation of different platform chemicals.

The separation of nonhydroxy fatty acids is advantageous as it gives pure ricinoleic acid which can be used for further processing to produce value-added products that have multiple applications [22]. The process of separation of methyl ricinoleate from other fatty acid methyl esters of castor oil is shown in Fig. 6. Castor oil is initially converted to its fatty acid methyl esters, and these castor oil methyl esters are further separated into nonhydroxy fatty acid methyl esters and ricinoleic acid methyl ester by solvent-based partition method using hexane and aqueous methanol system to obtain pure methyl ricinoleate [23].

The oil's adaptability in the manufacture of a wide range of castor oil-based items depends on these three functionalities. For example, the carboxylic group can produce a variety of esterification and amidation products [24, 25] whereas the single point of unsaturation can be altered via hydrogenation, epoxidation or vulcanization. To increase the oil's potential, the hydroxy functional group at carbon-12 can be acetylated, alkoxylated or removed by dehydration. The different transformations of castor oil and its fatty acids produce modified chemical entities that can have many

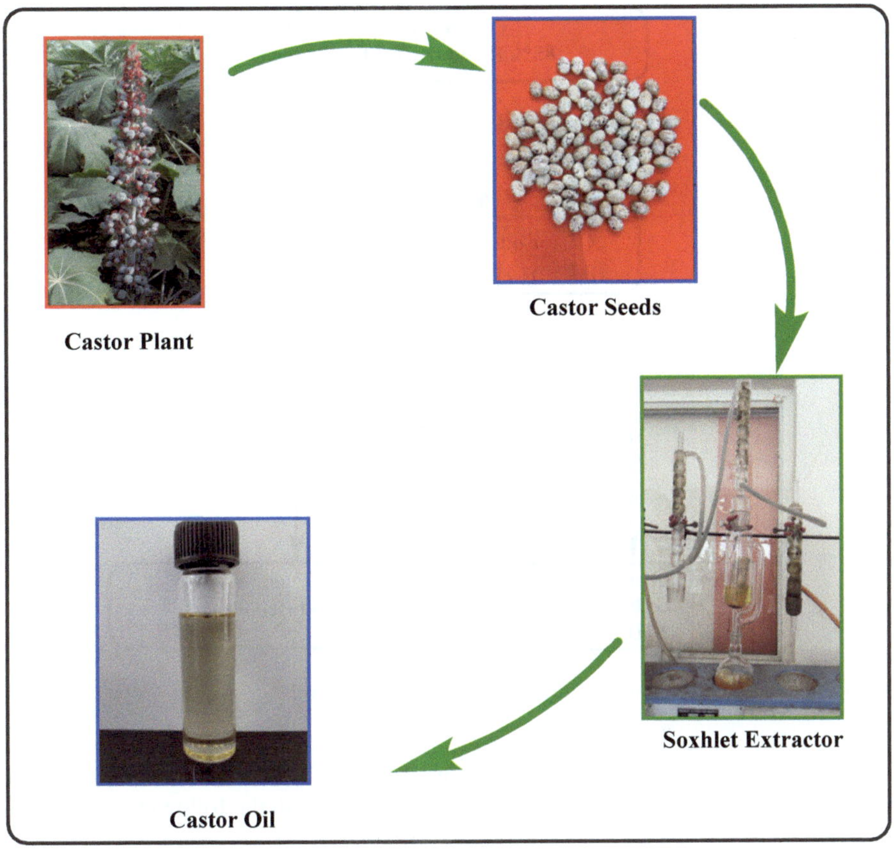

Fig. 4 Castor plant, seeds and oil extraction

Table 1 Fatty acid composition of castor oil

Fatty acid	%
Palmitic acid (C16:0)	1–1.5
Stearic acid (C18:0)	1–1.6
Oleic acid (C18:1)	3.0
Linoleic acid (C18:2)	3.5
Ricinoleic acid (18:1; OH)	85–90

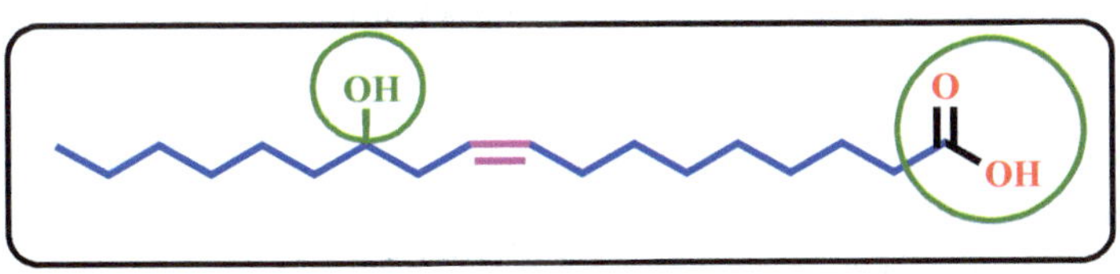

Fig. 5 Structure of ricinoleic acid showing the three functionalities

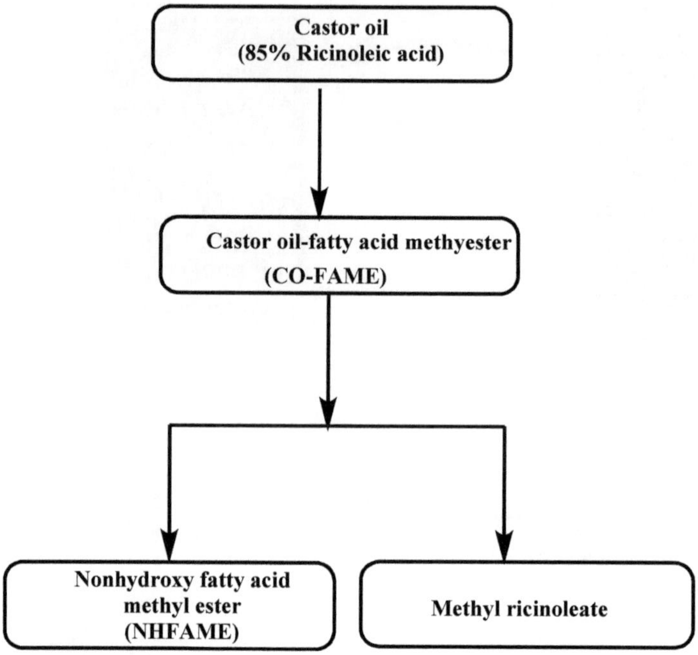

Fig. 6 Process for separation of methyl ricinoleate from castor oil

properties. In the present chapter, the focus is on the biolubricant and biomaterial aspects of castor-based compounds which are discussed in the following sections.

4 Biolubricant Applications of Castor

Lubricants are one of the energy-related materials that are extensively used in our daily lives. Lubricants are used to reduce the friction between components that are in contact and the base oil constitutes about 75–90% of the lubricant [26]. Due to their high demand and applications in both industry and our day-to-day activities, lubricants are released into the environment for several reasons causing pollution of air, water and soil. This detrimental effect can be addressed by using lubricants which are biodegradable and can reduce the impact of environmental pollution caused by the use of fossil fuel-derived lubricants. As discussed earlier, the need of the hour is the protection of environment from the petroleum-based oil pollution because almost 50% of fossil fuel-derived oils end up in the environment. This can be addressed either by recycling the petroleum-derived oils or by using biobased lubricants which are environmentally friendly or derived from natural materials like vegetable oils and fats [27]. Biobased lubricant oil or a biolubricant should have the characteristics of being biodegradable and non-toxic to the environment, with renewable nature and have reduced or no emissions of greenhouse gases [28]. It is reported that the current use of such biolubricants is limited, and the demand is found to be increasing due to environmental regulations and also due to the disposal-related issues of fossil fuel lubricants [29]. In this context, lubricants derived from vegetable oils due to their renewable nature and other positive fuel properties can be one of the top choices as alternate lubricant sources with sustainable potential. In addition to these environmentally friendly properties, vegetable oils have lubricity, low volatility and narrow viscosity ranges which help them to adhere to the surface of metals. The most commonly used biomass or plant sources for biolubricants are from soybean, palm, rapeseed, sunflower and coconut. However, the choice of source for biolubricants should not affect the food sector, and hence focus now is being shifted to used cooking oils, non-edible oils and other tree-borne seed oils [30, 31]. This is highly important in countries like India where the import of edible oils is more than 55% and hence edible oils cannot be used for other applications. In this context, castor oil is usually preferred for industrial applications over other edible oils. Castor oil, being a non-edible oil, was extensively studied for its lubricant applications including biodiesel. It is noteworthy to mention that castor oil is ideal for lubricant applications, and it has better low-temperature viscosity properties compared to other vegetable oils. The viscosity of castor oil is higher compared to other vegetable oils which is beneficial as it gives better lubrication, but on the other side it limits the application in higher speed-related operations. To make castor oil work in such conditions, it was mixed with vegetable oils like corn oils, soybean oil and rice bran oil to modify the rheological properties [32]. It was found that a blend of equal proportions of castor oil and soybean oil resulted in enhanced rheological properties for grinding application.

A detailed study on castor oil biodiesel and its blends with petrodiesel was carried out in order to know the potential of castor oil in biodiesel applications. The study showed that castor oil biodiesel exhibited high density and kinematic viscosity as per the American Society for Testing and Materials (ASTM) specifications which can limit the applications of castor oil biodiesel [33]. However, there are some limitations to using vegetable oils as lubricants in terms of oxidative and thermal stabilities which can be addressed by the modification of the substrate oils. For lubricant applications and to improve the physicochemical characteristics, vegetable oils are subjected to modifications like epoxidation, transesterification, esterification, hydrogenation, etc. [34]. In addition, the hydroxy functionality in ricinoleic acid makes castor oil more oxidatively stable compared to other common oils which is considered advantageous for the production of base oil for lubricant applications. The castor fatty acids were modified to acyloxy polyol ester derivatives with a variety of different alcohols followed by the epoxidation of the unsaturation to yield base oils with different acyl chain lengths. The chemistry in the preparation of these biolubricant base stocks is shown in Scheme 1. Briefly, castor oil is hydrolyzed to obtain the fatty acids, and in this case ricinoleic acid is the major fatty acid in castor oil and hence chemical modification is described on ricinoleic acid. Ricinoleic acid is acylated with acid anhydrides of different chain lengths to obtain the acyloxy derivatives. These acyloxy derivatives are then esterified with different alcohols to obtain the acylated polyol

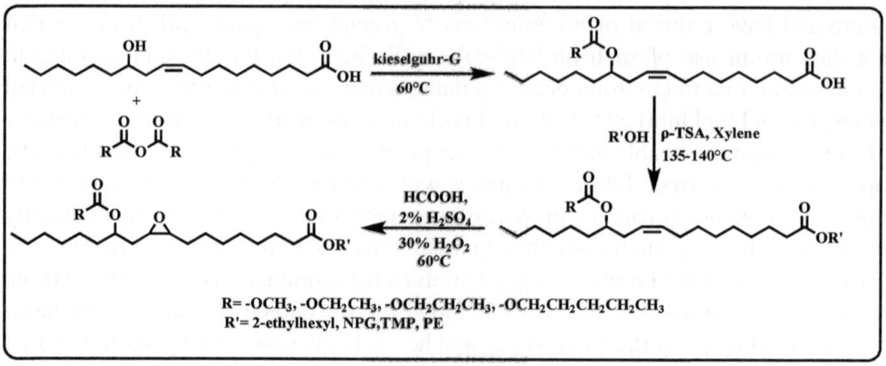

Scheme 1 Route for the preparation of epoxy acylated polyol esters of ricinoleic acid

esters of ricinoleic acid. Further, the acylated polyol esters were subjected to the epoxidation of the double bond to obtain the epoxy acylated polyol esters of ricinoleic acid as the final base oil components [35].

The generated products of different acyl chain lengths were found to exhibit lower pour points and very high load-carrying capacity. These castor base oils are anticipated to have potential applications for marine and aviation lubricant formulations.

5 Biomaterial Applications of Castor

Biomaterials are defined as materials produced or derived from plants, bacteria, animals, fungi and other life forms which can interact with biological systems and are intended to be used for biomedical applications. The seeds of plants serve as renewable biomass as they provide vegetable oils and fatty acids which are exploited as building blocks for the synthesis of a variety of biomaterials and polymers. Though fatty acids were known for such studies, they are not fully exploited for their role as potent biomolecules which shows that they can be explored in the fields of biomaterials [36]. The role of fatty acids for their antimicrobial activity was known from older times but was diminished by the invention of new antibiotics. However, the increasing antibacterial resistance of microbes to many of the existing antibiotics has directed the attention toward new lipid-based antimicrobials that have fewer side effects with low toxicity. Due to the emergence of antimicrobial resistance and the demand for biocompatible bioactives, research groups are now focusing on the design of biomaterials based on fatty acids as sustainable alternatives to other materials.

The vegetable oils from various crops like sesame, soybean, grape, olive and castor were reported to exhibit biomedical properties like antibacterial, anti-tumor and anti-inflammatory activities. These vegetable oils serve as potential substrates for producing biodegradable polymers and biomaterials that can have a reduced carbon footprint [37]. It is reported that fatty acids and their derivatives are receiving increased importance in developing biomaterials for applications in the biomedical field due to their renewable, biocompatible nature and non-specific mode of action [38]. However, edible oils like sunflower, coconut and ground oils cannot be utilized in the production of biomaterials as the demand for edible oil is considerably high for domestic consumption. Therefore, there is a need for moving to alternative sources for the production of biomaterials. In this regard, castor oil has a long history of being a viable source for the production of various biomaterials. In recent years, castor-derived fatty acid-based materials for biomedical applications have been increasingly reported. Especially ricinoleic acid which can undergo various transformations has been studied for its derivatives for biological applications [39]. Ricinoleic acid derivatives were reported to exhibit biological activities like antimicrobial, anticancer, analgesic and anti-inflammatory properties [40]. The castor-based polymeric materials and their composites gained a lot of interest as they are biocompatible and are anticipated to have other desired biological activities like antimicrobial activity, promising cell-biomaterial interactions and processability [41]. They are also finding extensive applications in the biomedical field as wound healing agents, bioadhesives and scaffolds for cardiac and bone tissue engineering.

Polyurethanes are one of the important classes of materials that are extensively studied in recent years as they find extraordinary applications in tissue engineering and biomedical engineering.

However, the practical application of polyurethanes is a major threat to producers as well as to consumers as they are composed of volatile organic compounds and precarious air pollutants.

To overcome this issue, waterborne polyurethane dispersions were created by dispersing polyurethanes in an aqueous medium as a stable phase in order to partially or entirely replace solvent-based polyurethanes for various industrial and biomedical applications [42]. Recent studies demonstrated that the majority of the raw materials (polyols, isocyanates and chain extenders) used in polyurethane dispersions (PUDs) come from petrochemical feedstock, which is viewed as a major source of hazardous air pollutants and a non-renewable resource [43]. In this regard, polyurethanes derived from castor oil are regarded as biodegradable materials which are reported to show improvised mechanical, chemical and thermal properties. In addition, they are also reported to exhibit high wear resistance and show excellent compatibility with blood and tissue [44]. They are also reported to be used as scaffolds which are employed in seeding the isolated cells and allow the cell to grow and proliferate where it forms a tissue construct and further this tissue construct will be implanted in the patients [45]. In recent studies, castor oil triglyceride was employed for the synthesis of cationic PU dispersions (Scheme 2), and it was identified that the hydrophobic nature of triglycerides and long fatty acid chains offered the resulting PU films with enhanced physicochemical properties. It is also reported that the synthesized PU dispersions have demonstrated promising activity, and these castor oil-based cationic waterborne polyurethanes with good mechanical and antibacterial properties can be added to paints for pathogen control or utilized as coatings for medical equipment and food processing [46].

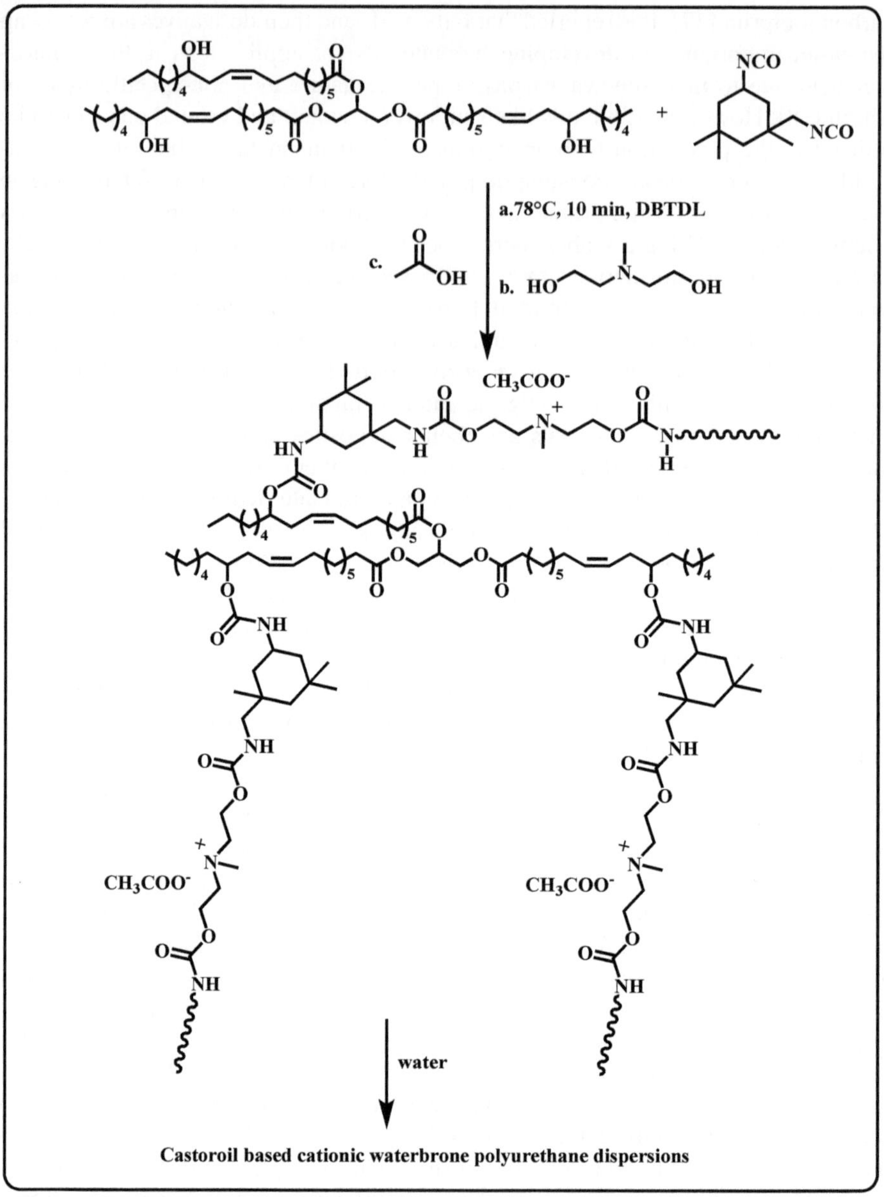

Scheme 2 Synthetic route for castor oil-based cationic waterborne polyurethane dispersions

Similarly, cyclic carbonates are being studied extensively both in industry and academia due to their versatile nature as green solvents and sustainable features [47]. They are reported as potential substitutes for polyurethanes that are used in coating, paint and biomaterial applications. Due to their green and sustainable nature and growing interest, an attempt was made to prepare new biobased cyclic carbonates from castor-derived ricinoleic acid with Lewis acids and without the use of carbon dioxide for potential applications in biodegradable polymers [48].

Likewise, the use of polyamides in biomedical and environmental applications has been increasingly reported due to their excellent biodegradability and biocompatibility. In general, the polycondensation of N,N-bis(2-hydroxy ethyl) fatty amides generated by base-catalyzed aminolysis of different oils is employed for the synthesis of vegetable oil-based poly(ester amide) resin. Many reports are available in the literature where different oils like linseed, *Pongamia glabra* and *Mesua ferrea* were utilized for the synthesis of poly(ester amide) resin, but none of them exhibited a better performance characteristic [49]. In this regard, castor oil-based biodegradable poly(ester amide) resin was synthesized as shown in Scheme 3 where it was reported to enhance biocompatibility and biodegradability properties [50]. The in vitro studies revealed the microbial specificity of the aforementioned synthetic poly(ester amide) resin where Gram-negative bacteria were found to be more effective than Gram-positive bacteria at degrading synthetic polymers through bacterial mediation.

As mentioned earlier, this part of the chapter also discusses the usage of two important castor oil-derived fatty acids, namely ricinoleic acid and 10-undecenoic acid and their derivatives for the preparation of new chemical entities with biological applications that can have the potential for biomaterial preparation and applications. These compounds can act as potential bioactive lipidic substrates that can have applications as ingredients for making biomaterials. Ricinoleic acid is well-explored for the synthesis of various polymeric materials like polyesters, polyamides and polyurethane. Polymers composed of ricinoleic acid find their applications as drug delivery agents. They are extensively studied for delivering various therapeutic drugs including Rotenone, Domperidone and Cisplatin. For example, a novel copolymer prepared by melt polycondensation of poly(ethylene glycol) and castor-derived fatty acids like ricinoleic and sebacic acids was evaluated for controlled release of anticancer drug, Paclitaxel. The synthetic route for the copolymer preparation is shown in Scheme 4. It was observed that the drug-release rate can be controlled by tuning the ratio of ricinoleic acid and sebacic acid in the amphiphilic polymer, which consequently affects the crystallinity of the micellar core [51].

In addition, other bioactives like phenolic lipids were synthesized from ricinoleic acid and its derivatives for potential applications in the cosmetic and food industries. Especially, a variety of antioxidants were prepared using methyl ricinoleate and 12-hydroxystearic acid methyl ester to obtain new molecules with antioxidant activity [52]. The hydroxy group of methyl ricinoleate was modified to obtain (Z)-methyl-12-aminooctadec-9-enoate that was employed as a substrate for the synthesis of phenolic lipid derivatives for potential bioactive applications [53].

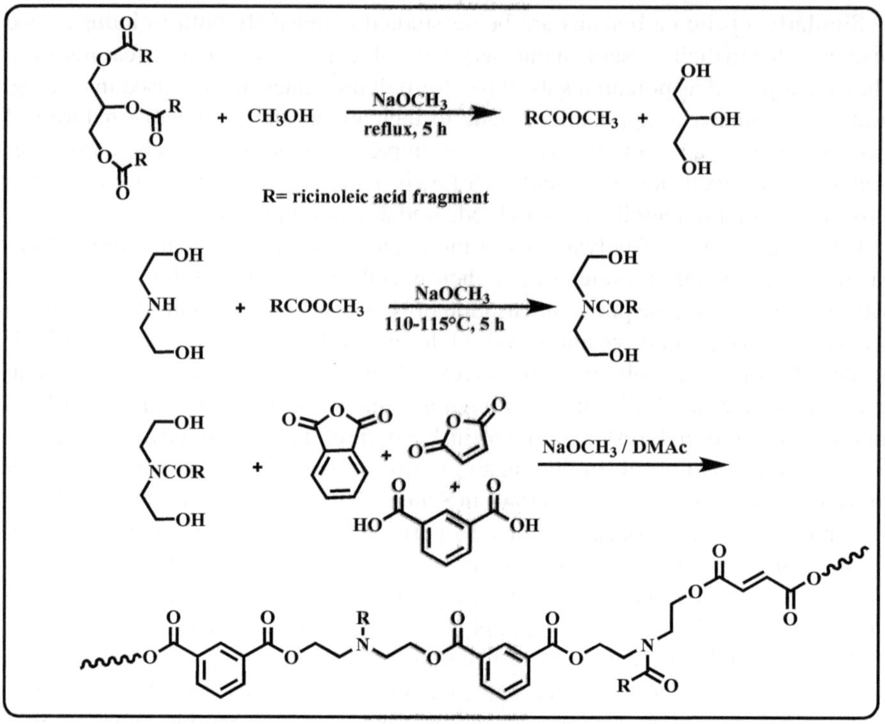

Scheme 3 Synthesis of biodegradable poly(ester amide) resin from castor oil

10-undecenoic acid or undecylenic is another important fatty acid derived from castor oil and is known for its application in many fields like polymers, as an intermediate in organic synthesis, antimicrobial compounds, etc. [54]. Especially the polymers obtained from 10-undecenoic acid like poly(thioether-ester) are reported to have potential biomedical applications. Poly(thioether-ester) with lower levels of toxicity was synthesized by employing renewable sources like 10-undecenoic acid and isosorbide (Scheme 5) and evaluated for its drug delivery potential. The biocompatibility nature of the poly(thioether-ester) nanoparticles makes them ideal candidates that can be employed as biomedical nanomaterials for drug delivery applications [55].

10-undecenoic acid is widely known for its antifungal properties and is reported to be used in various nonspecific topical antifungal agents which are majorly employed in treating superficial infections. Studies have shown that undecylenate-based creams which contain zinc undecylenate and undecylenic acid were more efficient than other creams at treating fungal infections. The non-irritating nature combined with potent antifungal activity makes it superior to the other available antifungal agents [56]. Many new derivatives were prepared from 10-undecenoic acid with a view to develop new potential antimicrobial and anti-biofilm agents [57]. 10-undecenoic acid is also reported to be employed in the synthesis of various insecticides (Fig. 7)

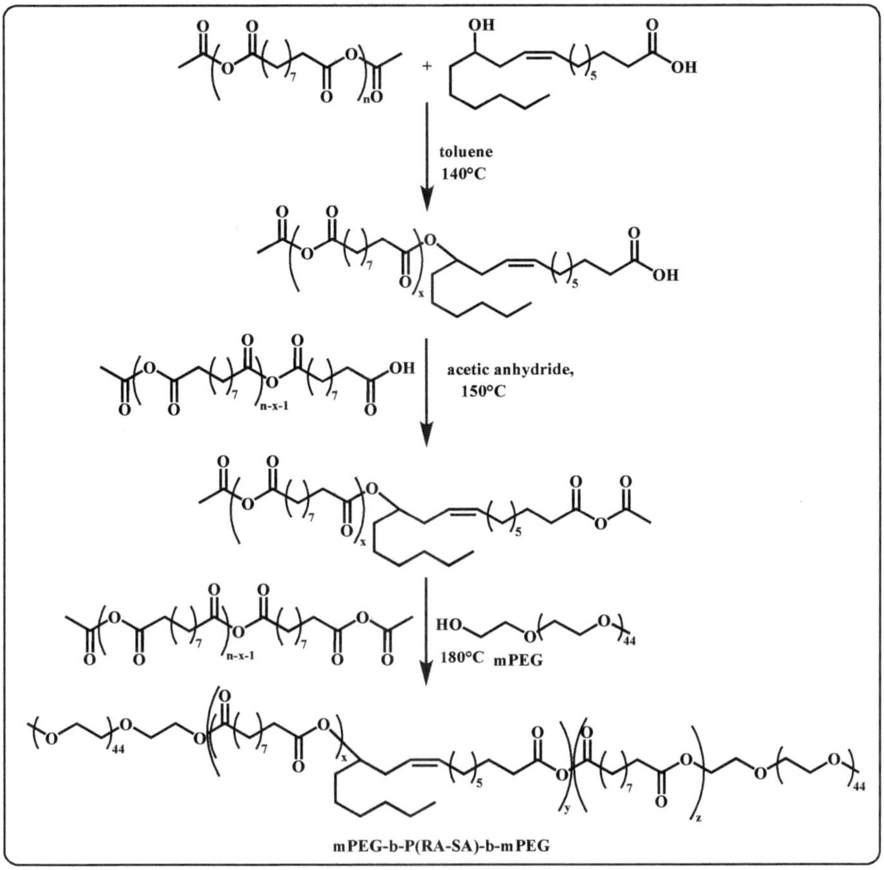

Scheme 4 Synthesis of Poly(ethylene glycol)-b-P(ricinoleic acid-sebacic acid)-b- Poly(ethylene glycol)

where it serves as the starting material in the synthesis of various epoxide hydrolase inhibitors [58].

10-undecenoic acid derivatives were also reported to show antifungal and antibacterial activities on plant pathogens. In a recent report, it was shown that 10-undecenoic acid-derived triazoles (Scheme 6) demonstrated antifungal and antibacterial activities against plant pathogens such as *Ralstonia solanacearum* and *Fusarium oxysporum* [59].

Similarly, 11-bromoundecanoic acid which is a derivative of 10-undecenoic acid is reported to serve as a crucial intermediate for the synthesis of various bioactive molecules. Various reports are available in the literature where 11-bromoundecanoic acid was valorized by employing it as a substrate for the synthesis of bioactive quaternary ammonium compounds, 1-monoacylglycerol and many other polymeric materials. In a recent report, it was shown that monoacylglycerols derived from

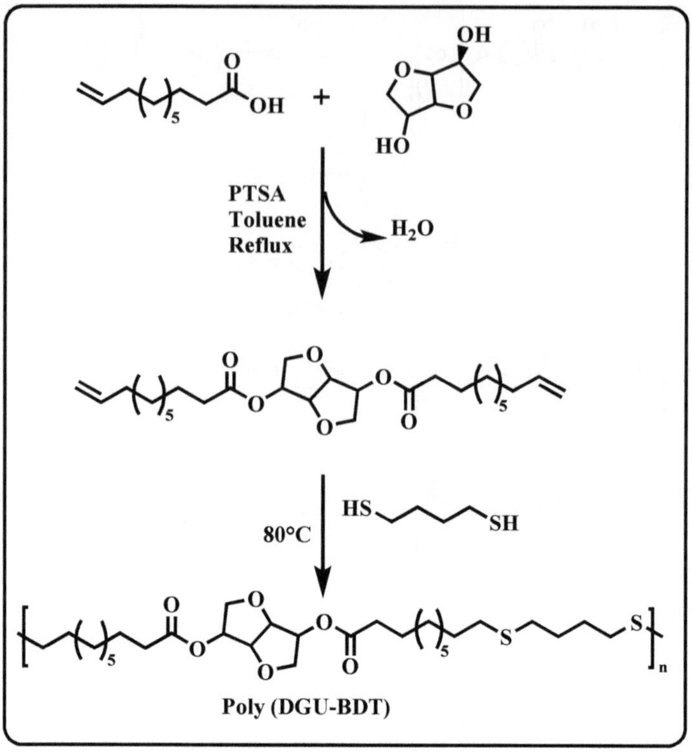

Scheme 5 Synthesis of poly(dianhydro-d-glucityl diundec-10-enoate-butanedithiol)

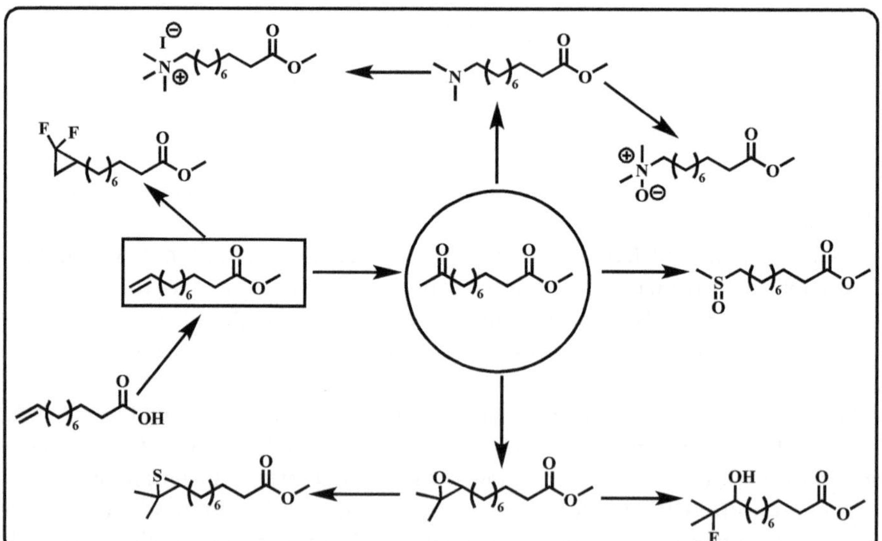

Fig. 7 Synthesis of epoxide hydrolase inhibitors

Scheme 6 Synthesis 10-undecenoic acid-based triazoles

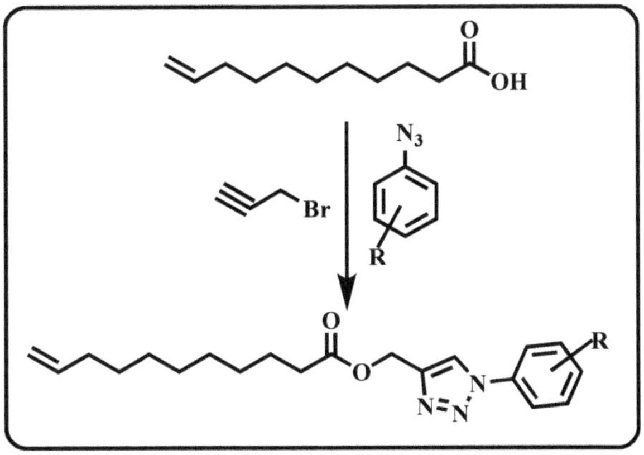

11-bromoundecanoic with phenolic acids (Fig. 8) exhibited potent antioxidant, antimicrobial and cytotoxic potential [60].

Similarly, 11-bromoundecanic acid is also employed as the starting material for the synthesis of oxadiazole containing 1-monoacylglycerols. The reported 11-bromoundecanic acid-based 1-monoacylglycerols exhibited potential antimicrobial activity, and they were also reported to show promising cytotoxic potential [61].

The topics covered in this chapter provide an overview of some areas of potential applications of derivatives obtained by the modification of castor-based fatty acids. Castor oil remains the most promising choice as a renewable biomass source for developing many new biomaterials in the future. It is the most valuable and sustainable resource for many industries like energy, chemical and pharmaceutical, agrochemical and biofuel sectors that can benefit from castor seed oil.

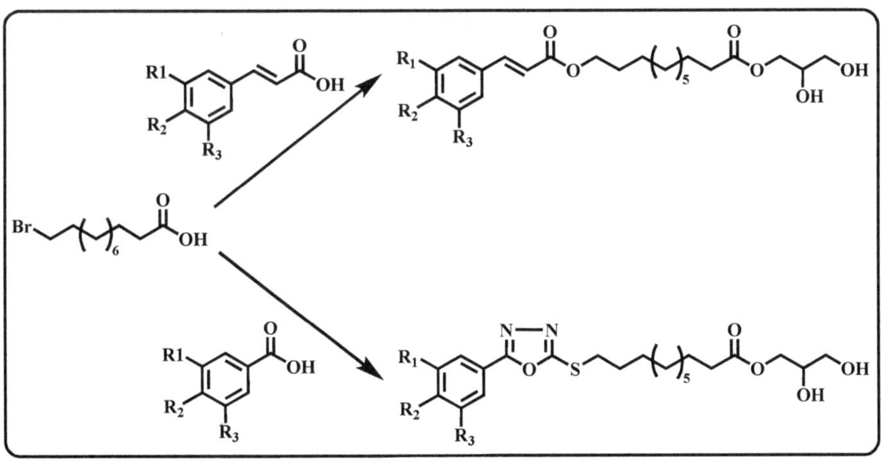

Fig. 8 New bioactive derivatives from 11-bromoundecanoic acid

6 Conclusion

The single point strategy to mitigate the problems of depriving fossil fuels and to minimize the emissions of greenhouse gases is by the usage of bio-derived materials for fuels, energy and material applications. It can be achieved by replacing the substrates that are presently being used for the production of fuels and biomaterials. By doing so, at least some part of the burden is borne by these sustainable alternative sources which can address the emission of undesirable gases which can result in a cleaner environment compared to fossil fuels. In this regard, vegetable oils not only serve as the substrate for biobased fuels and their derivatives but also as a good source for the development of biomaterials that have applications in polymers and biomedical fields. The use of vegetable oils is likely to grow as they can be the green materials that have the potential to replace the non-renewable sources with castor oil having a promising future for making sustainable biomaterials.

Acknowledgements The authors express sincere thanks to the Director, CSIR-IICT (Ms. No. IICT/Pubs./2023/262), for supporting to contributing to this chapter and providing all the required facilities.

References

1. Sumathi S, Chai SP, Mohamed AR (2008) Utilization of oil palm as a source of renewable energy in Malaysia. Renew Sustain Energy Rev 12:2404–2421
2. Awasthi MK, Sarsaiya S, Patel A, Juneja A, Singh RP, Yan B, Awasthi SK, Jain A, Liu T, Duan Y, Pandey A (2020) Refining biomass residues for sustainable energy and bio-products: an assessment of technology, its importance, and strategic applications in circular bio-economy. Renew Sustain Energy Rev 127:109876
3. https://www.eia.gov/energyexplained/biomass/
4. Biermann U, Bornscheuer U, Meier MA, Metzger JO, Schäfer HJ (2011) Oils and fats as renewable raw materials in chemistry. Angew Chem Int Ed Engl 50:3854–3871
5. Tongnuanchan P, Benjakul S (2014) Essential oils: extraction, bioactivities, and their uses for food preservation. J Food Sci 79:1231–1249
6. Kant R, Kumar A (2022) Review on essential oil extraction from aromatic and medicinal plants: techniques, performance and economic analysis. Sustain Chem Pharm. 30:100829
7. Mares EKL, Goncalves MA, da Luz PTS, da Rocha Filho GN, Zamian JR, da Conceição LRV (2021) Acai seed ash as a novel basic heterogeneous catalyst for biodiesel synthesis: optimization of the biodiesel production process. Fuel 299:120887
8. Anjaneyulu B, Kaki SS, Kanjilal S, Reddy JRC, Ravinder T, Prasad RBN, Rao BVSK (2017) Physico-chemical characterization and biodiesel preparation from Ailanthus excelsa seed oil. Energy Sour A Recovery Util Environ Eff 39:811–816
9. Callegari A, Bolognesi S, Cecconet D, Capodaglio AG (2020) Production technologies, current role, and future prospects of biofuels feedstocks: a state-of-the-art review. Crit Rev Environ Sci Technol 50:384–436
10. Chen SY, Lao-Ubol S, Mochizuki T, Abe Y, Toba M, Yoshimura Y (2014) Production of Jatropha biodiesel fuel over sulfonic acid-based solid acids. Bioresour Technol 157:46–350
11. Kaki SS, Prasad RBN (2014) Base catalyzed transesterification for biodiesel: novel catalyst options. In: Raghavan KV, Reddy BM (eds) Industrial catalysis and separations. Taylor & Francis Group, p 437

12. Osorio-González CS, Gómez-Falcon N, Sandoval-Salas F, Saini R, Brar SK, Ramírez AA (2020) Production of biodiesel from castor oil: a review. Energies 13:2467
13. No SY (2014) Application of hydrotreated vegetable oil from triglyceride based biomass to CI engines–a review. Fuel 115:88–96
14. Mikulec J, Cvengroš J, Joríková Ľ, Banič M, Kleinová A (2010) Second generation diesel fuel from renewable sources. J Clean Prod 18:917–926
15. Zeman P, Hönig V, Kotek M, Táborský J, Obergruber M, Mařík J, Hartová V, Pechout M (2019) Hydrotreated vegetable oil as a fuel from waste materials. Catalysts 9:337
16. Górnaś P, Rudzińska M (2016) Seeds recovered from industry by-products of nine fruit species with a high potential utility as a source of unconventional oil for biodiesel and cosmetic and pharmaceutical sectors. Ind Crops Prod 83:329–338
17. Shridhar BS, Beena KV, Anita MV, Paramjeet KB (2010) Optimization and characterization of castor seed oil. Leornard J Sci 17:59–70
18. Kemper TG (2005) Edible oil and fat products: processing technologies. In: Shahidi F (eds) Bailey's industrial oil and fat products. Wiley & Sons, pp 57–98
19. Mutlu H, Meier MA (2010) Castor oil as a renewable resource for the chemical industry. Eur J Lipid Sci Technol 2010(112):10–30
20. Scarpa A, Guerci A (1982) Various uses of the castor oil plant (Ricinus communis L.) a review. J Ethnopharmacol 5:117–137
21. Tunaru S, Althoff TF, Nüsing RM, Diener M, Offermanns S (2012) Castor oil induces laxation and uterus contraction via ricinoleic acid activating prostaglandin EP3 receptors. Proc Natl Acad Sci 109:9179–9184
22. Bhukya G, Kaki SS (2022) Design and synthesis of sebacic acid from castor oil by new alternate route. Eur J Lipid Sci 124:2100244
23. Berdeaux O, Christie WW, Gunstone FD, Sebedio JL (1997) Large-scale synthesis of methyl cis-9, trans-11-octadecadienoate from methyl ricinoleate. J Am Oil Chem Soc 74:1011–1015
24. G. R. O'Shea Company. www.groshea.com, August 2015
25. Deshpande DP, Haral SS, Gandhi SS, Ganvir VN (2012) Transesterification of castor oil. ISCA J Eng Sci 1:2–7
26. Asuquo JE, Anusiem ACI, Etim EE (2012) Extraction and characterization of rubber seed oil. Int. J. Mod. Chem. 1:109–115
27. Erhan SZ, Sharma BK, Liu Z, Adhvaryu A (2008) Lubricant base stock potential of chemically modified vegetable oils. J Agric Food Chem 56:8919–8925
28. Mobarak HM, Mohamad EN, Masjuki HH, Kalam MA, Al Mahmud KAH, Habibullah M, Ashraful AM (2014) The prospects of biolubricants as alternatives in automotive applications. Renew Sustain Energy Rev 33:34–43
29. Cecilia JA, Ballesteros Plata D, AlvesSaboya RM, Tavares de Luna FM, CavalcanteJr CL, Rodríguez-Castellón E (2020) An overview of the biolubricant production process: challenges and future perspectives. Processes 8:257
30. Singh Y, Farooq A, Raza A, Mahmood MA, Jain S (2017) Sustainability of a non-edible vegetable oil based bio-lubricant for automotive applications: a review. Process Saf Environ Prot 111:701–713
31. Correa DF, Beyer HL, Possingham HP, Thomas-Hall SR, Schenk PM (2017) Biodiversity impacts of bioenergy production: microalgae vs. first generation biofuels. Renew Sustain Energy Rev 74:1131–1146
32. Kannan IS, Karthikeyan S, Ravikumar K (2020) Lubrication performance of castor oil blended with other vegetable oils in grinding of Inconel 625. In: IOP conference series: materials science and engineering, vol 923. IOP, p 012027
33. Berman P, Nizri S, Wiesman Z (2011) Castor oil biodiesel and its blends as alternative fuel. Biomass Bioenergy 35:2861–2866
34. Kamalakar K, Mahesh G, Prasad RB, Karuna MS (2015) A novel methodology for the synthesis of acyloxy castor polyol esters: low pour point lubricant base stocks. J Oleo Sci 64:1283–1295
35. Azmeera T, Kotte K, Vidavalur S, Karuna MSL, Madhu D (2022) Epoxy acyloxy castor polyol esters: multifunctional base oil for aviation to marine formulations. Pet Chem 62:1273–1282

36. Yosief HO, Sarker MI (2021) Naturally derived fatty acid based antibacterial agents. In: Conversion of renewable biomass into bioproducts. ACS, pp 91–117
37. Ribeiro AR, Silva SS, Reis RL (2022) Challenges and opportunities on vegetable oils derived systems for biomedical applications. Biomater Adv 134:112720
38. Kanjilal S, Kaki SS (2019) Antimicrobial activities of fatty acids and their derivatives. Antimicrob Mater Biomed Appl 5:457
39. Kula J (2016) Synthesis and bioactivity of (R)-ricinoleic acid derivatives: a review. Curr Med Chem 23:4037–4056
40. Matysiak S, Kula J, Błaszczyk A (2019) Chiral amide derivatives of ricinoleic acid and 3-hydroxynonanoic acid synthesis and cytotoxic activity. Med Chem Res 28:948–958
41. Kanagaraj P, Huang W, Liu C (2020) Noncovalently functionalized sulfated castor oil–graphene oxide-strengthened polyetherimide composite membranes for superior separation of organic pollutants and their fouling mitigation. ACS Appl Mater Interfaces 12:37054–37066
42. Gogoi S, Karak N (2014) Biobased biodegradable waterborne hyperbranched polyurethane as an ecofriendly sustainable material. ACS Sustain Chem Eng 2:2730–2738
43. Alam M, Akram D, Sharmin E, Zafar F, Ahmad S (2014) Vegetable oil based eco-friendly coating materials: a review article. Arab J Chem 7:469–479
44. Madra H, Tantekin-Ersolmaz SB, Guner FS (2009) Monitoring of oil-based polyurethane synthesis by FTIR-ATR. Polym Test 28:773–779
45. Laschke MW, Strohe A, Scheuer C, Eglin D, Verrier S, Alini M, Pohlemann T, Menger MD (2009) In vivo biocompatibility and vascularization of biodegradable porous polyurethane scaffolds for tissue engineering. Acta Biomater 5:1991–2001
46. Liang H, Liu L, Lu J, Chen M, Zhang C (2018) Castor oil-based cationic waterborne polyurethane dispersions: storage stability, thermo-physical properties and antibacterial properties. Ind Crops Prod 117:169–178
47. Pescarmona PP (2021) Cyclic carbonates synthesized from CO_2: applications, challenges and recent research trends. Curr Opin Green Sustain Chem 29:100457
48. Narra N, Rachapudi BNP, Vemulapalli SPB, Korlipara PV (2016) Lewis-acid catalyzed synthesis and characterization of novel castor fatty acid-based cyclic carbonates. RSC Adv 6:25703–25712
49. Shende PG, Jadhav AB, Dabhade SB (2002) Polyesteramide resins from dehydrated castor oil and various dibasic acids. Pigm Resin Technol 31:310–314
50. Pramanik S, Sagar K, Konwar BK, Karak N (2012) Synthesis, characterization and properties of a castor oil modified biodegradable poly (ester amide) resin. Prog Org Coat 75:569–578
51. Zhou S, Sun W, Zhai Y (2018) Amphiphilic block copolymer NPs obtained by coupling ricinoleic acid/sebacic acids and mPEG: synthesis, characterization, and controlled release of paclitaxel. J Biomater Sci Polym Ed 29:2201–2217
52. Reddy KK, Ravinder T, Kanjilal S (2012) Synthesis and evaluation of antioxidant and antifungal activities of novel ricinoleate-based lipoconjugates of phenolic acids. Food Chem 134:2201–2207
53. Yarra M, Kaki SS, Prasad RB, Mallampalli KS, Yedla P, Chityala GK (2016) Synthesis of novel (Z)-methyl-12-aminooctadec-9-enoate-based phenolipids as potential antioxidants and chemotherapeutic agents. Eur J Lipid Sci Technol 118:622–630
54. Van der Steen M, Stevens CV (2009) Undecylenic acid: a valuable and physiologically active renewable building block from castor oil. Chemsuschem 2:692–713
55. Machado TO, Cardoso PB, Feuser PE, Sayer C, Araújo PH (2017) Thiol-enemini emulsion polymerization of a biobased monomer for biomedical applications. Colloids Surf B 159:509–517
56. Shapiro AL, Rothman S (1945) Undecylenic acid in the treatment of dermatomycosis. Arch Derm Syphilol 52:166–171
57. Reddy TVK, Jyotsna A, Devi BP, Prasad RBN, Poornachandra Y, Kumar CG (2016) Design, synthesis and in vitro biological evaluation of short-chain C12-sphinganine and its 1, 2, 3-triazole analogs as potential antimicrobial and anti-biofilm agents. Eur J Med Chem 118:98–106

58. Roe RM, Kallapur V, Linderman RJ, Viviani F (2005) Organic synthesis and bioassay of novel inhibitors of JH III epoxide hydrolase activity from fifth stadium cabbage loopers. Trichoplusiani PesticBiochemPhys 83:140–154
59. Gandhi B, Greeshma K, Ruvulapalli DP, Kaki SS (2022) Design, synthesis, and antimicrobial evaluation of novel 10-Undecenoic acid-based lipidic triazoles. Med Chem Res 31:1558–1570
60. Gandhi B, Juliya J, Dileep V, Uma Rajeswari B, Misra S, Kaki SS (2021) Antioxidant and biological activities of novel structured monoacylglycerol derivatives with phenolic acids. Eur J Lipid Sci Technol 123:2100055
61. Gandhi B, Jhansi M, Deshpande SS, Vinay T, Misra S, Kaki SS (2023) Design, synthesis and biological activity of novel oxadiazole containing monoacylglycerols as potential bioactive lipids. J Mol Struct 1284:135424

Optimization of Co-digestion of Fish Waste for Methane Generation and Utilization of Slurry for Organic Farming

K. M. PrasannaKumaran, M. Karthikeyan, C. Sanjay Kumar, N. Kamali Meera, and V. Kirubakaran

Abstract India is bestowed with a diverse coastal region that secured 3rd place in Global Fisheries and 2nd place in Aquaculture. Fisheries contribute 1.07% of GDP and 5.15% of Agriculture GDP. If it is utilized efficiently, it will contribute greatly to the economy and pollution management system. But there is no correct method of disposal of waste which leads to environmental issues. Using fish waste directly as a fertilizer has some limitations, because they will become malodorous to attract rodents and other animals. Tablets and oils derived from these waste products may not align with cultural and religious preferences. Some countries restrict meat by-products considering food safety and quality. Hence, the present paper analyses the fish waste management, processing, and value addition in the sense of feeding to a floating drum biogas plant with various agro residues as a co-digesting medium. In this sense, we discuss the measures to improve its performance as a sustainable biomass. The slurry from the bio-digester has been analyzed with respect to N, P, K content and treated properly for utilization as manure for eco-farming.

Highlights of the Study:

- Number of fish shops in Dindigul—approximately 30–35 shops.
- Waste generation per week—6–8 Tons.
- Current utilization—mostly dumped in open spaces and water bodies. Some wastes are collected and used for composting.

Keywords Fish wastes · Biogas plant · Co-digestion · Eco-farming

---

K. M. PrasannaKumaran (✉) · M. Karthikeyan · C. Sanjay Kumar · N. Kamali Meera · V. Kirubakaran
Centre for Rural Energy, The Gandhigram Rural Institute – Deemed to be University Gandhigram, Chinnalapatti, Tamil Nadu 624 302, India
e-mail: prasanna7raina@gmail.com

S. Kumar et al. (eds.), *Clean Energy Transition-via-Biomass Resource Utilization*, Green Energy and Technology, https://doi.org/10.1007/978-981-97-6321-4_5

1 Introduction

Fossil fuels are a form of energy that takes millions of years to become so. The wide-scale extraction of fossil fuels over these years led to a serious impact on the climate and the depletion of an incredible amount of them. Therefore, an alternative to the dwindling fossil fuel must be adopted. An effective and more feasible option is renewable energy sources. Biogas, a renewable and clean source of energy, is generated through bio-digestion. Since the organic and municipal waste produced contributes to environmental pollution, it can be utilized for biogas production. Fish and organic waste are rich sources of high-value organic carbon, ideal for methane production [1]. Nutrient recovery is possible from the product of the biogas as it has a higher concentration of plant nutrients. Bio-digested slurry is an excellent plant fertilizer rich in micro- and macro-nutrients.

Fish is a source of various nutrients such as protein, diet fats along with iodine, and omega-3 fatty acids. Most often, the fish is consumed fresh rather than canned. The annual fish waste generated is nearly 70 million tones. These wastes are still unutilized and contribute to pollution in both cities and village regions. Unutilized fish wastes cause emissions such as landfill gas, smell, and leachate when disposed of in a landfill. The acceleration in fish processing wastes will expand the renewable energy sector that targets fish waste and might play a significant role in the footprints of biofuel.

1.1 Global Seafood Consumption and By-product Utilization

Global seafood utilization has more than doubled over the past 50 years. There are large variations across countries and regions of the world in the amount of total fish supply for human consumption, reflecting different eating habits and traditions, availability of fish and other foods, prices, socio-economic levels, and seasons [2]. Per person per day per month apparent fish utilization can vary from less than 1 kg per capita in one country to more than 100 kg in another. Differences are also evident within countries, with consumption usually higher in coastal areas.

According to estimations based on 2011 baseline data, worldwide demand for seafood meant for human utilization is 143.8 million tons per year, and the whole usage footprint, which includes the inclusion uses of seafood, is 154 million tones. Since 1961, the mean yearly hike in worldwide apparent food fish consumption (3.2%) has outperformed population growth (1.6%) and exceeded utilization of meat from all terrestrial animals combined (2.8%) and separately (bovine, ovine, pig, and others), excluding poultry (4.9%). Fish food usage per person per day per year has increased from 9.0 kg in 1961 to 20.2 kg in 2015, at a pace of roughly 1.5% per year on average. A primary projection for 2016 and 2017 indicates an inclusive rise to around 20.3 and 20.5 kg, respectively [3]. Utilization growth has been driven not just by increasing production, but also by a combination of many other variables, including decreased wastage, better use, improved distribution routes, and higher demand, which is connected to population explosion, rural development, and rise in income of the people. Food sector waste is a significant cause of pollution in the environment. Research has been conducted in an attempt to create strategies for converting these wastes into usable goods. More than half of the residual material from the total fish harvest is unlikely to be utilized as food, amounting to about 32 million tons of garbage.

1.2 Fish Processing By-products

Contains:

Extracted Proteins
Extraction of Ether extract or fat
Minerals
Palmitic acid, Stearicacid, Oleicacid
Docosa hexaenoic acid.

1.3 By-product Utilization

In the year 2015, the raw components required to produce fishmeal and fish oil saw significant utilization. The total fish production, amounting to 13.9 million metric tons, combined with the by-products from wild catches at 3.75 million metric tons and aquaculture by-products at 1.94 million metric tons, collectively positioned fish waste as an exceptionally compelling and valuable source of nutrients. By 2025, fish meal generated from fish waste will account for 38 percent of global fish meal output, up from 29% in the 2013–2015 average [3]. In 2015, 0.856 million tons of fish oil were generated.

1.4 Environmental Issues for Dumping and Processing Fish Wastes

Fish waste can comprise, but is not limited to, meat, skin, bones, intestines, shells, and liquid stick water. Warm temperatures cause fish excrement to deteriorate quickly. As a result of bacterial breakdown, fish wastes cause aesthetic issues and unpleasant smells if not properly kept or controlled. The organic components of the trash have a high biological oxygen demand, which can cause environmental and health hazards if not treated effectively.

1.4.1 Disposal at Oceans

It was and still is normal practice to dispose of fish feces in a lake or ocean. The problem is that dumping everything in one spot might overburden the ecosystem, hence dumping is prohibited to safeguard our water bodies, and animal and human health. This might result in a lot of waste for fish processing businesses. Some fish waste is transferred to the sea for disposal. Concerns about the environmental impact of dumping fish feces into ocean waterways include decreased oxygen levels in seawaters at the ocean's bottom; burying or suffocation of living organisms; and establishment of illness or non-native and invasive species to the seafloor environment [4]. Organic waste emissions containing fish oils and fish products are nutrient-dense. This nutrient-rich biomass has the potential to result in a fast rise in the number of organisms feeding on this new food source. Because some of these species require oxygen to consume the nutrients, they can quickly exhaust the available dissolved oxygen in the saltwater, resulting in anoxic conditions. When this happens, the animals or creatures that require oxygen to thrive simply cannot live since the oxygen necessary for the flora and fauna is depleted. Furthermore, some of the wastes are converted into gases. Carbonic acid is produced when carbon dioxide is generated and dissolves in water. This increases the pH or acidity of saltwater, which has an impact on marine life.

1.5 Processing of Fish Waste

The impact of fish processing bio-waste on neighboring surroundings, habitats, and creatures is hotly debated. Although it is a natural contaminant, fish bio-waste has the potential to impact oxygen levels, salinity, temperature, pH levels, and overall organism abundance in marine water. The intrusion of this garbage also has an impact on the food webs of neighboring marine ecosystems, especially when considerable volumes are present [4]. The seafood industry generates a huge amount of fish waste both locally and worldwide. Improper disposal of fish wastes in the surroundings may cause environmental hazards due to the liberation of toxic gas at the time of bacterial decomposition. This inedible fish waste can be utilized rather than disposed of by processing.

1.6 Production of Biogas

Biogas production is one of the most basic and rapidly growing technologies. Biogas is a sort of biofuel derived from biogenic material and is commonly defined as a gas created by bacteria fermentation of organic material under anaerobic conditions (in the absence of oxygen). It may be made from a wide range of organic materials and wastes, such as sewage sludge, animal manure, and municipal organic waste. Materials such as biodegradable garbage, straw, manure, sugarcane, and by-products of agricultural and industrial operations, as well as specifically produced energy crops, can be utilized to generate energy [2]. Biogas production is enabled by processing biodegradable and organic wastes using a simple technology, anaerobic digestion. Anaerobic digestion involves the breakdown of organic material under airless, i.e., oxygen-free. It encompasses four steps they are Hydrolysis, Acidogenesis, Acetogenesis, and Methanogenesis (Fig. 1).

2 Materials and Methods

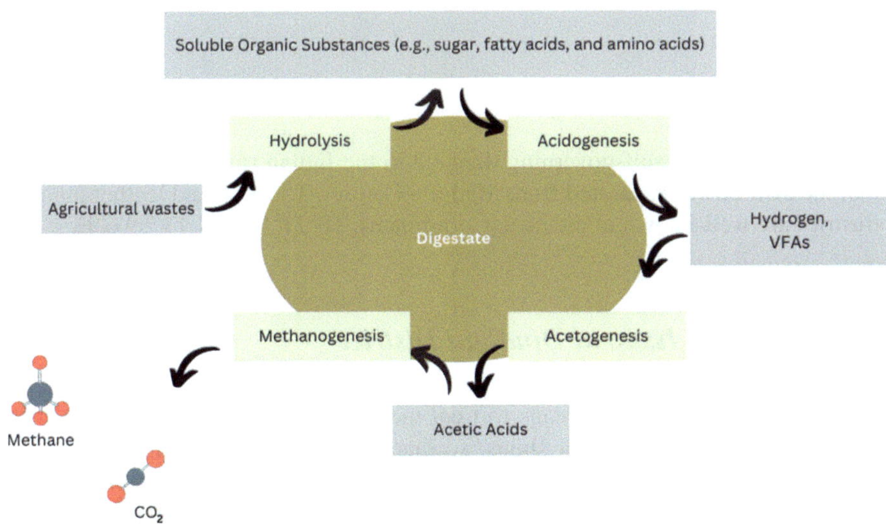

Fig. 1 Stages of biogas production

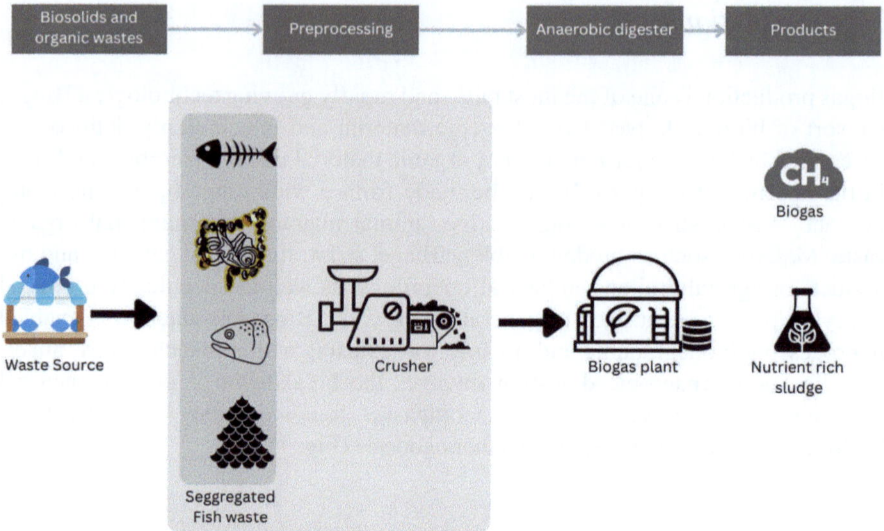

Fig. 2 Process flow diagram of fish processing and methane generation

2.1 Outline of Fish Waste

Fish wastes are disposed of in seas and oceans. This may lead to the burial of living organisms and the emergence of non-native diseases. Fish wastes are disposed of in certain permitted landfills. But the problems with landfills may include careful management to reduce the odors, pathogens, and contamination of water resources.

2.2 Study Area

Dindigul District is a self-governing division in the Indian state of Tamil Nadu. In 1985, the district was separated from Madurai District. The lock and leather tanning industries are well-known in the area. Coordinates: 10°21′14.4″N 77°59′6″E.

2.3 Fish Waste Issue in Dindigul District

The organization has created plans to treat meat and fish waste generated by fish markets, and chicken and meat shops, so that it may be properly disposed of. A sanitary supervisor from the civic authority told The Hindu that the firm had offered to take the rubbish for free. They have provided us with around 150 bins for rubbish

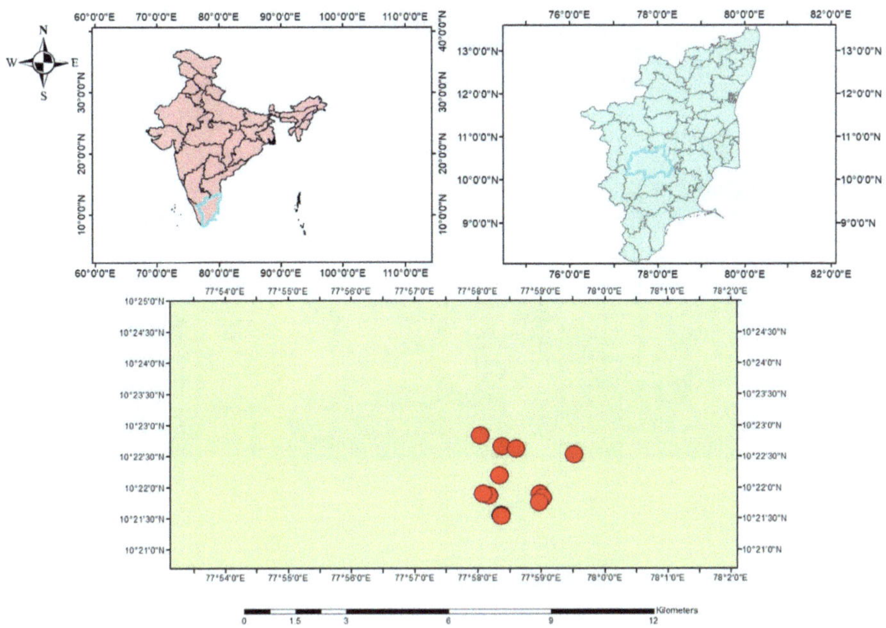

Fig. 3 Maps & waste resource mapping of fish wastes in dindigul district

collection across the civic body's 65 wards. "We've installed them at the 31 micro-composting yards and told sellers at the fish market and butcher stalls to bring their garbage and dispose of it in these bins," they explained. The city creates roughly six to eight tons of fish waste every week, with the busiest days being Sundays, Mondays, and Wednesdays. The waste is gathered at a single location in the city, from which a business based in Dindigul collects it every evening. Vendors have been instructed to deposit their waste in the designated bins at the nearest micro-composting yard. The Corporation Commissioner has issued a warning, noting that vendors who are detected dumping meat and fish waste in open areas and bodies of water may face penalties. Fish markets are not scattered throughout the Dindigul town. Most of the fish shops are in the Dindigul town market.

2.4 Technology Value Chain of Fish Wastes

Environmental hazards due to improper disposal of fish wastes can be limited by means of utilizing those wastes for biogas production which are a good source of high-valued organic carbon (Fig. 4).

The nutrients in fish market wastes include Magnesium (Mg), Phosphorous (P), Potassium (K), Calcium(Ca), and Nitrogen (N).

Fig. 4 Fish waste

2.5 *Treatment of Fish Waste*

The inedible fish waste includes skin and viscera (internal organs in the body's major cavities, particularly those in the belly, such as the intestines). Carcasses (fishbones) are segregated from these wastes since decomposing is difficult. The wastes are ground and heated in an oven to remove the odor from them (Fig. 5).

Fig. 5 Grinded fish waste

2.6 *Proximate Analysis of Fish Waste*

Proximate Analysis provides the entire formation of the biomass and so is critical for biomass energy utilization. Proximate analysis of a fuel yields the proportion of material that burns in a gaseous form (volatile matter), in a solid state (fixed carbon), and the amount of inorganic waste material (ash).

2.6.1 Moisture

The moisture content of biomass leftovers has a direct impact on their heating properties. The standard method for determining moisture in biomass includes heating the sample in a hot air oven: 105 ± 50 C.

$$\%\text{ of Moisture} = [(\text{M2} - \text{M3})/(\text{M2} - \text{M1})] * 100 \quad (1)$$

Where

M1—Mass of the empty crucible in grams, M2—Mass of the crucible with biomass in grams, and M3—Mass of the crucible and oven-dried sample in grams.

2.6.2 Volatile Matter

To determine volatile matter, the fuel is heated to a specified temperature and rate in a controlled atmosphere. The standard technique for determining volatiles in biomass includes heating the sample in a muffle furnace to 400 °C.

$$\%\text{ of Volatile Matter} = [(\text{M3} - \text{M4})/(\text{M2} - \text{M1})] * 100 \quad (2)$$

Here,

M4—Weight of the crucible and hot air oven-opened sample in grams.

2.6.3 Ash

The weight of the residue remaining after full combustion of the biomass at 600 C is defined as ash.

$$\%\text{ of Ash} = [(\text{M5} - \text{M1})/(\text{M2} - \text{M1})] * 100 \quad (3)$$

where M5—Weight of the crucible and left over in grams.

Table 1 Composition of fish wastes

	Criteria	Percentage level
1	% of moisture	46.432
2	% of volatile matter	41.99
3	% of ash	5.568
4	% of fixed carbon	6.01

2.6.4 Fixed Carbon (fc)

The fixed carbon content is calculated by subtracting the sum of A percent, VM percent, and percent M from the overall composition of 100% (Table 1 and Fig. 6).

$$\text{Fixed Carbon} = 100 - (\%\text{Moisture} + \%\text{Volatile matter} + \%\text{Ash}) \tag{4}$$

Calculation

M1 = 37.530 g; M2 = 56.870 g; M3 = 47.890 g; M4 = 39.769 g; and M5 = 38.607 g

$$\begin{aligned} \%\text{ of Moisture} &= [(M2 - M3)/(M2 - M1)] * 100 \\ \text{(i)} \quad &= [(56.870 - 47.890)/(56.870 - 37.530)] * 100 \\ &= 46.432\% \end{aligned}$$

Fig. 6 Ash content of fish wastes

$$\text{(ii)}\quad \begin{aligned}\%\text{ of Volatile Matter} &= [(M3 - M4)/(M2 - M1)] * 100\\ &= [(47.890 - 39.769)/(56.870 - 37.530)] * 100\\ &= 41.990\%\end{aligned}$$

$$\text{(iii)}\quad \begin{aligned}\%\text{ of Ash} &= [(M5 - M1)/(M2 - M1)] * 100\\ &= [(38.607 - 37.530)/(56.870 - 37.530)] * 100\\ &= 5.568\%\end{aligned}$$

$$\text{(iv)}\quad \begin{aligned}\%\text{ of Fixed Carbon} &= 100 - (\%M + \%VM + \%A)\\ &= 100 - (46.432 + 41.990 + 5.568)\\ &= 6.01\%\end{aligned}$$

2.7 *Biogas Slurry*

The biogas slurry contains 93% water and 7% dry matter, of which 4.5% is organic and 2.5% is inorganic. The digested biogas slurry also contains phosphate, potassium, zinc, iron, manganese, and copper, all of which have been lost from the soil because of intensive farming operations. Bio-slurry can also be used to improve the fertility of soil for agricultural production. Bio-slurry has more nutrients and micronutrients than composted manure and FYM and provides more conveniently available plant nutrients. The effects of applying bio-slurry are comparable to the effects of applying synthetic fertilizers. As a result, digested bio-slurry may be a valuable alternative to synthetic fertilizers. Biogas slurry is regarded as a good organic fertilizer source since it includes significant levels of both macronutrients (N, P, K) and micronutrients (Zn, Mn, B) required for plant growth [5]. The use of biogas slurry provides a sustainable solution for agriculture, the environment, and rural communities.

2.8 *Nutrient Content of Bio-slurry*

The use of biogas slurry as fertilizer is a significant aspect of the cyclic usage of biomass resources, and it has a relatively high commercial interest, as well as environmental and social benefits. Owners of biogas plants seek to sell biogas sludge

Table 2 Nutrient content of bio-slurry

S. no	Nutrients	Values (%)
1	Potassium	1.00
2	Nitrogen	0.22
3	Phosphorous	0.08
4	Organic substances	4.25
5	Solid substances	9.81

as a commodity. However, due to constraints such as high transportation costs and discomfort, uncontrolled fertilizer requirements, and low consistency of biogas slurry fertilizer efficiency, reaching a deal with dispersed farmers is difficult.

3 Results and Discussion

Initially, at the lag phase physical conversion of the cell catabolism is attributed to growth. At the exponential phase, the cell density increases. The concentration of the gas increases gradually. At the stationary phase, the cell population is relatively constant. At the dead phase, the growth begins to limit due to physical factors such as light, retention time, and atmospheric temperature. The chemical factors include pH, CO_2, CN ratio, and so on. Cell density due to the incapability of sustaining growth and nutrients are also depleted to a certain level. Figure 7 shows the concentration of CO_2 and CH_3 with respect to number of days. Figure 8 shows the concentration of H_2S with respect to the number of days.

3.1 Calibration of Gas Mixture in Online Biogas Analyzer

As a degree of measurement, calibration of the gas mixture is performed (Fig. 9).

The concentration of gases in analyzing the filling aluminum cylinder is as follows (Fig. 10):

$CH_4 = 67.2\%$, $O_2 = 0.4\%$, $CO_2 = 31.9\%$, $H_2S = 350$ ppm.

Biogas analysis from biogas produced from fish waste:

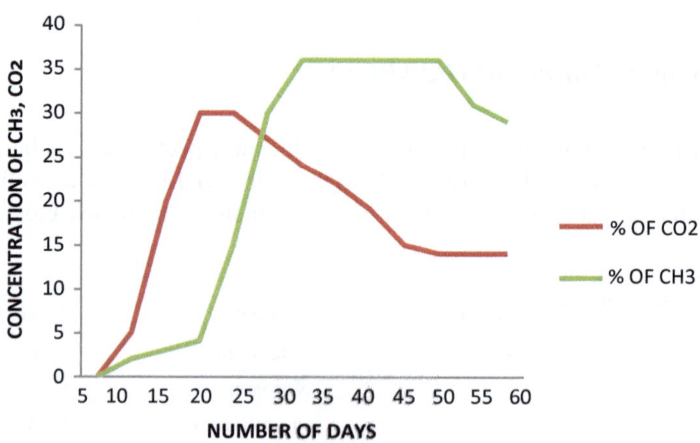

Fig. 7 Number of days versus concentration of CH_4, CO_2 (%)

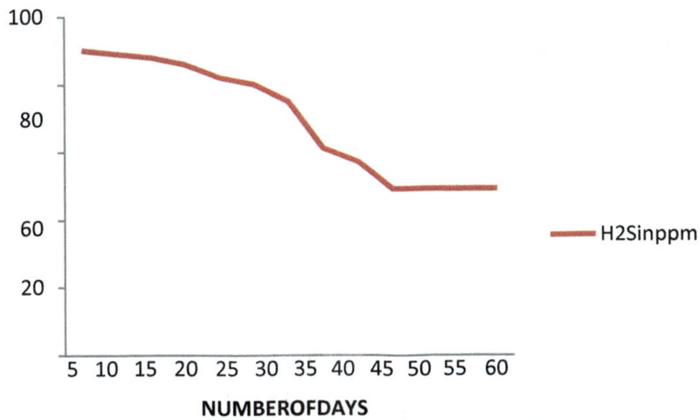

Fig. 8 Number of days versus concentration of H_2S (PPM)

After the calibration of the gas mixture process, by using the Online Biogas Analyzer, the concentration of the gases produced from the fish wastes after 60 days is as follows (Fig. 11):

$CH_4 = 36.4\%$; $O_2 = 4.1\%$; $CO_2 = 24.9\%$; $H_2S = 1$PPM.

Thus, the following observations were obtained:

- Retention time required is more.
- Proper utilization of fish waste for methane production and hence management of organic waste.
- The slurry obtained can be utilized as fertilizer for farming.

Fig. 9 Calibration of gas mixing in refilling aluminum cylinder

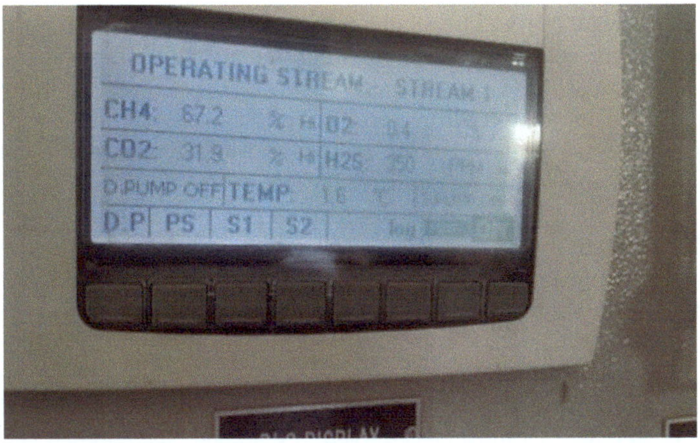

Fig. 10 Concentration of gases in aluminum cylinder

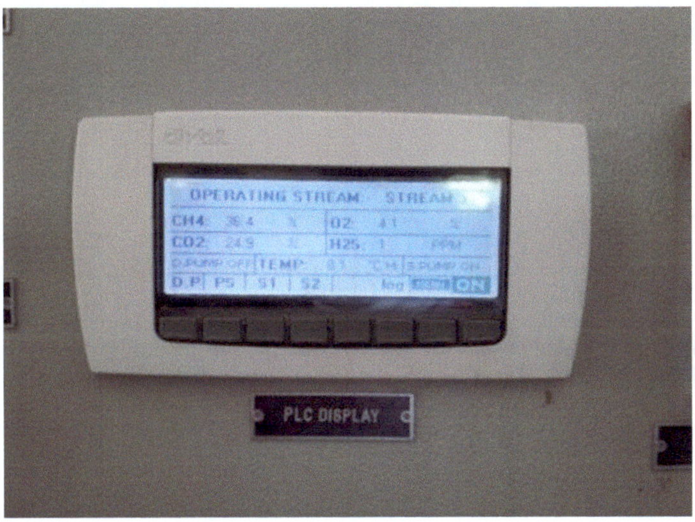

Fig. 11 Concentration of gases in biogas plant

4 Conclusion

Waste Management is essential as an increase in waste generation is a growing problem. The characterization "Not in my Backyard" highlights the necessity of solid waste management and what the way to resolve the issue should be. World fish consumption has been steadily increasing in the past four years as a result; there are more fish processing wastes. These wastes have greater potential for energy production. Also, the natural environment is protected by reducing the carbon footprints as recovery of methane from biogas plants is possible. In the case of landfills, the recovery of methane is tedious. Thus, the biogas plants reduce the emissions which contribute to the greenhouse effect. Anaerobic digestion of these wastes will give a solution for decreasing both environmental pollution and fossil fuel usage.

Although fish processing bio-waste is a natural contaminant, it is nonetheless pollution. Fish processing bio-waste, when present in high quantities or in a concentrated region, has the potential to impact neighboring marine habitats and ecosystems by altering natural chemical levels and disrupting food chain linkages. When fish processing plants dump their trash into the ocean, two things happen. First, an increase in nutrients generates an increase in algal blooms, which isn't always a bad thing. The rise in phytoplankton is then followed by an increase in zooplankton. This can result in anoxia in inert waters if there are too many organic materials present at one time, alternatively, a larger food source for an increasing fish population. In any case, it is an issue that must be observed, researched, and handled using developing technology wherever possible. It's a pity that we're still throwing nutrient-rich garbage into the ocean. We do, after all, have the technology to convert this garbage into something useful. Even if it requires a bit more time and money, it should be worthwhile.

References

1. LudwigSasse (1998) "BiogasPlants" A Publication of Doutsches Zentrumfur Entwicklung techologies–GATE
2. Mag dalena et al (2014) Potential for producing biogas from agricultural waste in rural plants in Polands. Sustainability 6:5065–5074. ISSN 2071-1050
3. Tomczah R et al (2012) Biogas production from fish wastes in co-digestion with sewage sludge
4. Ajip et al (2016) Perspective of waste utilization in sea food industry. Int J Manage Appl Sci 2(9). ISSN 2394-7926
5. Islam MS (2006, September) Use of bioslurry as organic fertilizer in Bangladesh agriculture. Prepared for the presentation at the international workshop on the use of bioslurry domestic biogas programme, Bangkok, Thailand, pp 3–16

Efficient Valorization of Biomass: A Perspective on Overcoming Critical Skepticisms Surrounding the Commissioning of Functional 2G Bio-refineries

Navnit Kumar Ramamoorthy, Revanth Babu Pallam, Shovan Rakshit, Pon Yazhine Tamilselvan, Sahadevan Renganathan, and Vemuri Venkateswara Sarma

Abstract By the year 2050, the world's human population, which is estimated to attain a massive 9.7 billion, is also hastily equipping itself to confront the complete exhaustion of fossil fuels. While ~$710 billion USD is the tentative, collective global investment towards sustainable, green bio-energy production, skepticisms related to the feedstock supply chain, and real-time commercial viability of 2G bio-refineries continue to baffle policymakers. In an attempt to provide a suggestive approach, the present chapter, using preliminary data analytics to model two mutually exclusive hypothetical scenarios, elucidates that conventional feedstocks fall short by 1.57×10^3-fold to cater to the anticipated quantities of bio-ethanol production and face a deficit of 2.68×10^9-fold to meet the estimated global livestock fodder demand. Focusing on circular economical, zero-waste bio-refineries, the chapter provides perspectives on industrial symbiosis, biomass banking, and a government-funded co-operative bio-energy society to promote indigenous bio-fuel production and improve a nation's gross domestic product (GDP) value.

Keywords Bio-energy · Policymakers · Hypothetical scenario model · Zero-waste bio-refineries · Circular economy · Indigenous bio-fuel production

---

Equal Contribution of Authors: Navnit Kumar Ramamoorthy and Revanth Babu Pallam have equally contributed towards conceptualizing the content of the book chapter.

N. K. Ramamoorthy · R. Babu Pallam · S. Rakshit · P. Y. Tamilselvan · V. Venkateswara Sarma (✉)
Fungal Biotechnology Laboratory, Department of Biotechnology, Pondicherry University, Kalapet, Pondicherry 605014, India
e-mail: sarmavv@yahoo.com

S. Renganathan
Biofuel Laboratory, Centre for Biotechnology, Anna University Chennai, Chennai 600025, India

S. Kumar et al. (eds.), *Clean Energy Transition-via-Biomass Resource Utilization*, Green Energy and Technology, https://doi.org/10.1007/978-981-97-6321-4_6

1 The Fuel Crisis, and Global Investments in Bio-energy

The year 2040 would record a ~23% increase (close to 115 million barrels) in global fossil fuel consumption per day; the existing fossil fuel reserves would hardly last for the next 4–5 decades [1]. Worldwide investments towards energy production are expected to rise by 8% to attain a ~$2.4 trillion USD in the near future [2], majorly being diverted towards renewable energies and bio-energy, which are key to outmaneuvering the accelerating energy crisis, lessening the burden on human population, and invigorating economic revival [2]. According to the IEA Sustainable Recovery Tracker, the cumulative investments towards sustainable, clean energy would amount to $710 billion USD [2]. Some significant alternate fuels, their production techniques, salient features, and their limitations are presented in Table 1. In 2021, China, the USA, and the EU had invested (in USD) $380 billion, $215 billion, $260 billion, respectively on biofuels. Around 80 countries including Canada, Lithuania, Argentina, have set targets towards cleaner bio-energy transitions [3]. Venture capitalists, the governments of nations, and policymakers are confronted with several bio-energy alternatives to focus on. However, the deciding criteria include the turnaround times of such investments/operations [2], market viability [4], process economy [5], and sustained availability [6] of the desired product, which invariably influence the inflow of funds and investments [2]. Among the available energy options, the present-day bio-energy accounts for ~55% of renewable energy [3]. Insurmountable pressures in the feedstock/fuel supply chain, tight market in sourcing man-power and related services, and rising establishment costs have further influenced policymakers' decision of investing in bio-fuels [2]. In addition to advanced, commercial-scale bio-fuel productions in the EU (bio-methane), the USA (aviation bio-fuels, bio-fertilizers), India (solid and gaseous biogas), and Australia (biomass-based electricity) [3], as per the renewable energy market update in the year 2023, its (bio-fuels') demand would reach significant highs in India, Brazil, and Indonesia [3].

2 Various Nations Have Been Banking on Bio-ethanol Production

Bio-ethanol, the prominent biomass-derived bio-fuel [38], with a world-wide market of $58 million USD, and employed in the transportation sector [39], showed a 110% increase in its global production in the year 2021 (3.78 billion liters), in comparison to its recorded production in 2007 [38]; it is expected to reach a demand of 164 billion liters by 2030 [1]. Among the bio-fuel alternatives available in the industry/market, bio-ethanol production has reached a comparatively mature stage in terms of its production, with several demonstrated yields and performance outputs [40]. In the year 2018, from the produced, cumulative total of 108 billion liters of bio-ethanol, ~90.7 billion liters were utilized by the transportation sector [1]. The blending of bio-ethanol and gasoline to create a fuel blend, termed "gasohol," possesses an

Table 1 Various alternate fuels, their production techniques, energy densities, potential applicability, and salient features and limitations

	Alternate fuel	Reported production techniques	Energy density	Demonstrated/ potential applicability	Salient features and limitations
Transitional-phase fuels	*Natural gas*	Stage-wise compression, cooling, dehydration, and separation [7]	53.6 MJ/kg [8]	T*, $\mathrm{I}^{\$}$, $\mathrm{C}^{@}$, $\mathrm{R}^{\&}$ [9]	**Salient features:** Cheap substrates; inexpensive establishment costs; technology available at scale-able levels [10]; easier transportation **Limitations:** Non-renewable in nature; excessive methane emission; varying prices in global market; highly inflammable [11]; tedious pipeline commissioning for delivery
	Compressed natural gas (CNG)				
	Liquefied natural gas (LNG)		22 MJ/L [8]	T*, $\mathrm{I}^{\$}$, $\mathrm{C}^{@}$, $\mathrm{R}^{\&}$ [12]	
	Liquefied petroleum gas (LPG)/Propane		25.3 MJ/L [8]	T*, $\mathrm{I}^{\$}$, $\mathrm{C}^{@}$, $\mathrm{R}^{\&}$	
	Other synthetic fuels from non-renewable sources		Paraffin oil—42 MJ/kg	T* and $\mathrm{I}^{\$}$	**Salient features and limitations** are the same as that for conventionally used fossil fuels
Zero-emission fuels	Ammonia	Majorly through Haber–Bosch process [13, 14; catalytic steam reformation [15]	15.6 MJ/L [16]	T*, $\mathrm{I}^{\$}$, $\mathrm{C}^{@}$, $\mathrm{R}^{\&}$ [17]	**Salient features**: Octane rating of 110–130; available infrastructure; high fuel-power efficiency; not-so flammable **Limitations:** High toxicity; release of emission; low-reactivity [18]
	Electricity	Majorly coal-firing [19]; chemical to electrical energy conversion as is the case with batteries	Coal-derived—8.1 kWh/kg [20]; Li-ion batteries—300 Wh/kg [21]	T*, $\mathrm{I}^{\$}$, $\mathrm{C}^{@}$, $\mathrm{R}^{\&}$ [22]	**Salient features:** Easier transportation; no emission **Limitations:** Vast expense in setting up power plants; costlier than fossil fuels; can be lethal during leakage [23]

(continued)

Table 1 (continued)

	Alternate fuel	Reported production techniques	Energy density	Demonstrated/ potential applicability	Salient features and limitations
	Hydrogen	Steam methane reformation [24]; gasification of biomass [25]; biomass-derived liquid gasification [26]; photochemical and photobiological splitting of water; electrolysis [27]	141.9 MJ/kg [28]	T*, $I^{\$}$, $C^{@}$, $R^{\&}$ [29]	**Salient features**: Clean combusting fuel; high energy density; usage of renewable substrates **Limitations**: Expensive storage and transportation; high-reactivity; high production costs; emissions are based on its method of production [30]
Renewable fuels	Bio-alcohols and biomass-derived solid, liquid, and gaseous fuels	Bio-refinery operation [31]; trans-esterification; thermal decomposition; pyrolysis [32]	Bio-diesel—38 MJ/kg; bio-ethanol—19.6 MJ/L; bio-butanol—29.2 MJ/L; bio-gas—20 MJ/m^3; bio-oil—36.9 MJ/kg; bio-hydrogen—141.86 MJ/kg [28]	T*, $I^{\$}$, $C^{@}$, $R^{\&}$ [33]	**Salient features**: Potential fuel blend or a stand-alone fuel; energy density on par with fossil fuels; lesser emission [6] **Limitation**: Bio-refinery unit-operations are cost-intensive; variable product yields; requires blending agents during addition to fossil fuel [31]
Non-renewable fuel	Nuclear energy	Controlled nuclear fission [34]	^{233}U—79,420,000 MJ/kg; ^{235}U—MJ/kg 79,390,000 MJ/kg; ^{238}U—80,620,000 MJ/kg [35]	T*, $I^{\$}$, $C^{@}$, $R^{\&}$ [36]	**Salient feature**: Highest energy density **Limitation**: Production techniques prone to disasters; radioactive waste generation; high capital/operational costs [37]

Energy consuming sectors: T*—Transportation sector; $I^{\$}$—Industrial sector; $C^{@}$—Commercial sector; $R^{\&}$—Residential sector improved octane number, while bio-ethanol, when blended with diesel fuel, significantly reduces emissions [41]. Every year, ~140 million tons of ethanol is diverted towards the synthesis of ethylene [1]. In a broad perspective, bio-ethanol is the alcohol produced by fermentation of simple and soluble sugars [6]. Based on the feedstock/raw material employed in a bio-refinery, the production processes categorize themselves as first-, second-, and third-generation bio-fuels [31]. Table 2 shows the various generations of bio-fuels, their production techniques, salient features, and limitations. The USA dominates the bio-ethanol production scene with a 56% contribution, seconded by Brazil contributing to 28%, followed by nations, such as the European Union—5%, China—4%, Canada—2%, Thailand, India, Argentina—1%, and other countries—2% [42]. Currently, 60% of the globally produced bio-ethanol utilizes corn as the feedstock, while ~25% utilizes sugar cane, followed by molasses and wheat-derived ethanol contributing to 2% and 3%, respectively [42]. Considering aspects of global food security, and expensive substrates [43], researchers have been working to explore newer sources of feedstock [44].

A bio-refinery, analogous to a petroleum refinery in its operations [53] and processes, sustainably valorizes various feedstocks to value-added products, such as food, platform chemicals, and bio-fuels [54]. Biomass, a renewable carbon source, abundant in agricultural residues and less in energy density, has been applied in second-generation bio-refineries (2G bio-ethanol) [54]. Developing nations require government subsidies and tax exemptions in order to implement bio-fuel stations/ bio-refineries on a commercial-scale [55].

3 2G Bio-ethanol, and the Cloud of Controversies Surrounding It

Several controversial remarks are being made against the production and usage of bio-fuels; but, according to Buyx, a senior research associate at the School of Public Policy at University College, London, the raised skepticisms are based on conglomerated perspectives of individual scenarios with respect to the dogma of "biomass-to-bio-fuel-conversion" existing in different nations, and they are pertinent only to the present-generation and production impediments [40]. In the long term, they may actually be "sustainable" [40]. Certain perceived drawbacks have been stated, pertaining to the production of bio-fuels, majorly bio-ethanol, such as: (i) depletion of bio-diversity, due to excessive biomass usage and mono-crop cultivation [56]; (ii) over-use of water resources, and unscrupulous use of fertilizers [56]; (iii) bio-ethanol-blended fuels reported to releasing higher quantities of carbon dioxide; (iv) a lower energy density, roughly ~34%, in comparison to petrol/diesel; (v) internal-combustion engines' requirements in internal component modification prior to bio-fuel usage; (vi) ethanol's hygroscopicity, resulting in its transportation to remain a tedium [57]; (vii) phase separation of blends at lower temperatures, and requirement of blending agents [58]. A section of the population claims that

Table 2 The various generations of bio-fuels, their salient features and limitations

Various generations of bio-fuels, their salient features and limitations			
Source	Salient features	Limitations	Bio-fuels capable of being produced from the mentioned sources
First-generation bio-fuels			
Edible feedstock	Easier pre-processing of grains, starch-rich substrates, and vegetable oils; non-requirement of biomass pre-treatment procedures [45]; economical investment and operational costs [6]	Demands monocrop cultivation and larger cultivable area [46]; requires larger quantities of fresh water; promotes unscrupulous usage of pesticides/fertilizers to promote higher crop yields; yield is heavily impacted by climatic variations; results in a stringent "food vs. fuel" competition [46]	Bio-diesel—38 MJ/kg; bio-ethanol—19.6 MJ/L; bio-butanol—29.2 MJ/L; bio-gas—20 MJ/m^3 [28]
Second-generation bio-fuels from conventional sources of lignocellulosic biomass			
Non-edible feedstock/ crop residues	No negative influence on human food security; feedstock can be harvested from forests too [47]; usage of crop weeds may provide meager income to agriculturalists; effective route for valorization of waste agricultural residues [4]	May promote deforestation; expensive, varying physical/chemical/ physio-chemical pre-treatments required for effective solubilization of fermentable components of biomass [6]; establishment and operational costs are comparatively higher than the first-generation bio-fuels; results in a rising "fuel vs. livestock fodder" competition [44]	Bio-diesel—38 MJ/kg; bio-ethanol—19.6 MJ/L; bio-butanol—29.2 MJ/L; bio-gas—20 MJ/m^3; bio-oil—36.9 MJ/kg; bio-hydrogen—141.86 MJ/ kg [28]
Third-generation bio-fuels from conventional sources of lignocellulosic biomass			
Algal biomass, majorly micro-algae	No threat to human and animal food security [48]; algal biomass can be cultivated on brackish water [49], effluent water, contaminated water, etc., thereby decreasing the requirement of processed/fresh water	Requires larger surface areas of water bodies for algal cultivation; the establishment costs are relatively higher in comparison to the first- and second-generation bio-fuels [48]; eutrophication of water bodies may endanger aquatic lives [50]	Bio-diesel—38 MJ/kg; bio-ethanol—19.6 MJ/L; bio-butanol—29.2 MJ/L; bio-gas—20 MJ/m^3; bio-oil—36.9 MJ/kg; bio-hydrogen—141.86 MJ/ kg [28]

(continued)

Table 2 (continued)

Various generations of bio-fuels, their salient features and limitations			
Source	Salient features	Limitations	Bio-fuels capable of being produced from the mentioned sources
Fourth-generation bio-fuels			
Genetically manipulated algae/ microbes	No potential influence on food availability and security [6]; no particular specification for the cultivation media; provides avenues for bio-catalytic upgradation of fuels, in vitro metabolic engineering of microbes to replicate fuel pathways [51]	Environmental impacts of the genetic manipulations need to be addressed [52] prior to large-scale cultivation; high costs of research and implementation [31]	Bio-diesel—38 MJ/kg; bio-ethanol—19.6 MJ/L; bio-butanol—29.2 MJ/L; bio-gas–20 MJ/m^3; bio-oil—36.9 MJ/kg; bio-hydrogen–141.86 MJ/ kg [28]

Energy densities of each of the mentioned bio-fuels are represented in MJ/kg

such targeted crop cultivation may, in fact, increase CO_2 emissions by taking up most of the space in arable lands, rather than the assumed decrease while substituting fossil fuels with bio-fuels [59]; certain others also speculate, based on preliminary researches, that bio-ethanol blending may not significantly reduce emissions of greenhouse gases [60]. Skepticism among the public strengthens further, when they learn that they may not be able to use bio-fuel-blended fossil fuels in older versions of automobiles [61]; contradicting the above-claims, reports suggest that 85% (v/v) of bio-ethanol-containing gasoline can readily be used in internal-combustion engines [62]. Additionally, claims stating a reduction in fuel economy have added momentum to the controversies surrounding real-time usage of bio-fuels; cost-efficiency of such fuels have triggered criticism in the global market; a typical bio-alcohol-producing process would involve stages of pre-treatment, enzymatic saccharification, fermentation, distillation, dehydration, and blending, which the public comprehend to be tough to implement in a cost-efficient manner towards commercialization [63]. In a study, it has been observed that "sustainable aviation bio-fuels" would not decrease greenhouse gas emissions, and it can remain to be a clean-burning fuel only when actually sustainable sources, such as crop residues, municipal sludge, and waste water are used as feedstocks [64].

4 Bio-fuels Suffer from a Dearth of Due Recognition

Bio-refinery processes integrate the fundamentals of bio-chemical engineering's unit-operations, chemistry, bio-chemistry, biology, molecular biology, recombinant DNA technology, bio-engineering [65], process control, mechanical engineering, fluid thermodynamics [5], and material sciences [54]. However, often bio-refinery processes and bio-energy-related researches do not get the due recognition [66] in terms of proposal grants for doctoral/post-doctoral research; additionally, certain scientific journals and their editorial teams are prejudiced, and with a pre-conceived notion, outrightly reject bio-fuel-related research works. An imbalance between the encouragement provided by various nations in terms of tax subsidies and incentives [55], and the number of sanctioned academic research projects in bio-fuels would, in the longer run, reduce the number of aspiring bio-energy researchers. While other researches related to health-care and pharmaceutical compounds receive acclaim in the world of academic publishing, it still has to be understood that, if not for fuels and post the extinction of fossil fuels, the healthcare industry/pharmaceutical industries, heavily reliant on the transportation [67] and aviation sectors [68] on a day-to-day basis, would stall and come to a complete stand-still.

Amid several advantages, backlashes, and criticisms, related to bio-ethanol production, food security, and cost-efficiency, one major understanding is with regard to the appropriate, sustainably available, cost-effective, and lesser pre-processing-requiring feedstock's usage for bio-refinery operations. The present chapter, with certain assumptions based on data collected from sources, analyzes two independent hypothetical scenarios, applying a preliminary multi-variate analysis to validate the bio-fuel vs. livestock fodder debate. As a potential solution to the raising concern, certain significant researches, with respect to waste stream-derived biomass valorization, have been compiled and presented. Additionally, approaches such as zero-waste and circular bio-economical models for cost-efficient and scalable bio-fuel production have been discussed as well.

5 A Hypothetical Scenario Analysis Explicates an Escalating 2G Bio-ethanol Versus Livestock Fodder Controversy

This section of the chapter, based on previously compiled and reported statistical data for the 30 most-populated countries of the world [69], assesses the potential bio-ethanol yields during two disjoint/mutually exclusive hypothetical scenarios. The first scenario considers complete utilization of agricultural residues from the 30 countries for bio-ethanol production, while the second scenario evaluates the annual generation of crop residues/livestock fodder against its (fodder's) requirement for the global livestock population.

5.1 Scenario 1: What if the Entire Global Agricultural Residues Were Used as Bio-refinery Feedstock?

The country-wise fossil fuel requirement in liters/day (L/d) are listed in Fig. 1 [70]; the United States of America shows the highest fossil fuel requirement of 3.16×10^{12} L/d, while Congo (Dem. Republic) recorded the least requirement of 1.84×10^9 L/d (Fig. 1). On the basis of the data reported for cultivated land [71, 72] and average crops' yields per year [73], the country-wise crop residues' quantities were estimated (Fig. 2). Myanmar recorded the highest crop residues' yield of 3.76×10^{12} kg per year, followed by the USA, China, and India; the least annual generation of 4.68×10^8 kg of crop residues was recorded by Congo (Fig. 2). 86% of the harvested crops' components are not consumable by humans [74], and this factor was used for evaluating the global, annual generation of crop residues (Fig. 2). Assuming: (i) an average of 35% (w/w) of cellulose in the crop residues; (ii) a 100% saccharification of the crop residues to glucose entities; (iii) a maximum achievable stoichiometric conversion of 0.5 g ethanol per g of sugar [6], the annual 2G bio-ethanol yields for each of the 30 countries were ascertained (Fig. 3). Myanmar showed the highest possible ethanol yield of 4.68×10^{11} L per year, followed by the USA, China, and India; Congo (Dem. Republic) recorded the least-anticipated, yearly ethanol yield of 5.82×10^7 L (Fig. 3). Based on data from the above-stated estimation (Fig. 3) and the recorded fossil fuel requirements (Fig. 1), the consequential country-wise reductions in the annual dependence on fossil fuel was estimated (Fig. 4); it was observed that Tanzania would record a maximum of 32.81% reduction, while South Korea would record a minimum of 0.30% reduction in their fossil fuel requirements (Fig. 4). China recorded the highest crude oil import of 1.73×10^9 L/d followed by the USA and India; the least crude oil import of 6.36×10^3 L/d was recorded by Myanmar (Fig. 5). During the end of refinery operations, 84% of crude oil would result in commercially available fossil fuels [75]. Though all the mentioned countries except Kenya and Tanzania possess their own crude oil reserves, in the present hypothetical estimation, to avoid ambiguity, it was assumed that all the mentioned countries depend on imported crude oil for their fossil fuel requirements. Accordingly, based on data recorded for the country-wise annual crude oil requirement (Fig. 5) and their possible bio-ethanol productions (Fig. 3), the potential savings on annual forex towards crude oil import has been shown in Fig. 6; the USA would record a highest saving of \$62.4 billion USD (United States Dollars) per year, while the least forex savings on crude oil import would be recorded by Congo, showing a \$0.024 billion USD per year (Fig. 6). The savings were calculated as on the 23rd of July2023, and the price per liter of crude oil was \$0.5 USD on the date of estimation [76]. However, despite diversion of the entire agricultural residues towards bio-refineries, it could be observed that the actual world-wide bio-ethanol requirement of 8.25×10^{14} L/year for ~20% (v/v) blending to fossil fuels is 1.57×10^3-fold greater than its anticipated global production of 5.26×10^{11} L/year.

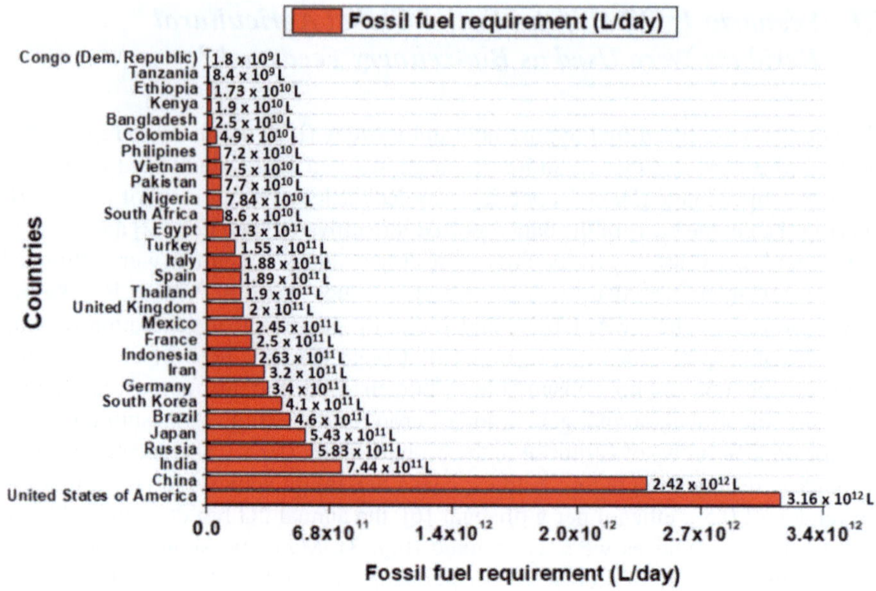

The data related to the fossil fuel requirement of Myanmar could not be ascertained

Fig. 1 Fossil fuel requirements of the highly populated countries of the world

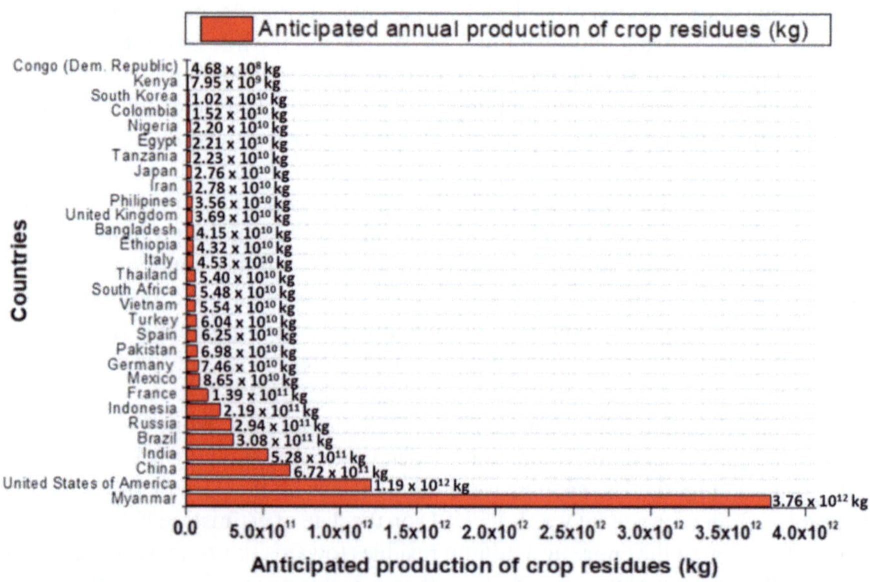

Fig. 2 Anticipated generation of crop residues from the 30 highly populated countries of the world

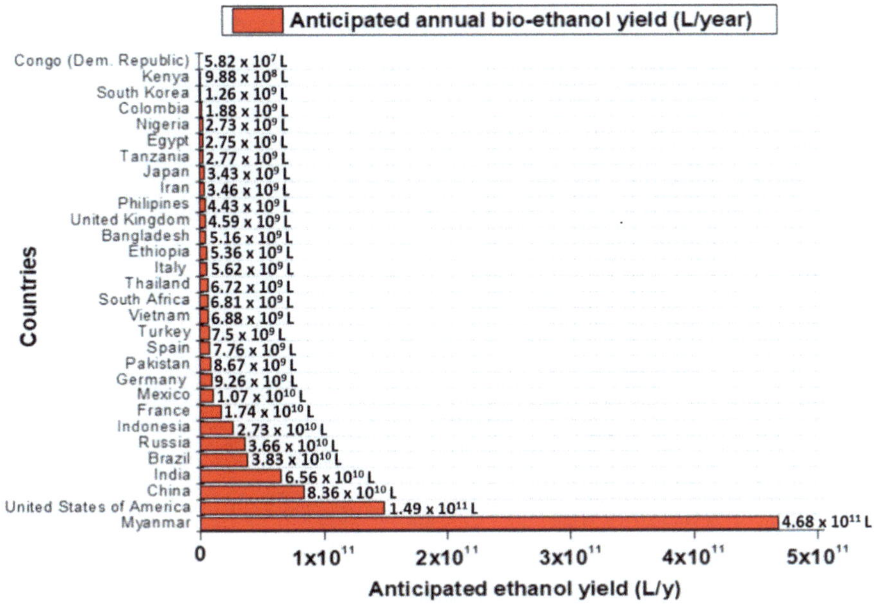

Fig. 3 Possible, cumulative bio-ethanol yield from the 30 major countries of the world

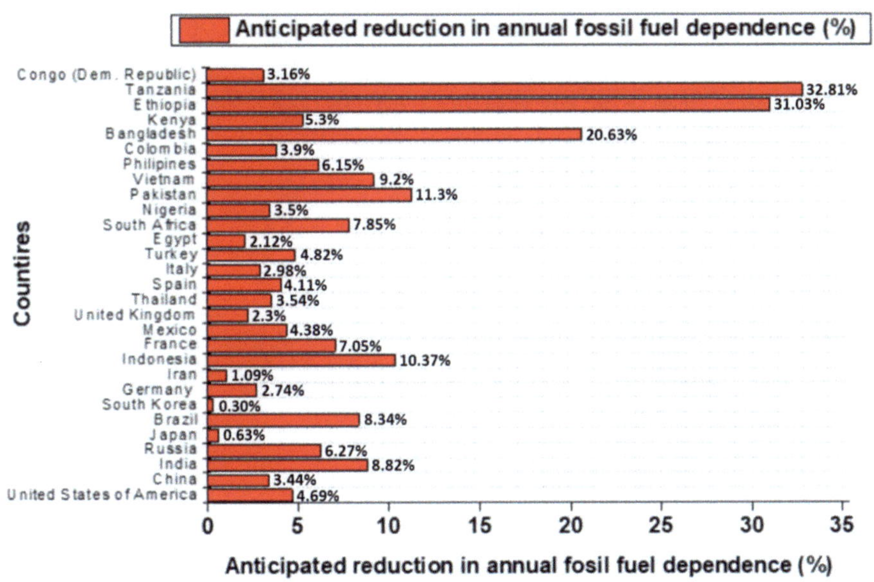

Fig. 4 The anticipated percentages of reduction in fossil fuel dependence of the 30 major countries of the world

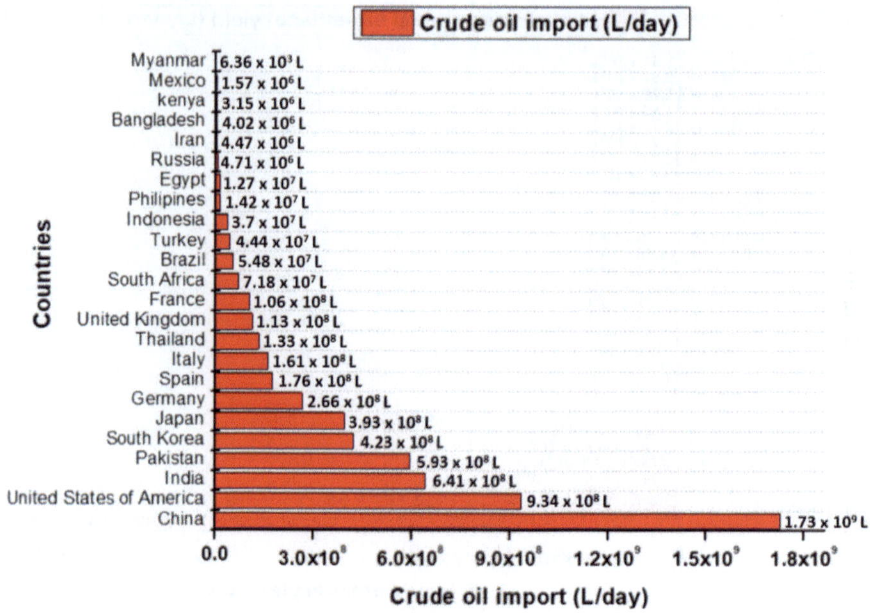

The crude oil import data for Colombia, Tanzania, Congo (Dem. Republic), Vietnam, Ethiopia, and Nigeria could not be ascertained.

Fig. 5 Quantities of crude oil imported by the most-populated countries of the world

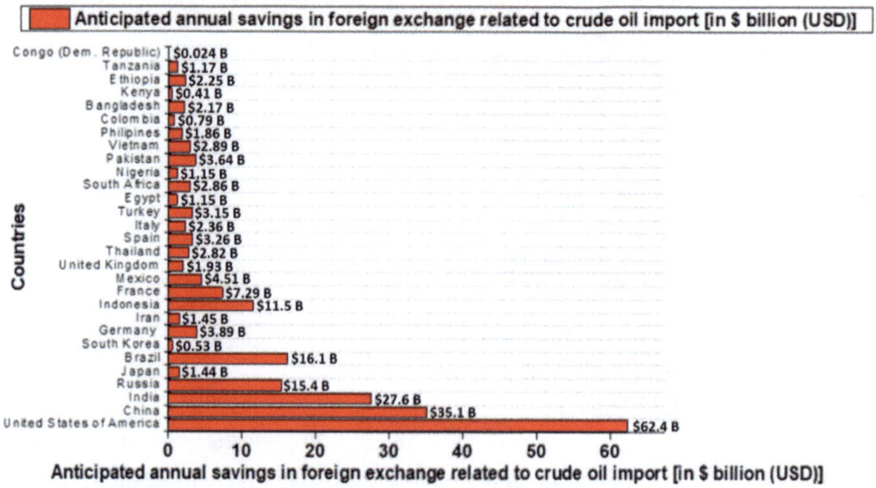

Fig. 6 The anticipated annual savings on crude oil import as a result of indigenous bio-ethanol production

5.2 *Scenario 2: What if the Globally Produced Agricultural Crop Residues Were Completely Utilized as Livestock Fodder?*

During evaluation of the second scenario, it was observed that the global livestock population amounts to roughly 20 quintillion individual animals [77]. Based on the distribution of mammalian species population across the world [78], the approximate livestock count has been estimated and listed in Fig. 7. Since several countries have regulatory measures against utilization of forest resources/lignocelluloses derived from forests for fuel purposes, the present estimation includes only the livestock population and agricultural residues from lands in which agriculture is practiced. The global, annual fodder requirement for: (i) cattle is 1.15×10^{22} kg/y [79]; (ii) pigs is 5.26×10^{21} kg/y [80]; (iii) buffaloes is 2.19×10^{21} kg/y [81]; (iv) camels is 3.29×10^{20} kg/y [82]; (v) asses is 2.31×10^{20} kg/y [83]; (vi) goats is 3.46×10^{20} kg/y [84]; (vii) horses is 1.39×10^{21} kg/y [85]; (viii) sheep is 1.75×10^{20} kg/y [86]; (ix) domesticated elephants is 2.63×10^{6} kg/y [87], as also shown in Fig. 8. On the basis of data obtained for annual crop residue generation (Fig. 2), around 8×10^{12} kg of agricultural residues could potentially be generated per year. On the contrary, the possible global, annual livestock fodder requirement is 2.14×10^{22} kg/year, which is roughly 2.68×10^{9}-fold higher than its anticipated production. In addition to this, phenomena, such as urban sprawl [88], reduction of cultivable land due to urbanization [88], infertility of cultivable lands to due excessive synthetic fertilizer/pesticide usages [89], may further reduce the generation of crop residues. Furthermore, agriculturalists have been levying heavy prices on post-harvest residues, and have been resorting to marketing them for a higher revenue to bio-refineries [90].

The currently generated quantities of crop residues are several folds lesser than their actual requirements to support livestock; they also fall short of the stipulated requirement for bio-fuel production purposes, and it is explicit that a bio-fuel versus fodder competition has already begun; this is on the lines of the "fuel versus food competition," which the first-generation bio-fuels had resulted in [31]. Another shortcoming of the entire dependence on agricultural residues is its seasonal availability [91], perhaps limited to the selected harvest seasons across a year, along with a huge variability in the produce [44]. Hence, there is a pressing need for the exploration of alternate sources of biomass for bio-fuel production [5].

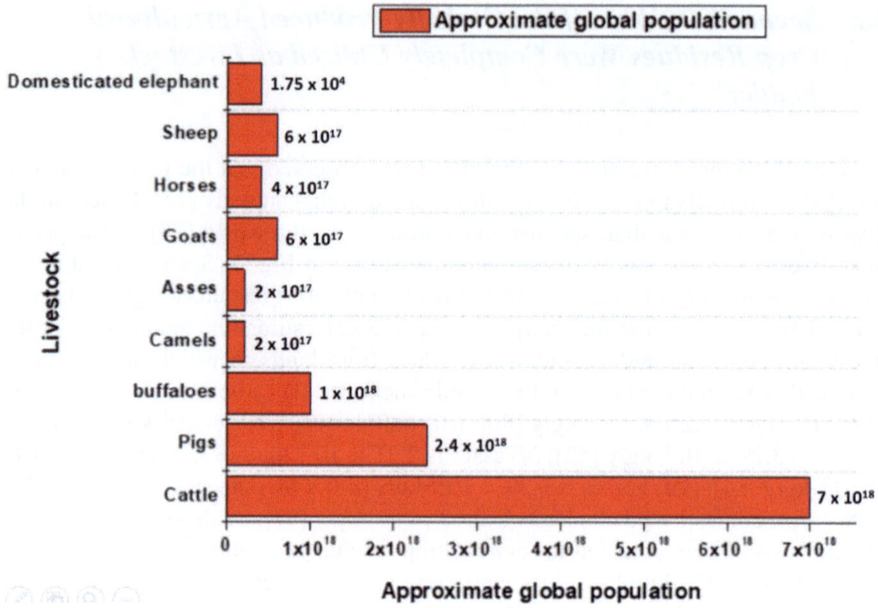

Fig. 7 An estimate of the global livestock population

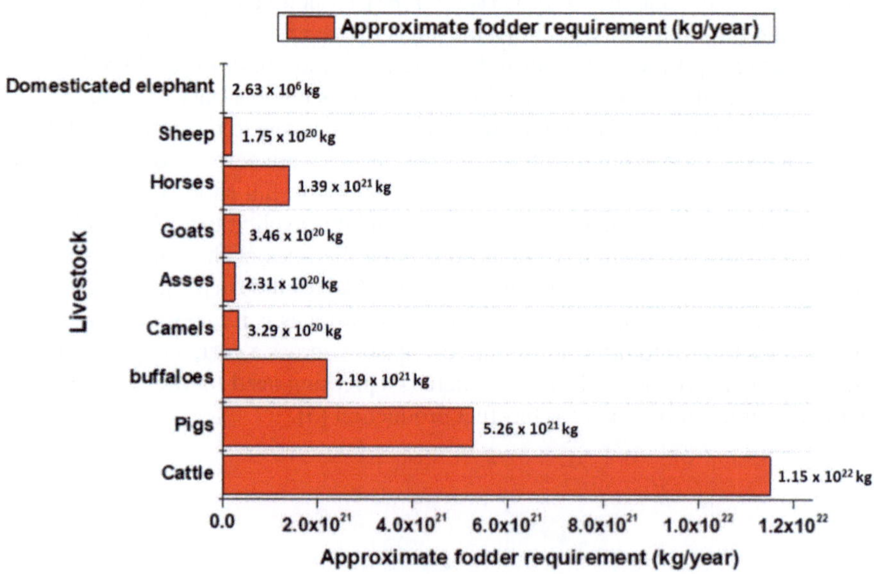

Fig. 8 An estimate of the fodder requirement for the global livestock population

6 Unconventional/Municipal Waste Stream-derived Lignocelluloses as Alternate Feedstocks Mitigate Capital Investments, and Improve Bio-refineries' Operational Bio-economy

A keen attention to the scopes of bio-process economics, along with meticulous cost–benefit analyses [5], is a prime requisite prior to large-scale commissioning of alternate biomass valorization routes, while also considering the low-value end-product—biofuel. One of the foremost factors having a significant influence on the final marketing price of bio-fuels is the cost incurred in feedstock procurement, which amounts to 75–80% of the total production costs [55]; in a country like India, the logistical challenges faced during the transport of biomass to bio-refineries contributes to roughly $0.05 USD [54]. Hence, opting for unconventional/waste stream-derived, highly energy-dense sources of bio-refinery feedstocks could be an alternative, which may improve its competitive investment pricing, attract governmental funds and venture capitalists alike [92]. But, scale-up-related challenges, and varying compositions of such alternate biomass sources impose a restriction on efficient quantification, and direct industrial-level implementation of waste-based bio-refinery models [42].

The biomass to bio-fuel nexus primarily dwells on establishing a balance between the harvested energy and the techno-economy of the entire operation [93] in a sustainable fashion [94]. Collection and transportation of biomass to appropriate pre-processing centers is a major challenge to be addressed [95] while considering the utilization of waste stream-derived lignocellulosic biomass. Such a process, apart from creating a recycle chain [96] and substantiating a country's overall socio-economic status [94], also creates avenues for economic development in rural areas of a nation [95]. Crop residues are usually transported to a centralized location as pressed briquettes/pellets, and supplied to a bio-refinery [95]; the initial accumulation of crop residues occurs in grain-processing locations, given the food grains' economic value. However, while considering large-scale requirement of waste-derived lignocelluloses, the labor and transportation costs may incur a heavy pricing on the produced bio-fuel [44]. In contrast to the described logistic operations concerning crop residues, the transportation of non-hazardous medical wastes and municipal wastes to landfill sites or disposal/incineration zones is performed using trucks [44]. Instead, as suggested in a recent work, if such trucks/transport vehicles are directed towards bio-refineries, a major cost involved in pre-processing of the waste biomass would considerably be reduced [44]. Another alternate, suggestive approach to utilize waste biomass such as vegetable/fruit waste, kitchen waste [97], vegetative waste, cellulose-rich hospital wastes as biomass [58], is the establishment of bio-refineries in closer proximity to such waste generation zones [95], thereby imparting a lesser burden on the geo-spatial cost component of a bio-alcohol plant too [92]. While considering urban vegetative wastes, which includes tree pruning, fresh/dried leaves, twigs, fallen branches, etc., [98] of a community, a through mapping, inclusive of rates of waste generation, geographical proximity and supply chain assessment, could be performed [99] to design a perennially operable 2G ethanol plant. Co-location of bio-refineries, with prior validation of the "repurpose potential," within biomass pre-processing centres' proximity would reduce the capital investments [92]; such strategic planning may further improve collaboration of industries and nurture industrial symbiosis [100]. If necessary, the policy-frameworks could be suitably altered in such a manner that such energy-dense food wastes are segregated at the point of collection [38]. A monitoring database, connected by a secure biometric system, as already proposed to be implemented in well-developed nations, would further aid in channelizing the smart collection and segregation of wastes [101], inculcating a socio-economic civic responsibility in every citizen. The housekeeping laborers, who at present are educated about the segregation of collected wastes, would also double as barefoot bio-refinery engineers [4]. Such employees would be provided with an orientation program to acquaint them with the processes involving minimal chemical usage and meager investments [90]. As a result, efficient, indigenous ethanol supply from recycled waste using a self-sustainable approach would become a reality [4]. In line with governmental legislations pertaining to food security, concerned with minimizing the generation of waste/or its effective valorization, such bio-energy harvesting techniques are potent circular and bio-economical sustainable options [38]. In terms of the operational economy, while conventional lignocelluloses demand a stage of pre-treatment to enhance the accessibility of the cellulose architecture to cellulolytic enzymes [5], waste-derived lignocelluloses typically include pre-processed and previously utilized commodities, such as fabric, cotton, bandages, papers, packaging cartons, cooked vegetables and fruit, municipal sludge, etc., which do not essentially demand a mandatory stage of ~40% cost-incurring pre-treatment [90]. Additionally, they may require lower enzyme loadings [5], owing to their easier potential to be de-crystallized/de-polymerized, consequentially releasing higher quantities of soluble saccharide entities [102]. Subsequently, fermentation times are decreased and the overall turnaround time for the end-product, bio-ethanol, would be lessened [102], adding to significant reduction in processes' economies [4].

6.1 Food Wastes as a Potential Feedstock for Bio-energy Plants

Surplus rice produce and cooked, spoilt food, majorly from households, cafeterias, restaurants, food-processing facilities, retail sectors, etc., [103] beyond the standards of human consumption or re-use in poultry/cattle farms [96] could be subjected to the cost-efficient dry milling, autoclave-based mashing (90–100 °C for 20 min), coupled with sequential enzymatic saccharification using α-amylase and glucoamylase [96] to result in a mash, which is fermentable by *Saccharomyces* spp. to bio-alcohols [96]. According to a report released by the European Commission, ~126 Mt of food waste was estimated to be accumulated towards the end of the year 2020, and it is expected to reach ~416 Mt by the year 2025 [103]. For instance, in the European Union, ~60% of all bio-waste is comprised of waste food, which bears a large possibility to be utilized as a bio-refinery feedstock [38]. Furthermore, the nutrient-rich nature of food wastes, which includes lipids, minerals, amino acids, apart from carbohydrates, proves to be a potential substrate for the production of a myriad of enzymes such as cellulases, pectinases, amylases, proteases, lipases, etc. [103]. Ramamoorthy et al. used a mixture of market-rejected rice grains and spoilt rice, containing a total reducing sugar content of 70% (w/w), to produce in-house amylolytic enzymes, employing *Aspergillus* spp. as the amylase-producing fungi [96]. Following stages of sequential amylolytic saccharification using α-amylase and glucoamylase, respectively, a bio-ethanol yield of 59.2 g/L has been reported [96]. An effective recycle loop for unconsumed sugars may be used to completely valorize the starch-rich substrate. Care has to be taken by establishing proper regulatory standards to assess the quality of grains prior to implementation of this technique, as it may recommence the fuel vs. food debate, which had made this age-old approach non-existent [96]. According to the Food and Agriculture Organization of the United Nations, ~1.3 billion tons of food wastes are generated across the world [93]. The following quantities of food-related waste generations have been recorded: beverage—26%; dairy—21%; fruit and vegetable processing—14.8%; cereal and grain processing—12.9% [104]. Matsakas and his research group employed a cellulolytic enzyme complex, produced using *Myceliopthora thermophila* to hydrolyze a 30% (w/w) dry material-containing food wastes, to record an ethanol titer of 19.27 g/L [103]. With fairly lesser establishment costs and self-sustainability, with respect to the hydrolytic enzyme production [105], such food/waste food-based bio-refineries can readily be employed for large-scale bio-alcohol production [96]. Throughout the European countries and other developed countries of the world, bread, being a staple food, is produced at quantities greater than 100 million tons per year, which is several-fold higher than its actual demand, in turn amounting to its wastage each day; for instance, annually ~2.92 k tons of bread is wasted in the United Kingdom [1]. Waste bread/unconsumed bread, which had crossed its date of expiry, with an average starch content of ~50–70% (w/w), was subjected to hydrolysis and enzymatic saccharification using a commercial enzyme formulation, followed by an *S. cerevisiae* KL17-based fermentation to result in bio-ethanol titers of 75 g/L (for the acid-hydrolyzed sample) and ~98 g/L (for the enzymatically hydrolyzed sample). The solid residues obtained after the acid and enzyme-based hydrolysis, and after the fermentation, resulted in a methanation potential of 345 mL and 379 mL of methane per gram of volatile solids, respectively [1]. The team of researchers further claimed that ethanol produced using bread waste may amount to a reduction of ~50% or more in the usage of fossil fuels [1].

6.2 Vegetable and Fruit Wastes, and Kitchen Wastes are Proven to be Rich Sources of Fermentable Sugars

Peels of fruits such as apricot, avocado, banana, raspberry, kiwi, sapodilla [97], watermelon, grapefruit, etc., roughly contain 2–14% (w/w) of galactose/maltose/fructose/sucrose/glucose. Kitchen wastes, including vegetable peels, spoilt portion and fibrous husks of coconuts [97], are typically rich sources of lignocelluloses and more than ~50% carbohydrates [38]; they also contain significant quantities of lipids, amino acids, minerals [106], and pentoses [97] in addition to being a feedstock with a lower carbon footprint [38]. These wastes, which usually form a part of household/kitchen wastes, are dumped in landfills [107]. As is the case with conventional sources of biomass, such vegetable and fruit peels also comprise cellulose, encased within networked layers of lignin and hemicellulose [39]. Post a dilute acid/alkali-mediated pre-treatment [97], following enzymatic saccharification and fermentation, considerably high titers of bio-ethanol have been reported [39].

6.3 Segregated Municipal Solid Wastes are Consistently Available

Urban vegetative waste accounts for several millions of tons of lignocellulose-rich residues each day in every country [99]. For instance, in a university's campus, sprawling ~780 acres, roughly 2 tons of such vegetative wastes are generated [4]. In a lower-middle-income, highest population-bearing country such as India, 0.1–0.6 kg of waste is generated each day [39]. While tier-1 cities generate around 50 k tons of wastes every day, the sub-urbs, and cumulative total from its 23 metropolitan cities amounts to 30 k tons/day [39]. The municipal solid waste comprises 40–45% of organic wastes, 8% of paper-based waste, 20% of recyclable waste, and 22% inert wastes [39]. Such biomasses, typically contain cellulose within the range of 23–36% (w/w), hemicellulose in the range of 6–8.45% (w/w), and lignin in the range of 27.74–11.16% (w/w) [99]. The biomasses undergo cellulolytic degradation to result in sugars of ~154.5 g/kg, and subsequent fermentation to ethanol to result in yields close to ~62.5 g/kg; as a stage of pre-processing, the wastes are subjected to pre-treatment using dilute acid/alkali [99]. As a real-time example, Raud et al. claimed that bio-ethanol, produced thus, has the potential to replace 60% of the commercially available diesel fuel usage in a city's public transportation [99]. In a pilot-scale bio-process, employing a shredded and autoclaved mixture of urban vegetative waste and e-commerce packaging wastes (3:1 ratio on weight-basis), Ramamoorthy and his colleagues performed a ~1-month-long oyster mushroom (*Pleurotus* sp.) cultivation, which resulted in 18.25% (w/w) of readily marketable mushrooms, cellulase yields of 0.88 g/d.s and 13.25% soluble sugars, and 78% delignification and complete hemicellulose consumption, thereby doubling as a stage of pre-treatment. During enzymatic saccharification, it was observed that the spent mushroom substrates required a reduced enzyme loading of 12 FPU/g, elucidating the efficacy of the process. Subsequent fermentation yielded ethanol titers of ~63.8 g/L. The researchers further reported that around ~100 kg of biomass could be processed in the self-constructed, low-capital bio-refinery hut; the additional revenue of $3.38 USD/kg of mushrooms (conversion from Indian National Rupees to United States Dollars—date of assessment 28 July 2023) is an added salient feature of the proposed work. In the above-elaborated approach, a typically cost-incurring pre-treatment process conversely yielded revenue instead [90]. Annually, ~6.72×10^7 kg of cellulose-rich packaging wastes and e-commerce-related packaging waste are generated, while ~21×10^8 tons of urban vegetative wastes are accumulated, globally, per year [90], thereby justifying their surplus generation rate and the year-long availability as a potential feedstock.

In a recent 100-L pilot-scale work employing a source-segregated bio-waste-based feedstock, Panagiota et al., in a ~24-h-long fermentation, obtained a bio-ethanol yield of $86.6 \pm 4.9\%$, recorded a cellulose degradation of $76.8 \pm 5.2\%$ and a higher starch decomposition of $94.6 \pm 2.4\%$ [38]. The researchers had collected ~3 tons of source-segregated bio-wastes, which comprised (in % w/w) 7.7 ± 3.4 celluloses, 9.4 ± 4.9 hemicelluloses, $12.5 \pm 5\%$ acid soluble lignin. With an optimized solid loading of 25%, the pilot trials of the dried feedstock were performed in a solid-state fermentor using treatments with commercial formulation of starch-degrading (40 μL/g) and cellulose-degrading enzymes (175 μL/g), respectively [38].

In an attempt to valorize sewage sludge, following mechanical milling to ~3 mm, moistening using 40% (v/v) water and autoclaving, the sludge material was subjected to a 48-h-long hydrolytic action of *Bacillus flexus* at 30 °C. The obtained, hydrolyzed sludge, during a ~10–70-h-long fermentation process, performed between 20 and 30 °C, at a pH of 6.5, and a yeast inoculum loading of 6% (w/v), resulted in a bio-ethanol titer of ~0.035% (v/v) [108].

6.4 Waste Papers are Cellulose-rich Wastes Seldom Entering the Recycle Chain

Newspapers contribute to ~35–40% percentage of municipal solid waste generation, and their disposal has become a menace in several developing countries, owing to decreased landfill capacities [109]. Shruti et al., following a 1.5% (v/v) sulfuric acid pre-treatment at 121 °C for 45 min, subjected the newspapers to a ~120-h long bacterial cellulase-catalyzed deconstruction to attain a maximum soluble sugar release of 0.14 g/g of substrate; subsequently, using baker's yeast (5% (v/v)) in a 144-h long fermentation, the researchers reported a bio-ethanol yield of ~6.12–6.91% (v/v) [109]. In a similar work, utilizing waste paper, Egbe Zimife et al. had reported bio-ethanol production after individual 6-h-long pre-treatments using 5% (v/v), 10% (v/v), and 50% (v/v) sulfuric acid and sodium hydroxide to 40 g waste paper [110].

6.5 Wastes from the Textile Industry Contain the Highest Quantity of Cellulose Per Unit Weight

On a daily basis, a typical cotton ginning industry, while processing every 227-kg bale of cotton, generates ~37–147 kg of cellulose-rich wastes, which are potential allergens causing occupational hazards and respiratory ailments [111]. Cotton fibers, the hairs of the seeds of *Gossypium* spp., after preliminary steps of mechanical processing, scouring, and bleaching, result in usable cotton, which is ~99–100% rich in cellulose [102], thereby making such biomasses readily amenable to cellulolytic hydrolysis, evading a step of pre-treatment [102]. Typically, cotton wastes/textile industry wastes are dumped in landfills or incinerated [102]. Decades earlier, in the year 1981, Brink had hypothesized that a cotton-based bio-refinery can amount to ~2–4 million gallons of ethanol annually [112]. Furthermore, amounting to a production rate of ~23 million tons/y, cotton accounts for one-third of the world-wide textile industries, elucidating a consistent perennial availability of the biomass source [113]. Mahalakshmi and colleagues, in a 24-h-long fermentation, performed at 35 °C for 24-h, using 5% (v/v) sulfuric acid pre-treated cotton waste as substrates and 1% (w/v) yeast inoculum loading, resulted in a bio-ethanol yield of 9.3% (v/v) [111]. Svetlana et al. utilized de-sized, bleached, 2-h-long corona-treated cotton fabrics and waste cotton as a substrate, followed by a commercial cellulolytic blend-based saccharification at 50 °C, in a18-h long *S. cervisiae* var. *ellipsoideus*-based fermentation (inoculum load of ~1.5×10^6 CFU/mL), at 30 °C, to yield bio-ethanol titers close to 0.9 g/L.h [102]. Cotton lint, pre-treated with 3% sulfuric acid, when subjected to a *S. cerevisiae*-based fermentation, resulted in a bio-ethanol yield of ~3–16.5 g/L; a 3.5-fold increase in ethanol titer has been recorded while using a 5% H_2SO_4-treated lint waste [114]. A mixture of cotton linter and denim material, after a 12% NaOH-based pre-treatment at 45 °C, when subjected to an enzymatic hydrolysis at 3% (w/v) substrate loading with a cellulase-cellobiase blend (20:30 IU/g) for 24–96-h, resulted in a 99% soluble sugar release. When subjected to baker yeast-based fermentation for ~4 days, at 37 °C, the same resulted in bio-ethanol titers of 0.48 g/g of glucose [113].

6.6 Biomedical Wastes, Comprising Topical Dressing Wastes from Hospitals are Cellulose-dense, and They are Available All Year-long

Ramamoorthy and his colleagues have demonstrated the usage of a mixture of surgical waste cotton and waste packaging cardboard for bio-ethanol production. Annually, around 0.65 and 0.7 k kg of waste surgical cotton and waste packaging card-boards are generated in India. These wastes are dumped in landfills and they do not find an established recycle route [58]. Post a step of pre-processing, comprising autoclaving, shredding, and pulverization, and 10% (w/v) *Saccharomyces cerevisiae* RW-143 inoculum-loaded 36-h-long fermentation in three different works, the research group reported bio-ethanol yields of: ~60 g/L while using 15% (v/v) liquor ammonia-treated biomass [115]; ~51.24 g/L after a *Paceilomyces* sp.-based biologically pre-treated biomass [116]; ~48 g/L after a cold atmospheric plasma-pre-treated biomass mixture [58]. Given its sustained availability, the stated source of biomass could be employed for cellulase complex production in batch [117] and fed-batch modes [118] of operation to enhance enzyme titers obtained from wild *Trichoderma* spp., which further would confer a bio-refinery process self-sustainable, circumventing the need for external procurement of enzyme complexes [102].

In a recent work, Ramamoorthy et al. have reported the usage of disposed COVID-19 PPE, majorly comprising cotton fabric-based masks, N-95 masks, lab aprons, hospital bedspreads, etc., for the production of a cellulolytic enzyme complex using *Trichoderma citrinoviride* VN-15. Since the onset of the pandemic, roughly 87×10^3 tons of PPE wastes have gathered, owing to improper disposal, contributing to 45% solid waste accumulation [44]. Currently, such PPE are disposed in landfills or incinerated [5], and these wastes do not have an established recycle route. Globally, around 52 k hospitals are existential, while in a lower-middle-income country such as India, around ~69 k hospitals exist [44]. After a stage of shredding and pulverization to an average particle size of ~0.625 μm and autoclaving, the segregated PPE were subjected to cyclic-fed-batches [44] and auxostat modes [5] of enzyme production in two parallel, individual works. Approximately, a 43% increase in cellulase titers were observed during cyclic fed-batches, and parameter-balancing auxostats resulted in consistent cellulase titers. Furthermore, the researchers hypothesized a bio-ethanol yield of ~143 g/L while utilizing COVID-19 PPE-based substrates [44].

6.7 Algal Biomasses are a Potential Feedstock to Cater to the Production of Bio-energies

Phototrophic and heterotrophic algae, majorly cultivated cost-efficiently in open pond systems, synthesize and store carbohydrates utilizing CO_2 from the atmosphere or simple organic sources, respectively [91]. While unicellular microalgae, capable of growing in typically lower incubation periods, are high in lipid content, the eukaryotic macroalgae comprise a lesser quantity of lipids in comparison, and possess other substances, such as carrageenan, sugar alcohols, alginate, etc. [91]. The high adaptability, an innate extremophilic nature, pesticide-free cultivation, and the requirement of moderate physiological factors deem *Arthrospira platensis*, commonly known as Spirulina, a cyanobacteriaum, is a potential feedstock candidate for bio-refineries [119]. Nahian et al., collected algae, dried, ground it, and subjected to a 2% (v/v) autoclave-based acid hydrolysis to release 15.2 g/L of fermentable sugars, and subsequently subjected the hydrolysate to ethanol production, resulting in a 0.9% (w/w) ethanol yield [119]. Algal cultivation, based on the nature of the species of algae chosen, is a cost-efficient process of wastewater management, and it also helps in reclaiming fresh water sources. Microalgae, consisting of ~11–15% soluble saccharides, and devoid of inhibitory quantities of lignin, can readily be implemented as a raw material for 2G ethanol bio-refineries [41]. Samar et al. harvested the microalgae, *Microcystis* sp., from a high-rate algal pond, treated it with 0.5 N sulfuric acid, at 120 °C for 4-h, neutralized it, and subjected the 45% sugar-containing saccharified hydrolysate to a 43.6-h-long anaerobic fermentation, at 30 °C, to result in a bio-ethanol yield of 18.6 g/L using a 15.09% (w/v) culture of *S. cerevisiae,* which was previously immobilized in sodium-alginate beads [41].

7 Full-scale Circular Economical Bio-refineries Promise Socio-economic Sustainability, Quick, Secure, and Improved Market Returns

Depletion of renewable energy resources and increasing pollution, in turn suppress the socio-economic and cultural facets of sustainability in energy production and environmental management; several governmental policymakers have been devising analytical data-driven models, frameworks, and integrated bio-processes towards improving the existing linear economy found in energy production alternatives [120]. Of late, bio-energy researchers have been adapting to a paradigm shift from the linear economy of bio-energy production systems [4], which comprises a single end-product, usually a result of several multi-stage unit-operation-involving processes [93]. Circular economy, a fairly new approach, stems out from the fundamentals of industrial ecology and symbiosis [120], to propose value-added, and commercializable/scalable bio-energy-related business propositions [4]. A full-fledged circular economical bio-refinery model [4], operating at a regio-specific/protypical eco-techno park scales, more or less aspiring to stay on the lines of natural bio-conversions, thrives on consolidating three essential principles: (i) recycling/regeneration of abundant wastes; (ii) establishment of a closed, cascaded loop [42] for maintaining substrate/energy flow and balance; (iii) amelioration and elimination of wastes/effluents [120] as value-added by-products [4]. However, tangibility of such propositions has been a major concern since their inception [120]. The prime variables to be considered during determination of the optimal parameters for improved biomass valorization, returns, and functionality, include: (i) a thorough estimation of circularity indices [120], right from the point of biomass collection until exit of the bio-refinery effluent [4]; (ii) empirical modeling of material balances in order to improve the energy/by-product harvest [4] while reducing the energy incurred in running the operation [120]; (iii) improving the shelf-life [120] and devising recycling, and for consistent generation rates of the products [4]. When considering the concept of circular energy utilization/by-product harvest/efficient recycle, such bio-process routes must prove to be consistent, economically viable, feasible, and meet the general principles of process economics [5].

A recent, intellectual property-secured, commercial-scale, circular economical, zero-waste bio-refinery model demonstrated by Ramamoorthy et al. presents a 126-day-long operation, where in 96 kg of a pre-defined ratio of vegetative waste and cellulose-dense e-commerce waste, when subjected to various individual unit-operations of the model, yielded a range of value-added and profitable products (Fig. 9) [121].

The different stages of this model are as follows: (i) The first stage of mushroom cultivation yielded 36 kg of oyster mushrooms, and parallelly resulted in 73.8% lignin degradation; (ii) the remnant biomass, during a subsequent solid-state fermentation using a species of *Trichoderma* sp., resulted in a cellulase complex with 124 k filter paper units; (iii) the spores harvested from the solid-state fermentation resulted in ~1.2×10^{13} bio-priming agents; (iv) post saccharification and fermentation of the remnant biomass from the cellulase production process, ~1.1 L of bio-ethanol was produced; (v) the biomass residues, obtained after the fermentation process, resulted in ~6.8 kg of organic bio-fertilizer, post subjection to a cow dung-based aerobic windrow composting [4]. The research group further reported, based on a cost–benefit analysis, that the model is ~70% profitable while evaluating the market

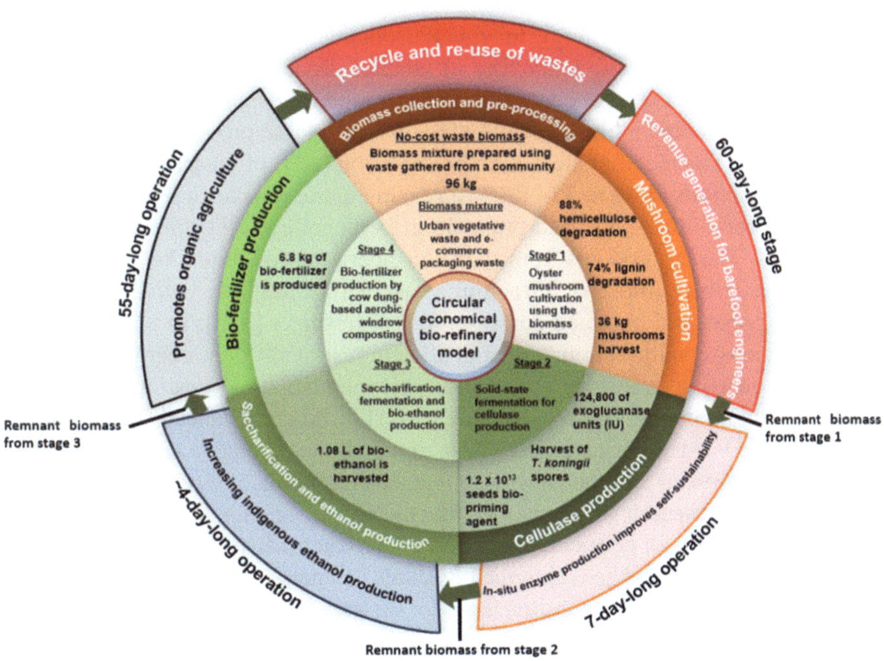

Fig. 9 An illustrative representation of a zero-waste circular economical bio-refinery model proposed by Ramamoorthy et al. [4, 5], Pallam et al. [121]

value of the products against expenses incurred during the bio-process [4]. The cascading biomass-utilization model is in line with the concept of the stage III bio-refineries proposed by Clark and Deswarte, where several raw materials, subjected to multiple, cost-effective unit-operations, result in commercially viable end products and by-products [54]. Following the principles of the Circular Economy Action Plan (CEAP), the bio-refinery model dwells on the dogma of reducing wastes, re-using or valorizing the same, and recycling the process effluents [54]. In principle, a zero-waste bio-refinery is regarded commendable, inclusive of continuous cycles of energy harvest [54] and effluent valorization [100].

Caroline et al. postulated and demonstrated a circular economical bio-refinery approach using rotting/rotten sweet potatoes' wastes as substrates for alcohol-based sanitizer production, wherein they observed a breakeven point during the production of 34% bio-alcohol; while a 39% rate of production showed marginally increased profitability, an 80% increased rate of production resulted in a net present value of ~$9.13 k USD, an internal rate of return of 64%, and a discounted payback period of 0.7 year [93]. Starchy tubers account for an annual food wastage of 45%, while, in countries such as Brazil, sweet potatoes individually account for ~350 k tons of wastage [93]. In the mentioned work, sweet potato wastes were collected, subjected to manual size reduction, steaming at 76 °C, mashing, sequential saccharification using commercial formulations of α-amylase and glucoamylase, and a 19-h-long *S. cerevisiase*-based fermentation at 34 °C to produce bio-ethanol [93].

In a recent review, Julio and colleagues highlighted the potential of valorization of the 11% residual energy in vinasse, obtained from a typical cane-based ethanol production effluent; every liter of ethanol production amounts to 15 L of vinasse [122]. With high concentrations of nitrogen, phosphorous, sulfate, potassium, melanoidins, and a highly acidic pH (~4–6), Vinasse proves to be a potential pollutant [122]. The fact that the anaerobic digestion (AD) of vinasse could result in methane productions close to 324 L/kg of organic matter, makes it a potential, scalable, circular economical, business proposition to integrate an AD plant in cane-based bio-refineries. The by-products from such unit-operations include filter cakes, straw, cane bagasse, hemicellulose-rich fraction, yeast extract, and ash; with a ~30% solid content, the press mud obtained could serve as a fertilizer [123]. In a similar work focused on vinasse valorization, Victor et al. demonstrated water, energy, and fertilizer production using an aerobic and an anaerobic membrane bio-reactor. While, with respect to organic matter, the anaerobic reactor showed an enhanced 97% removal, and the aerobic bio-reactor recorded a 92% removal, the aerobic reactor recorded an 80% nitrogen removal and the anaerobic reactor recorded a 56% nitrogen removal. Furthermore, the research group claimed that the aerobic reactor would be capable of recovering ~527 cu.m of permeates per hour, which can readily be used as a precursor for fertilizer production. The technique possesses an energy generating capability of 26.1 kW/m^3 of vinasse; the operational cost of the aerobic bio-reactor is ~14% lesser than that recorded for the anerobic bio-reactor [122].

Post a stage of biomass pre-treatment, a significant fraction of hemicelluloses (~20–50%), comprised of its fundamental unit xylan is harvested, and it finds wide applicability in the production of xylo-oligosaccharides, which serves as a valuable precursor in food and energy sectors [53]. In the food sector, while xylose itself serves as a source of an artificial sweetener, xylo-oligosaccharides are used in the production of pre-biotics, such as xylobiose and xylotriose, to improve the population of intestinal *Bifidobacterium*, which offers several health benefits ranging from relieving constipation, improved intestinal absorption of nutrients, and controlling diarrhea, related to gastro-intestinal disorders and cancers [53].

In spite of several circular bio-economical models being proposed and being adapted on a larger scale, as stated earlier, its quantification/tangibility in terms of by-products, energy flows, and harvests is an ordeal. A thorough framework for its assessment, based on multi-criteria analysis dependent on reliable process indicators, and subsequent study of its feasibility in real-time circular economical bio-refineries could prove to be a potential solution [100]. With emerging real-time data analytics and artificial intelligence-driven models, such multi-variate analyses would become comparatively less tedious [44]. In a recent work, taking into account 8 key areas and 18 process indicators, Linda et al. assessed 6 different approaches for valorization of stillage by-product from a wheat-utilizing bio-refinery [100]. They ascertained the utilization of the stillage by-product as fodder, fertilizer, for incineration, as feedstock for distant biogas plants, as feedstocks for closely situated biogas plants, and as feedstock for local biogas for heat and power. Among all the studied alternatives, the research group zeroed-in on the by-product's usage for animal fodder or as a feedstock for a nearby biogas station [100].

8 Biomass Banks and Bio-refinery Co-operatives Provide a Futuristic Outlook

A suggestive and a newer approach towards effective utilization of agricultural, non-agricultural residues, and waste-based lignocellulose is the postulation of regulatory directives to regularize their collection, pre-processing/logistic management, and transportation to bio-refineries. A standing example is the establishment of 20 "biomass banks" in rural districts of the state of Maharashtra of the Indian sub-continent. The banks would install biomass pellet makers, which after systematic storage of pelleted biomass, would source them to industries requiring the same as feedstocks [124].

Akin to the government-regulated and financed dairy co-operatives existing in several countries, the nations' governments [125], apart from incentivizing bio-alcohol production and its connected service loops, can also establish well-regulated and monitored bio-refineries, which may be self-financed or partly government-funded [126]. As proposed in a recent circular economical bio-refinery model [4, 121], and a pilot-scale mushroom cultivation-based pre-treatment work [90], the low-income marginal community farmers, and house-keeping laborers of the society, who are also referred to as barefoot engineers, when imparted with basic techniques of unit-operations, can run their own co-operative bio-refineries [127]. Smaller farmers can supply their produce/collected wastes to a mid-sized co-operative, where bio-alcohols would be produced [44]. Ethanol, produced thus, would be supplied to a larger regio-specific central facility, where post assessment of the bio-alcohol's quality using specific gravity quantification, would be transported to commercial fuel-filling stations of the state. Such a systematic functioning, if implemented, can greatly contribute to a nation's: (i) indigenous ethanol production, (ii) self-sustainability; (iii) independence on fossil fuel/crude oil import; (iv) increase in GDP; (v) waste management and reduction in environmental pollution; (vi) rural empowerment by providing a supply chain to generate revenue for the economically weak barefoot engineers.

9 Conclusions and Future Directions

Three-fourth of bio-refineries' end-product's cost-determining factor is the price incurred in sourcing conventional crop-based lignocellulosic residues as feedstocks; the urgency in the need for waste-based bio-refineries has been sketched out in the chapter. However, another ~30–40% of the operational cost of bio-refineries are determined by the pre-treatment procedures employed. Mechanical, chemical, hydrothermal, ionic liquid-based, deep eutectic solvent-based pre-treatments, though are efficient in terms of their intended purposes, have been reported to be energy and cost-intensive; alternate techniques, such as biological pre-treatment using ligninolytic bacteria/fungi/microbial consortia, in vitro bio-catalyst-based pre-treatments, apart from not generating fermentation inhibitors, have been observed to be economical, while the only drawback being the varying efficiencies of processes and lengthier operational durations. Biological pre-treatments, implementing genetically manipulated microbes, though are being experimented at lab-scales, their pilot- and commercial-scale applicability has not been majorly reported. Mushroom cultivation using feedstocks has been reported to be a lucrative agripreneurial venture, wherein the growth of mushrooms doubles as a biomass's lignin degradation step; during extended operational cycles, mushroom cultivation over lignocelluloses has also been proven to secrete cellulases, simultaneously saccharify and solubilize a significant portion of its sugar polymers. Photocatalytic bio-reactors have been implemented towards direct saccharification of biomass, circumventing a step of pre-treatment. In-house/on-site/in situ cellulase production greatly reduces ~50% the expenses incurred in procuring commercial cellulase blends; several wild-type cellulolytic fungus-driven, alternate cellulose-rich substrates-employing, cost-effective scalable works have been reported. Integrated techniques such as simultaneous saccharification and fermentation (SSF), and simultaneous saccharification and co-fermentation (SScF) have been performed using a microbial co-culture or using genetically engineered microbes. Consolidated bio-processing (CBP), employing a single microbe to saccharify and ferment a source of biomass, though has several reported lab-scale investigations, has not been majorly reported to be working at industrial scales. Co-utilization/co-fermentation of hexose and pentose sugars, which potentially improves bio-alcohol titers and profit margins, has been attempted using wild microbial strains, and genetically modified microbial strains, and as separate unit-operations too. Upgradation of short-chain bio-alcohols to a higher energy density-possessing larger-chain alcohols is also being attempted. With several bio-energy research groups implementing techniques stemming out of various scientific domains, and concurrently working on futuristic sustainable energy production models, decision-makers have several options to invest. With the world's superpowers combining to invest in bio-energy alternatives, a consortium of several nations with a unanimous motto of "economic bio-energy for all," should begin commissioning ready-to-scale alternate energy production models. Furthermore, increased encouragement of bio-energy-related works in the academic research scene would draw more young minds towards addressing the conundrum, and jointly driving the policy-frameworks and government investments towards creating a robust future for the generations to come.

Statements and Declarations

Funding The first author is thankful to the University Grants Commission for the financial assistance received through the Dr. D. S. Kothari postdoctoral fellowship scheme.

Competing Interests The authors hereby declare that they have no competing interests.

References

1. Narisetty V, Nagarajan S, Gadkari S, Ranade V, Zhang J, Patchigolla K, Bhatnagar A, Awasthi MK, Pandey A, Kumar V (2022) Process optimization for recycling of bread waste into bioethanol and biomethane: a circular economy approach. Energy Convers Manage. https://doi.org/10.1016/j.enconman.2022.115784
2. Overview and key findings (2022) World energy investment. IEA. https://www.iea.org/reports/world-energy-investment-2022/overview-and-key-findings. Accessed 21 July 2023
3. Will energy security concerns drive biofuel growth in 2023 and 2024 (2023) Renewable energy market update—June 2023. IEA. https://www.iea.org/reports/renewable-energy-market-update-june-2023/will-energy-security-concerns-drive-biofuel-growth-in-2023-and-2024. Accessed 21 July '23
4. Ramamoorthy NK, Vengadesan V, Pallam RB, Krishnasamy V, Sadras SR, Sahadevan R, Vemuri VS (2023a) A commercial-scale, circular-economical bio-refinery model for sustainable yields of mushrooms, cellulase-complex, bio-priming agents, bio-ethanol, and bio-fertilizer. Kavaka 59:26–39. https://doi.org/10.36460/Kavaka/59/2/2023/26-39
5. Ramamoorthy NK, Pallam RB, Subash Chandrabose K, Sahadevan R, Vemuri VS (2023b) A critical process variable-regulated, parameter-balancing auxostat, performed using disposed COVID-19 personal protective equipment-based substrate mixture, yields sustained and improved endoglucanase titers. Prep Biochem Biotechnol. https://doi.org/10.1080/10826068.2023.2204479
6. Ramamoorthy NK, Ghosh P, Sahadevan R, Sarma VV (2022a) An insight into the applications of fungi in ethanol biorefinery operations. In: Deshmukh SK, Kandikere RS, Badalyan SM (eds) fungal biotechnology prospects and avenues, 1st edn. CRC Press, Boca Raton
7. Natural gas fuel basics. Alternative fuels data center. https://afdc.energy.gov/fuels/natural_gas_basics.html. Accessed 21 July 2023
8. Mazloomi K, Gomes C (2012) Hydrogen as an energy carrier: prospects and challenges. Renew Sustain Energy Rev 16(5):3024–3033. https://doi.org/10.1016/j.rser.2012.02.028

9. Have commenced dispensing CNG and PNG connections to households: IndianOil. https://economictimes.indiatimes.com/industry/energy/oil-gas/have-commenceddispensing-cng-and-png-connections-to-households-indianoil/articleshow/100214761.cms?from=mdr. Accessed 21 July 2023
10. Bhardwaj S, Das P (2017) A review: advantages and disadvantages of biogas. Res J Eng Technol (IRJET) 4(10):890–893
11. What are the advantages and disadvantages of natural gas. Energy Tracker Asia. https://energytracker.asia/advantages-and-disadvantages-of-natural-gas/. Accessed 3 July 2023
12. Liquefied natural gas: understanding the basic facts. https://www.energy.gov/fecm/articles/liquefied-natural-gas-understanding-basic-facts. Accessed 21 July 2023
13. Kandemir T, Schuster ME, Senyshyn A, Behrens M, Schlögl R (2013) The Haber-Bosch process revisited: on the real structure and stability of "ammonia iron" under working conditions. Angewandte Chemie (International ed. in English). 52(48):12723–12726. https://doi.org/10.1002/anie.201305812
14. Valera-Medina A, Amer-Hatem F, Azad AK, Dedoussi IC, de Joannon M, Fernandes RX, Glarborg P, Hashemi H, He X, Mashruk S, McGowan J, Mounaim-Rouselle C, Ortiz-Prado A, Ortiz-Valera A, Rossetti I, Shu B, Yehia M, Xiao H, Costa M (2021) Review on ammonia as a potential fuel: from synthesis to economics. Energy Fuels 35(9):6964–7029. https://doi.org/10.1021/acs.energyfuels.0c03685
15. Synthetic ammonia. United States Environmental Protection Agency. https://www3.epa.gov/ttnchie1/ap42/ch08/final/c08s01.pdf. Accessed 21 July 2023
16. Ammonia for power: a literature review. Ammonia Energy Association. https://www.ammoniaenergy.org/. Accessed on 3 July 2023
17. Yüzbaşıoğlu AE, Avşar C, Gezerman AO (2022) The current situation in the use of ammonia as a sustainable energy source and its industrial potential. Curr Res Green Sustain Chem 5. https://doi.org/10.1016/j.crgsc.2022.100307
18. Herbinet O, Bartocci P, Dana AG (2022) On the use of ammonia as a fuel—A perspective. Fuel Commun 11. https://doi.org/10.1016/j.jfueco.2022.100064
19. Coal & Electricity. World Coal Association. https://www.worldcoal.org/coal-facts/coal-electricity/. Accessed on 3 July 2023
20. Coal Equivalent. European Nuclear Society. https://www.euronuclear.org/glossary/coal-equivalent/. Accessed July 2023
21. Lithium-ion battery. University of Washington. https://www.cei.washington.edu/education/science-of-solar/battery-technology/. Accessed 5 July 2023
22. Probstein RF, Hicks RE (2003) Synthetic fuels. In: Meyers RA (ed) Encyclopedia of physical science and technology (Third Edition). Academic Press. https://doi.org/10.1016/B0-12-227410-5/00761-4
23. Advantages and disadvantages of electrical energy. Nuclear Energy. https://nuclear-energy.net/energy/electricity/advantages-and-disadvantages. Accessed 3 July 2023
24. Herraiz L, Lucquiaud M, Chalmers H and Gibbins J (2020) Sequential combustion in steam methane reformers for hydrogen and power production with CCUS in decarbonized industrial clusters. Front Energy Res 8(180). https://doi.org/10.3389/fenrg.2020.00180
25. Song H, Yang G, Xue P, Li Y, Zou J, Wang S, Yang H, Chen H (2022) Recent development of biomass gasification for H_2 rich gas production. Appl Energy Combust Sci 10. https://doi.org/10.1016/j.jaecs.2022.100059
26. Yang H, Chen H (2015) Biomass gasification for synthetic liquid fuel production. In: Luque R, Speight JG (eds) Gasification for synthetic fuel production. Woodhead Publishing. https://doi.org/10.1016/B978-0-85709-802-3.00011-4
27. Wang Z, Roberts RR, Naterer GF, Gabriel KS (2012) Comparison of thermochemical, electrolytic, photoelectrolytic and photochemical solar-to-hydrogen production technologies. Int J Hydrogen Energy 37(21):16287–16301. https://doi.org/10.1016/j.ijhydene.2012.03.057
28. Cavelius P, Engelhart-Straub S, Mehlmer N, Lercher J, Awad D, Brück T (2023) The potential of biofuels from first to fourth generation. PLoS Biol 21(3). https://doi.org/10.1371/journal.pbio.3002063

29. Knosala K. Langenberg L, Pflugradt N, Stenzel P, Kotzur L, Stolten D (2022) The role of hydrogen in German residential buildings. Energy Build 276. https://doi.org/10.1016/j.enbuild.2022.112480
30. Yue, Lambert H, Pahon E, Roche R, Jemei S, Hissel D (2021) Hydrogen energy systems: a critical review of technologies, applications, trends and challenges. Renew Sustain Energy Rev 146. https://doi.org/10.1016/j.rser.2021.111180
31. Ramamoorthy NK, Ghosh P, Sahadevan S, Sarma VV (2022b) Bioprocessing approaches for enzyme-based waste biomass saccharification. In: Varma P (ed) Enzymes in the valorization of waste, 1st edn. CRC Press
32. Nachenius RW, Ronsse F, Venderbosch RH, Prins W (2013) Biomass pyrolysis. In: Advances in chemical engineering, vol 42. Academic Press. https://doi.org/10.1016/B978-0-12-386505-2.00002-X.
33. Top applications of biofuels (2020) AZ big media. https://azbigmedia.com/business/top-applications-of-biofuels/. Accessed 3 July 2023
34. Nuclear energy. National geographic. https://education.nationalgeographic.org/resource/nuclear-energy/. Accessed 3 July 2023
35. Computing the energy density of nuclear fuel. https://whatisnuclear.com/energy-density.html. Accessed 3 July 2023
36. Nuclear power in a clean energy system. Fuel report IEA. https://www.iea.org/reports/nuclear-power-in-a-clean-energy-system. Accessed 3 July 2023
37. Advantages and challenges of nuclear energy (2021) Office of nuclear energy. https://www.energy.gov/ne/articles/advantages-and-challenges-nuclear-energy. Accessed 3 July 2023
38. Tsafara P, Passadis K, Christianides D, Chatziangelakis E, Bousoulas I, Malamis D, Mai S, Barampouti EM, Moustakas K (2022) Advanced bioethanol production from source-separated bio-waste in pilot scale. Sustainability. https://doi.org/10.3390/su141912127
39. Meenakshi, Sangwan, S, Shikha (2021) Biodegradable municipal solid waste as substrate for ethanol production: a review. Just Agric 2(1)
40. Roundtable on Environmental Health Sciences, Research, and Medicine; Board on Population Health and Public Health Practice; Institute of Medicine (2014) The Nexus of biofuels, climate change, and human health: workshop summary. National Academies Press (US), Washington (DC). Ethical and social issues. https://www.ncbi.nlm.nih.gov/books/NBK196458/. Accessed 31 July 2023
41. El-Mekkawi S, Abdo SM, Samhan FA, Ali GH (2019) Optimization of some fermentation conditions for bioethanol production from microalgae using response surface method. Bull Natl Res Centre 43. https://doi.org/10.1186/s42269-019-0205-8
42. Althuri A, Venkata Mohan S (2022) Emerging innovations for sustainable production of bioethanol and other mercantile products from circular economy perspective. Bioresour Technol 363. https://doi.org/10.1016/j.biortech.2022.128013
43. Deenanath ED, Iyuke S, Rumbold K (2012) The bioethanol industry in sub-Saharan Africa: history, challenges, and prospects. J Biomed Biotechnol. https://doi.org/10.1155/2012/416491
44. Ramamoorthy NK, Pallam RB, Sahadevan R, Sarma VV (2023) Cellulase production from disposed COVID-19 personal protective equipment (PPE) using cyclic fed-batch strategies. Process Biochem 127:112–126. https://doi.org/10.1016/j.procbio.2023.02.013
45. Dahman Y, Syed K, Begum S, Roy P, Mohtasebi B (2019) Biofuels: their characteristics and analysis. In: Verma D, Fortunati E, Jain S, Zhang X (eds) Biomass, biopolymer-based materials, and bioenergy. Woodhead Publishing Series in Composites science and engineering. Woodhead Publishing. https://doi.org/10.1016/B978-0-08-102426-3.00014-X.
46. Gasparatos A, Stromberg P, Takeuchi K (2013) Sustainability impacts of first-generation biofuels. Anim Front 3(2):12–26. https://doi.org/10.2527/af.2013-0011
47. Kumar R, Dhurandhar R, Chakrabortty S, Ghosh AK (2022) Downstream process: toward cost/energy effectiveness. In: Sahay S (ed) Handbook of biofuels. Academic Press. https://doi.org/10.1016/B978-0-12-822810-4.00012-9
48. Ullmann J, Grimm D (2021) Algae and their potential for a future bioeconomy, landless food production, and the socio-economic impact of an algae industry. Organic Agric 11. https://doi.org/10.1007/s13165-020-00337-9

49. Sahle-Demessie E, Hassan AA, El Badawy A (2019) Bio-desalination of brackish and seawater using halophytic algae. Desalination 465:104–113. https://doi.org/10.1016/j.desal.2019.05.002
50. Chislock MF, Doster E, Zitomer RA, Wilson AE (2013) Eutrophication: causes, consequences, and controls in aquatic ecosystems. Nat Educ Knowl 4(4):10
51. Taniguchi H, Okano K, Honda K (2017) Modules for in vitro metabolic engineering: pathway assembly for bio-based production of value-added chemicals. Synth Syst Biotechnol 2(2):65–74. https://doi.org/10.1016/j.synbio.2017.06.002
52. Brookes G, Barfoot P (2020) Environmental impacts of genetically modified (GM) crop use 1996–2018: impacts on pesticide use and carbon emissions. GM Crops Food 11(4):215–241. https://doi.org/10.1080/21645698.2020.1773198
53. Pinales-Márquez C, Rodriguez-Jasso RM, Araújo RG, Loredo-Treviño A, Nabarlatz D, Gullón B, Ruiz AH (2021) Circular bioeconomy and integrated biorefinery in the production of xylooligosaccharides from lignocellulosic biomass: a review. Ind Crops Prod 162. https://doi.org/10.1016/j.indcrop.2021.113274
54. Devi A, Bajar S, Kour H, Kothari R, Pant D, Singh A (2022) Lignocellulosic biomass valorization for bioethanol production: a circular bioeconomy approach. Bioenergy Res 15(4):1820–1841. https://doi.org/10.1007/s12155-022-10401-9
55. Adewuyi A (2022) Underutilized lignocellulosic waste as sources of feedstock for biofuel production in developing countries. Front Energy Res. https://doi.org/10.3389/fenrg.2022.741570
56. Various advantages and disadvantages of biofuels. Conserve energy future. https://www.conserve-energy-future.com/advantages-and-disadvantages-of-biofuels.php. Accessed 21 July 2023
57. Drawbacks of bioethanol. Biofuels. https://sites.lafayette.edu/egrs352-sp15-biofuels/drawbacks-of-bioethanol/. Accessed 21 July 2023
58. Ramamoorthy NK, Nagarajan R, Ravi S, Sahadevan R (2020) An innovative plasma pretreatment process for lignocellulosic bio-ethanol production. Energy Sour Part A Recovery Utilization Environ Effects. https://doi.org/10.1080/15567036.2020.1815900
59. The biofuel controversy. Eyes on environment. https://www.nature.com/scitable/blog/eyes-on-environment/the_biofuel_controversy/. Accessed 21 July 2023
60. What's the problem with biofuels, Emissions analytics. https://www.emissionsanalytics.com/news/whats-the-problem-with-biofuels. Accessed 21 July 2023
61. Sunday letters: biofuel skepticism (2013) Houston Chronicle, Chron. https://www.chron.com/opinion/letters/article/sunday-letters-biofuel-skepticism-4649465.php. Accessed 21 July 2023
62. Sindhu R, Binod P, Pandey A, Ankaram S, Duan Y, Awasthi MK (2019) Biofuel production from biomass: toward sustainable development. In: Kumar S, Kumar R, Pandey A (eds) Current developments in biotechnology and bioengineering, Elsevier. https://doi.org/10.1016/B978-0-444-64083-3.00005-1
63. Biofuels industry fighting rising tide of skepticism (2013) Advanced biofuels USA. https://advancedbiofuelsusa.info/biofuels-industry-fighting-rising-tide-of-skepticism/. Accessed 21 July 2023
64. The biofuels myth. Why 'Sustainable Aviation Fuels' won't power climate-safe air travel. https://biologicaldiversity.org/programs/climate_law_institute/pdfs/2022_The_Biofuels_Myth_Center_for_Biological_Diversity.pdf. Accessed 21 July '23
65. Pallam RB, University P, Rakshit S, Ramamoorthy NK, Sahadevan R, Sarma VV (2023) Ein hybride Bioreaktorsystem fur die Durchfuhrung von Fermentationen mit hoher Myzeldichte und quasistationarem Zustand zur Herstellung von Pilzenzymen. (A hybrid bio-reactor for high mycelial density, quasi-steady-state fungal enzyme production) German Patent and Trademark Office DE 20 2023 100 255
66. Biofuels pioneer vents fury at City as he quits (2008) The Guardian. https://www.theguardian.com/environment/2008/mar/11/energy.climatechange. Accessed 23 July 2023

67. The essentiality of transportation to the pharma industry (2020) Pharma's alamanac. https://www.pharmasalmanac.com/articles/the-essentiality-of-transportation-to-the-pharma-industry. Accessed 21 July 2023
68. Pharmaceutical logistics. Centre for excellence for independent validators. https://www.iata.org/contentassets/494bc14afd934b0193735e9a47091d72/pharma-ceiv-brochure.pdf. Accessed 23 July 2023
69. The largest countries of the world. Worlddata. https://www.worlddata.info/the-largest-countries.php/. Accessed 23 July 2023
70. Annual refined petroleum products consumption (2022) U.S. energy information administration. https://www.eia.gov/international/data/world/petroleum-and-other-liquids/annual-refined-petroleum-products-consumption. Accessed 5 July 2023
71. Field listing—land use. CIA. https://www.cia.gov/the-world-factbook/field/land-use/. Accessed 3 July 2023
72. The world factobook word. https://www.cia.gov/the-world-factbook/countries/world/. Accessed 3 July 2023
73. Energy investment is set to pick up by 8% in 2022 against the backdrop of the global energy crisis, but almost half of the increase in capital spending is linked to higher costs. https://www.iea.org/reports/world-energy-investment-2022/overview-and-key-findings. Accessed July 2023
74. Livestock turn food we can't eat into protein. Global food justice alliance. https://www.globalfoodjustice.org/environment/livestock-turn-food-we-cant-eat-into-protein. Accessed 3 July 2023
75. Fact #676: May 23, 2011 U.S. Refineries produce about 19 gallons of gasoline from a barrel of oil. Office of energy efficiency & renewable energy. https://www.energy.gov/eere/vehicles/fact-676-may-23-2011-us-refiners-produce-about-19-gallons-gasoline-barrel-oil. Accessed 23 July 2023
76. Crude oil price live. Commodities control. http://www.commoditiescontrol.com/live-crude-oil-price.html. Accessed July 2023
77. Number of animals on the earth. Pets on mom. https://animals.mom.com/number-animals-earth-3994.html. Accessed 5 July 2023
78. Wild mammals make up only a few percent of the world's mammals. Our world in data. https://ourworldindata.org/wild-mammals-birds-biomass. Accessed 3 July 2023
79. Scientific feeding of dairy animals. ICAR – Krishi Vigyan Kendra, North Goa. https://kvknorthgoa.icar.gov.in/litpub/Technical%20Folders/Animal%20Science/Scientific%20Feeding%20of%20Dairy%20Animals.pdf. Accessed 5 July 2023
80. Livestock::Pig::Feeding management. TNAU Agritech portal. Animal husbandry. https://agritech.tnau.ac.in/animal_husbandry/ani_pig_feeding%20mgt.html. Accessed 5 July 2023
81. Livestock: Buffalo. ICAR CCARI. https://ccari.icar.gov.in/dss/buffalo.html. Accessed 5 July 2023
82. Forage for camels. Douliere Hay France. https://doulierehayfrance.com/en/forage-for-camels/. Accessed 5 July 2023
83. How to safely feed donkeys. Tribune superior equine nutrition. https://tributeequinenutrition.com/blogs/news/how-to-safely-feed-donkeys. Accessed 5 July 2023
84. Goat fodder requirement for beginners. Goat farming. https://www.goatfarming.in/goat-fodder-information-for-beginners. Accessed 5 July 2023
85. Feed requirement of horses. Livestock and animals. https://agriculture.vic.gov.au/livestock-and-animals/horses/health-and-welfare/feed-requirements-of-horses. Accessed 5 July 2023
86. Feeding management and methods. Expert system for sheep and goat. http://www.agritech.tnau.ac.in/expert_system/sheepgoat/Feeding%20Management%20of%20Sheep%20and%20Goats.html. Accessed 5 July 2023
87. Priyadharshana M, Vadivel V (2022) Fodder plants and foraging behaviour of Asian elephants in Srivilliputhur Elephant Reserve, Tamil Nadu, India. Indian J Nat Prod Resour 13(3):391–397. https://doi.org/10.56042/ijnpr.v13i3.50452

88. Hatab AA, Cavinato ME, Lindemer A, Lagerkvist CJ (2019) Urban sprawl, food security and agricultural systems in developing countries: a systematic review of the literature. Cities. https://doi.org/10.1016/J.CITIES.2019.06.001
89. Joko T, Anggoro S, Sunoko HR, Rachmawati S (2017) Pesticides usage in the soil quality degradation potential in Wanasari Subdistrict, Brebes, Indonesia. Appl Environ Soil Sci. https://doi.org/10.1155/2017/5896191
90. Ramamoorthy NK, Vengadesan V, Pallam RB, Sadras SR, Sahadevan R, Sarma VV (2023) A pilot-scale sustainable biorefinery, integrating mushroom cultivation and in-situ pretreatment-cum-saccharification for ethanol production. Prep Biochem Biotechnol. https://doi.org/10.1080/10826068.2022.2162922
91. Özçimen D, Benan İnan (2014) An overview of bioethanol production from algae. In: Biernat K (ed) Biofuels status and perspectives. IntechOpen, Poland. https://doi.org/10.5772/59305
92. Martinkus N, Latta G, Brandt K, Wolcott M (2018) A multi-criteria decision analysis approach to facility siting in a wood-based depot-and-biorefinery supply chain model. Front Energy Res 6. https://doi.org/10.3389/fenrg.2018.00124
93. Trevisan Weber C, Ranzan L, Menz Liesegang LL, Ferreira Trierweiler L, Trierweiler JO (2020) A circular economy model for ethanol and alcohol-based hand sanitizer from sweet potato waste in the context of COVID-19. Braz J Oper Prod Manage 17(3):1–12. https://doi.org/10.14488/BJOPM.2020.028
94. Raina N, Sriariyanun M, Areeya S, Pathak AK, Slathia G, Saraswat H (2022) Trends in harnessing energy from waste biomass: pathways & future potential. In: ES3 web of conferences, p 355. https://doi.org/10.1051/e3sconf/202235501001
95. Lin T, Rodríguez LF, Davis S, Khanna M, Shastri Y, Grift T, Long S, Ting KC (2014) Biomass feedstock preprocessing and long-distance transportation logistics. GCB Bioenergy 8:160–170. https://doi.org/10.1111/gcbb.12241
96. Navnit Kumar R, Revanth Babu P, Vinoth V, Vignesh K, Sudha Rani S, Renganathan S, Venkateswara Sarma V (2023) Analyses of challenges faced during the production of amylolytic enzymes and bio-ethanol using left-over cooked-rice, and rice unfit for human consumption. Asian J Mycol (In-press)
97. TR S, Ramamoorthy NK, Sahadevan R (2019) Mixture of potato, sapodilla, kiwi peels and coir as a substrate for the production of cellulases using trichoderma atroviride ATCC 28043 by a solid-state cyclic fed-batch strategy and evaluation of its saccharification efficiency. Int J Sci Eng Manage (IJSEM) 4(6):130–133
98. Kudakasseril KJ, Nair GR, Hussain A, Raghavan GSV (2013) Feedstocks, logistics and pre-treatment processes for sustainable lignocellulosic biorefineries: a comprehensive review. Renew Sustain Energy Rev 25:205–219. https://doi.org/10.1016/j.rser.2013.04.019
99. Raud M, Kesperi R, Oja T, Olt J, Kikas T (2014) Utilization of urban waste in bioethanol production: potential and technical solutions. Agron Res 12:397–406
100. Hagman L, Feiz R (2021) Advancing the circular economy through organic by-product valorisation: a multi-criteria assessment of a wheat-based biorefinery. Waste Biomass Valor 12:6205–6217. https://doi.org/10.1007/s12649-021-01440-y
101. Pardini K, Rodrigues JJPC, Diallo O, Das AK, de Albuquerque VHC, Kozlov SA (2020) A smart waste management solution geared towards citizens. Sensors 20. https://doi.org/10.3390/s20082380
102. Nikolić S, Lazić V, Veljović Đ, Mojović L (2017) Production of bioethanol from pre-treated cotton fabrics and waste cotton materials. Carbohyd Polym 164:136–144. https://doi.org/10.1016/j.carbpol.2017.01.090
103. Matsakas L, Christakopoulos P (2015) Ethanol production from enzymatically treated dried food waste using enzymes produced on-site. Sustainability 7:1446–1458. https://doi.org/10.3390/su7021446
104. Tsegaye B, Jaiswal S, Jaiswal AK (2021) Food waste biorefinery: pathway towards circular bioeconomy. Foods 10. https://doi.org/10.3390/foods10061174
105. Ramamoorthy NK, Pallam RB, Rakshit S, Sahadevan R, Sarma VV (2022) Fungal α-amylase: various sources, commercial significance, bio-process challenges, and recent advancements

pertaining to its production. In: Gehlot P, Singh J (ed) Biomolecules of fungi nutraceutical, pharmaceutical, industrial and environmental protection, 1st edn. Om Publications, India
106. Pleissner D, Lam WC, Sun Z, Lin CS (2013) Food waste as nutrient source in heterotrophic microalgae cultivation. Biores Technol 137:139–146. https://doi.org/10.1016/j.biortech.2013.03.088
107. Khandaker MM, Abdullahi UA, Abdulrahman MD, Badaluddin NA, Mohd KS (2020) Bioethanol production from fruit and vegetable waste by using saccharomyces cerevisiae. In: Inambao F (ed) Bioethanol technologies. IntechOpen. https://doi.org/10.5772/intechopen.94358
108. Smoliński A, Karwot J, Bondaruk J, Bąk A (2019) The bioconversion of sewage sludge to bio-fuel: the environmental and economic benefits. Materials (Basel, Switzerland). 12(15). https://doi.org/10.3390/ma12152417
109. Byadgi SA, Kalburgi PB (2016) Production of bioethanol from waste newspaper. Proc Environ Sci 35:555–562. https://doi.org/10.1016/j.proenv.2016.07.040
110. Zimife EM, Obinna NC, Eze AF, Dango GZ (2022) Bioethanol production from waste paper: an alternative energy source. IOP Conf Ser Earth Environ Sci 1054(1). https://doi.org/10.1088/1755-1315/1054/1/012002
111. Mahalakshmi M, Angayarkanni J, Rajendran R, Rajesh R (2011) Bioconversion of cotton waste from textile mills to bioethanol by microbial saccharification and fermentation. Scholars Res Libr Ann Biol Res 2(3):380–388
112. Brink DL (1981) Making alcohol from cotton gin waste and cotton stalks. In: Proceedings of the symposium on: cotton gin trash utilization alternatives. University of California Cooperative Extension, Hanford, CA
113. Jeihanipour A, Taherzadeh MJ (2009) Ethanol production from cotton-based waste textiles. Biores Technol 100(2):1007–1010. https://doi.org/10.1016/j.biortech.2008.07.020
114. Kabir K, Batagarawa SM, Muhammad I, Nazifi A, Haruna I, Ahmed HI, Bamanga R, Bugaje I (2020) Bioethanol production from waste cotton lint using *Saccharomyces cerevisiae*. J Appl Sci 20:58–66. https://doi.org/10.3923/jas.2020.58.66
115. Ramamoorthy NK, Ravi S, Sahadevan R (2018) Production of bio-ethanol from an innovative mixture of surgical waste cotton and waste card board after ammonia pre-treatment. Energy Sour Part A Recovery Utilization Environ Effects 40:2451–2457. https://doi.org/10.1080/15567036.2018.1502843
116. Ramamoorthy NK, Sahadevan R (2020) Production of bioethanol by an innovative biological pre-treatment of a novel mixture of surgical waste cotton and waste card board. Energy Sour Part A Recovery Utilization Environ Effects 42:942–953. https://doi.org/10.1080/15567036.2019.1602208
117. Ramamoorthy NK, Sambavi TR, Sahadevan R (2019) A study on cellulase production from a mixture of lignocellulosic wastes. Process Biochem 83:148–158. https://doi.org/10.1016/j.procbio.2019.05.006
118. Ramamoorthy NK, Sambavi TR, Sahadevan R (2019) Assessment of fed-batch strategies for enhanced cellulase production from a waste lignocellulosic mixture. Biochem Eng J 152. https://doi.org/10.1016/j.bej.2019.107387
119. Hossain MNB, Basu JK, Mamun MD (2015) The production of ethanol from micro-algae spirulina. Proc Eng 105:733–738. https://doi.org/10.1016/j.proeng.2015.05.064
120. Priyadarshini P, Abhilash PC (2020) Circular economy practices within energy and waste management sectors of India: a meta-analysis. Bioresour Technol 304. https://doi.org/10.1016/j.biortech.2020.123018
121. Pallam RB, Ramamoorthy NK, Sadras SR, Sahadevan R, University P, Sarma VV, Vengadesan VK (2023) Ein konsolidietres, selbsttragendes, kreislaufwirtschaftliches Bioraffineriesystem. (A consolidated, self-sustainable, circular economical bio-refinery model) German Patent and Trademark Office DE 20 2022 107 273
122. Moreira VR, Carpanez TG, dos Santos FS, Santos LD, Fernandes DdS, Franca-Neta LS, Lange LC, Amaral MCS (2022) Circular economy in biorefineries: scale-up of anaerobic/aerobic membrane bioreactors for vinasse recycling. J Clean Prod 377. https://doi.org/10.1016/j/jclepro.2022.134448

123. de Carvalho JC, de Souza Vandenberghe LP, Sydney EB, Karp SG, Magalhães AI Jr, Martinez-Burgos WJ, Medeiros ABP, Thomaz-Soccol V, Vieira S, Letti LAJ et al (2023) Biomethane production from sugarcane vinasse in a circular economy: developments and innovations. Fermentation 9.https://doi.org/10.3390/fermentation9040349
124. Harnessing biomass in rural India for bioenergy and livelihood creation (2023) BAIF. https://baif.org.in/biomass-harnessing. Accessed 23 July 2023
125. What is a dairy cooperative. Cheese. https://mikehudsondist.com/dairy-cooperative/. Accessed 21 July 2023
126. Cooperatives in the Dairy Industry. United States Department of Agriculture. https://www.rd.usda.gov/files/cir1-16.pdf. Accessed 21 July 2023
127. Pondicherry University team gets bio-fuel patents (2023). https://www.thehindu.com/news/cities/puducherry/pondicherry-university-team-gets-bio-fuel-patents/article67002163.ece. Accessed 3 August 2023

Microalgae for the Sustainable Production of Biohydrogen

Revanth Babu Pallam, Navnit Kumar Ramamoorthy, Shovan Rakshit, Krishna Kumar Jaiswal, and V. V. Sarma

Abstract Due to its advantageous qualities, such as high gravimetric energy density and clean combustion, biohydrogen has emerged as a notable renewable energy fuel. As one of the most potential fuels of the future, it requires a scientific, ecological, and environmental research before it can be used on a large-scale. The technologies for H_2 gas production can rely on the use of certain renewable resources, but at present, large-scale methods rely primarily on the conversion of natural gas, heavy oils, naphtha, and coal, and only a small amount of H_2 is produced through alternative methods such as electrolysis and biomass processing. Hydrogen may be produced in a completely environmentally friendly manner using technologies based on some living organisms, such as microalgae and bacteria. Hydrogen can be produced from microalgae by photosynthesis, photorespiration, fermentation, and thermochemical processes for biomass conversion. Both marine and terrestrial ecosystems are home to microalgae, which can be either prokaryotic or eukaryotic. Green algae (Chlorophyta) and diatoms (Bacillariophyta) are two examples of eukaryotic microalgae, while cyanobacteria (Cyanophyceae) are examples of prokaryotic microbes. Present chapter focuses on the mechanisms of production of Biohydrogen by microalgae, microalgal cultivation methods, types of reactors used for cultivation of microalgae, parameters influencing the production of Biohydrogen by microalgae, metabolic limitation and bottlenecks and economic evaluation for the Biohydrogen produced by Microalgae.

Keywords Biofuel · Biohydrogen · Microalgae · Renewable energy · Third generation biofuel and sustainable environment

---

Equal Contribution of Authors: Revanth Babu Pallam and Navnit Kumar Ramamoorthy have equally contributed to the book chapter.

R. Babu Pallam · N. K. Ramamoorthy · S. Rakshit · K. K. Jaiswal · V. V. Sarma (✉)
Department of Biotechnology, Pondicherry University, Kalapet, Pondicherry 605014, India
e-mail: sarmavv@yahoo.com

K. K. Jaiswal
Department of Green Energy Technology, Pondicherry University, Puducherry, India

S. Kumar et al. (eds.), *Clean Energy Transition-via-Biomass Resource Utilization*, Green Energy and Technology, https://doi.org/10.1007/978-981-97-6321-4_7

1 Introduction

Nowadays, one of the greatest concerns for the worldwide research community is the hunt for renewable energy [69]. Population increases and industrial activities have resulted in increasing the demand for different types of fuel energy and it is the right time to take necessary action to meet the global demands of energy requirements [70]. It has been estimated that the Global population would reach 800–1000 crores by 2040. It is expected that to meet the Global energy requirements, 100–170 qBtu is required. However, as the socioeconomic status of countries differs from country to country, the forecast techniques of energy requirements of the world may not be accurate [60]. Globally, significant efforts have been taken to investigate the alternative sources of energy [71] in order to meet the raising energy demands, while at the same time making attempts to reduce the emissions. Hydrogen is an excellent alternative fuel as it produces a huge amount of energy in burning, while emitting only water as a byproduct. However, sustainability in the future is a big challenge for any source of bio-energy [41, 63, 72].

Biohydrogen generated by microbes, such as microalgae, has shown a significant promise as an environmentally friendly energy source. Microorganisms, in contrast to physicochemical processes, are self-repairing and self-replicating while using solar energy and organic substrates. The biochemical conversion of biohydrogen involves a variety of microorganisms, enzymes, biochemical pathways, and activities [95]. The process of hydrogenesis involves the transfer of electrons from the lysed water molecules to enzymes such as hydrogenase and nitrogenase [47]. The hydrogenase or nitrogenase enzyme is a protein that produces hydrogen and regulates the metabolic processes of prokaryotes and various eukaryotic microorganisms such as green algae and cyanobacteria. The first attempts for producing biohydrogen from microalgae were made in the 1930s using light through two processes namely, bio-photolysis and photofermentation [27].

Recent developments in light mediated catabolism revealed the potential of producing electrons from endogenous substrate breakdown for hydrogen generation, while developments in heterotrophic dark fermentation lead to the production of biohydrogen from anaerobic processes. Recent progress has concentrated on improving the photosynthetic capability of microalgae by molecular engineering, particularly through genetic modification of the microalgae. The long processing time of petroleum, coal, and natural gas and the unsustainable extractions of fossil fuels have resulted in a significant resource decrease. Moreover, the usage of fossil fuels leads to the emission of greenhouse gases (GHG) into the environment, specifically the release of carbon dioxide (CO_2), in the process of combustion [11]. Between 1998 and 2018, CO_2 emissions shot up by 48%, making carbon capture and storage (CCS) technology necessary to avoid severe negative environmental impacts. In this alarming situation, fossil fuels continue to meet over eighty percent of the world's energy needs, despite increasing investment in and use of sources of clean energy [23].

Among sustainable choices, hydrogen has the significant advantage of producing water as a byproduct while emitting no CO_2. Thermodynamic qualities, when compared to standard fuels, indicate the passion in its investigation efforts, while some areas of manufacturing and storage remain to be understood. Presently many industrial processes rely on hydrogen, although the majority is created using processes that involve steam reforming or electrolysis, which are not fully devoid of involvement of fossil fuels [22].

In the last few years, several ecological systems investigated have emphasized the biomass of algae as a source of energy from living matter. The photosynthetic mechanism allows microalgae to accumulate and utilize the energy of light. They utilize substances found in the environment, like water and CO_2, to sustain cell development and metabolic activity. These microorganisms may collect significant volumes of CO_2 and receive the nutrition required for growth through substrates or fluids derived from industrial waste. Simple and cheap algal growing techniques guarantee favorable rates of growth. In addition, these could be established on barren lands without interfering with farming areas where they can be used to produce food. Public interest in microalgae has increased substantially in recent years due to the diversity and adaptability of their metabolic products across a wide spectrum of species [83].

2 Microalgae

Microalgae, which were earlier considered as belonging to both prokaryotes and eukaryotes, are ubiquitous in both marine and terrestrial environments. Prokaryotic microorganisms like cyanobacteria (Cyanophyceae) and eukaryotic microalgae like green algae (Chlorophyta) and diatoms (Bacillariophyta) have become examples of microalgae [28, 46]. Because of their unicellular or basic multicellular structures, microalgae can survive under adverse environments and proliferate quickly. They have a short duplication time and are photoautotrophs. They can grow practically in any place with sunlight and a few ingredients because they have the ability for growth in hard conditions. The production of biofuels from microalgae offers numerous benefits. More than 50,000 species of microalgae are thought to exist, although only about 30,000 have been researched and analyzed [73]. They can develop quickly in a wide range of environments and can survive under hard situations. As a result, they can be cultivated on soil that is unsuitable for agriculture.

3 Mechanisms of Biohydrogen Production

3.1 Bio-photolysis

Microalgae can generate hydrogen via bio-photolysis or fermentation. Light-dependent and light-independent processes of fermentation can be clearly distinguished. Dark fermentation is an action that does not rely on light, whereas bio-photolysis or light fermentation is a photo-dependent activity. In the bio-photolysis pathway, microalgae use water and energy form sun light to produce hydrogen and oxygen. Bio-photolysis can either be direct or indirect. Further biohydrogen could be created from microalgae using electrochemical technologies such as microbial fuel cells (MFC), microbial photoelectrochemical cells (PEC), and microbial electrolysis cells (MEC). Considering their wide range of applications, they are frequently picked over traditional chemical, biological, and biochemical processes [101].

3.1.1 Direct Bio-photolysis

The bio-photolysis mechanism is analogous to photosynthesis. Green microalgae are the only living creatures capable of effectively undertaking bio-photolysis to make biohydrogen by breaking water in an anaerobic condition [7]. In the presence of light, photosystem II (PSII) oversees water breaking [24]. Direct bio-photolysis is a two-step response, which is depicted in Fig. 1. Photosystem II (PSII) initially converts the molecules of water into proton, electron, and oxygen. The hydrogen is subsequently produced using hydrogenase enzymes [57, 81].

Indeed, direct bio-photolysis is an intriguing technology since it uses sunlight to transform readily available water into oxygen and hydrogen molecules. Furthermore, it is more advantageous because no carbon-based substrates are required for the medium used for growing. Direct bio-photolysis, which uses microalgae as its primary photon source, produces no greenhouse emissions. Microalgae may convert

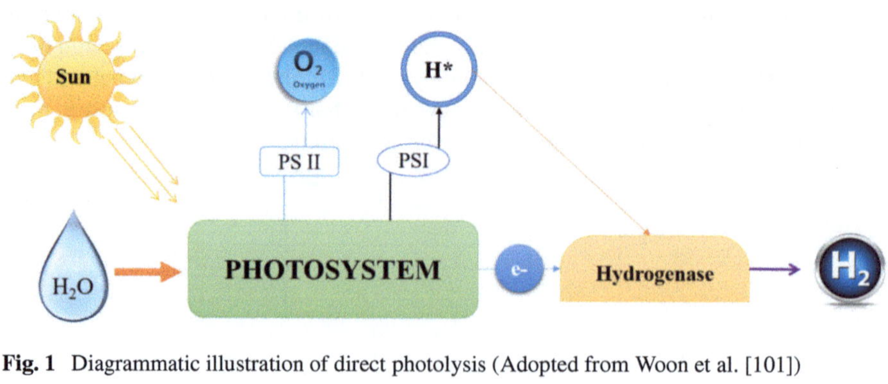

Fig. 1 Diagrammatic illustration of direct photolysis (Adopted from Woon et al. [101])

sunlight at a pace that is ten times faster than that of land plants. However, in environments where a certain concentration of oxygen is present, the hydrogenase machinery will limit biohydrogen production [59].

Microalgae's growth constraints and the low photosynthetic mechanism rate of conversion make the bio-photolysis approach for making biohydrogen ineffective. Efficiency in the conversion of photons is essential for the bio-photolysis process, which converts protons into hydrogen. The photon conversion efficiency of direct bio-photolysis was 80%, while that of green microalgae was just 12.5% [59]. In fact, efficiency was relatively low, with rates of 1% or less [84]. Furthermore, the efficiency of microalgal light absorption has a major impact on biohydrogen generation. Maximum light intensity can be used to boost microalgal growth and biomass. Microalgae can thrive in environments with low levels of incident light because they have evolved elaborate chlorophyll antennas in their photosystem. However, microalgae's photosynthetic efficiency drops when they are subjected to intense light. At high light intensities, surplus photons are lost through fluorescence or heat because the rate at which they are absorbed by the top layers of chlorophyll antennae molecules is much higher than the rate at which they are utilized for photosynthesis [56].

In the direct bio-photolysis process, electrons are generated through the oxidation of water via the PSII. Electrons are energized by photons from the sun via PSII after traveling along the electron transport chain. Photosystem I (PSI) and ferredoxin (Fd) both boost electron transport. The electron transfer product of plastoquinol, NADPH, transforms NADP+. Water is oxidized to produce oxygen and hydrogen, and the PSII then utilizes these electrons for photosynthesis. The water-oxidation cycle produces protons (hydrogen ions), which are then used to create a proton gradient for ATP synthesis by ATP synthase. In the chloroplasts of microalgae, protons serve as the ultimate acceptors of electrons during photosynthesis. As a result, the process generates equal amounts of hydrogen and oxygen gases [2, 81].

The [Fe–Fe] hydrogenase is extremely oxygen sensitive. Lower levels of oxygen will destroy the [4Fe–4S] cluster found in the site of activity of hydrogenase, rendering it ineffective [24, 59]. According to Stripp et al. [85] the reactive oxygen species formed initially at the distant Fe atom of the catalytic site's [Fe–Fe] sub-cluster. The reactive oxygen species would subsequently swiftly damage the [4Fe–4S] cluster and reduce enzymatic activity. As such hydrogenase actions were inhibited, the generation of molecular hydrogen dropped dramatically [85].

Hydrogen and oxygen productions can now be found in different parts of the world thanks to a two-phase photosynthetic approach based on Sulfur depletion. At the outset, the microalgae are cultivated in a Sulfur-rich medium that promotes cell division and photosynthesis. Under these conditions, normal photosynthesis can take place, with oxygen being produced from the oxidation of water. To further reduce photosynthesis and oxygen evolution, the microalgae are then moved to a Sulfur-deficient substrate. Sulfur-deficient cultures show significantly reduced activity of the oxygen-producing photosystem II (PSII). During PSII's water-splitting process, the sulfur-rich D1 proteins in the reaction center are effectively destroyed and must be replaced. However, when sulfur is scarce, D1 protein synthesis and the PSII repairing cycle are both inhibited, which decreases oxygen production [99]. When the rate of photosynthetic production of oxygen falls under the rate of cell respiration, the culture develops an anaerobic condition, which is beneficial for triggering the hydrogenase [6]. At sulfur deficiency conditions, hydrogen generation and hydrogenase activity are boosted. Wang et al. [98] report that *Chlorella pyrenoidosa*, under conditions of very low light (30 $E/m^2/s$), can produce hydrogen even in the absence of a Sulfur-deficient atmosphere. This is significantly less light than is required for photosynthetic production of oxygen to meet respiratory demands. The rate of hydrogen production was extremely low, at just 6 mol/L/h, when exposed to such dim light. Under sulfur deprivation circumstances, [94] studied the production of hydrogen in two *Chlamydomonas* strains: *Chlamydomonas reinhardtii* (CC425) and *Chlamydomonas moewusii* (SAG 24.91). Hydrogen production peaked at 2.69 mmol/L for *Chlamydomonas reinhardtii* (CC425) and 0.78 mmol/L for *Chlamydomonas moewusii* (SAG 24.91). Hydrogen was produced at rates of 17.02 mol/L/h by *Chlamydomonas reinhardtii* (CC425) and 5.12 mol/L/h by *Chlamydomonas moewusii* (SAG 24.91). Since the medium contains no sulfur, it is easier to produce hydrogen [56].

3.1.2 Indirect Bio-photolysis

Indirect bio-photolysis was developed as a solution to the problems that had been plaguing direct bio-photolysis. Hydrogen is produced by microalgae in an anaerobic environment from glycogen and starch stores through fermentation or respiration [7, 24] through indirect bio-photolysis. The two-stage indirect bio-photolysis method is necessary for hydrogen production. Indirect bio-photolysis is depicted in Fig. 2 on a more schematic level. First, microalgae use photosynthesis to increase the amount of endogenous substrate while also producing oxygen [59]. This process converts solar energy into chemical energy in the form of lipids, carbohydrates, and other organic molecules. In the second phase of metabolism, the fixed carbon present in the form of carbs is used to produce an enormous number of electrons. The Plastoquinone pool and PSII receive the transferred electrons almost immediately. The electron transport chain breaks down without enough oxygen. An anaerobic situation is created when atmospheric oxygen is depleted, which activates the oxygen-sensitive hydrogenase enzyme. PSI, ferrodoxin (Fd), and hydrogenase subsequently take the freed electrons and use them to generate hydrogen [82].

3.1.3 Dark Fermentation

Green algae and other anaerobes or microalgae can produce hydrogen anaerobically in the dark by transforming organic compounds, in contrast to photosynthesis and photofermentation, which generate energy from sunlight [39]. High hydrogen outputs in photobioreactors have contributed to the rise in the popularity of dark fermentation. Also, it has a cheap production cost, even when using complicated substrates like waste as feed [48]. Heterotrophic satellite cultures that can survive

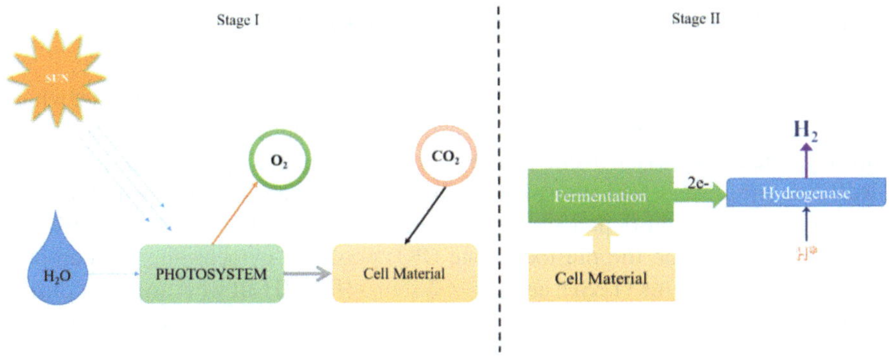

Fig. 2 Diagrammatic illustration of indirect bio-photolysis (Adopted from Woon et al. [101])

in the microalgal biomass's absence of oxygen are responsible for the production of hydrogen, which opens the door to the possibility of dark fermentative biohydrogen from microalgal culture. Hydrolysis, acidogenesis, acetogenesis, and methanogenesis are only a few of the reactions that take place during dark fermentation of complex organic molecules. Hydrolysis results in the production of hydrogen gas, fatty acids, carbon dioxide, and other metabolic intermediates via the acidogenesis process [42]. The produced fatty acids are subsequently fermented via acetogenesis to produce hydrogen and acetate. The subsequent step is for the acetoclastic methanogens to decarboxylate the acetate, which results in the formation of methane and carbon dioxide. Additionally, hydrogen-using methanogens reduce carbon dioxide to methane by using hydrogen gas as an electron source [81].

Hydrogen is just an intermediate metabolite formed during the acidogenesis and acetogenesis phases of dark fermentation. Hydrogen may be extracted from dark fermentation by blocking the processes that produce methane and use up hydrogen. However, microorganisms that use hydrogen in complex biohydrogen bacterial consortiums are notoriously difficult to regulate. Non-methanogenic bacteria were shown to decrease hydrogen production in colonies treated with BESA from the green algae *C. vulgaris* and *D. tertiolecta* [42].

$$C_6H_{12}O_6 + 2NAD^+ \rightarrow 2CH_3COCOOH + 2NADH + 2H^+$$

Hydrogenase, pyruvate-ferredoxin oxidoreductase, and NADH-ferredoxin oxidoreductase all have electron fluxes that can be affected by the amounts of NADH and acetyl-CoA, as well as the environment. This means that reduced molecules like ethanol, butanol, and lactate must be produced from NADH to keep the oxidative-reductive cycle going. Hydrogen production is thus reduced.

4 Photo Synthetic Electrons and Hydrogenase

The major mechanism of microalgal metabolism is oxygenic photosynthesis, which involves intricate redox processes at the surface of chloroplast thylakoid membranes. Carbohydrate reserves rich in energy are produced when atmospheric CO_2 is fixed by a RuBiSco (ribulose-1,5-bisphosphate carboxylase/oxygenase) enzyme in the dark, following an early light-dependent process that generates ATP, reduced NADH, and NADPH. In particular, the plastoquinone (PQ) pool, the cytochrome b6f complex (Cyt b6f), photosystems I (PSI), and photosystems II (PSII) all work together during the light phase to generate an electron transport chain. After passing through PSI, the electrons travel to ferredoxin (Fd), the cell's ultimate acceptor [36, 58].

In anaerobic conditions, Fd may provide electrons to the hydrogenase enzyme. Hydrogen can be split into electrons and protons with the help of this enzyme, which catalyzes the reversible reaction. *Chlamydomonas reinhardtii* and other algae containing hydrogenases have been the subject of extensive research. HydA1, which is particularly important under anaerobic conditions, and HydA2 are two of the Fe-hydrogenase genes encoded by this model organism [29, 53]. One of the many responses that micro algal organisms have to anoxia or stress is the function of hydrogenase, and this activity is strictly controlled at various levels. This protein is delivered to chloroplasts via an N-terminal transit peptide after being translated by ribosomes in the cytoplasm (Happe et al. 2013). The protein and the catalytic center, the H cluster, must combine into a 47–48 kDa enzyme. The mature proteins produced by the HydEF and HydG genes are necessary for its activation [66]. Cysteine links the [4Fe–4S] units that make up the H cluster to a di-iron unit that can accept electrons from Fd but is rapidly deactivated by oxygen [25, 74, 85, 102]. Multiple organisms have been shown to be exhibiting hydrogen metabolism,they include *Chlorella fusca* [100], *Chlamydomonas moewusii*, *Lobochlamys culleus* [52], *Chlorococcum littorale* [91], and *Tetraselmis subcordiformis* [106].

5 Approaches for Cultivation of Microalgae

There are two primary methods of cultivation: (i) open pond system and (ii) photobioreactors (PBR) for phototrophic microalgae. Multiple strategies for microalgal production in open ponds and PBR have been researched and documented, with varied outputs depending on the technique [28]. Microalgae are capable of being grown in open ponds using an average depth of 0.2–0.5 m by stirring it with a wheel of paddles in order to avoid settling. The PBR system is made up of a transparent row of pipes or plates, with culture being circulated from the center reservoir. Despite PBR systems providing more influence over the culture environment, they are a bit more costly than open ponds. Microalgae can be grown either in batch or continuous modes. Each method has its own pros. and cons.

5.1 Batch Cultivation for Microalgae

Microalgae are often grown in batches in sealed photobioreactors. The light source is housed in a clear container in these photobioreactors. Because the system is closed, every nutrient that is needed has to be fed into this system at the start of production. Throughout the cultivation process, no ingredients are added in between or withdrawn. For microalgal cultivation in batches, a greenhouse with an open pond could be used. However, compared to open systems, closed photobioreactors offer several advantages [67, 68]. In order to provide phototrophic microorganisms in liquid suspension with optimal illumination, nutrients, CO_2, mass transfer, and agitation, a closed batch photobioreactor has been designed. Therefore, they have more control over the environment for growth than do open ponds. It also allows for the species that need to be grown in the reactor to be controlled. The Photobioreactors require ten times higher light intensity when compared to open systems, which allows a higher cell density in cultivation. This reduces the amount of water that must be processed during the collection of cells, and hence the overall cost of operation [21]. In addition, a photobioreactor has much lower evaporation loss than a natural pond. In addition, in uncultivable land, microalgal culture can be performed in a closed photobioreactor. The expensive reactor building costs and the supply of power required to operate the reactors are some of the important limiting factors to operate a photobioreactor. As the cell density increases in a time-bound manner, the light and nutritional requirements cannot be maintained consistently in a batch cultivation technique. As a result, steady-state continuous processes that govern cell density are necessary.

5.2 Continuous Cultivation for Microalgae

Microalgal cultivation is often undertaken in open ponds on a continuous or semi-continuous basis. Shallow ponds, tanks, circular ponds, and raceway ponds are the most used continuous systems for mass microalgae cultivation [61, 78, 97]. In a semi-continuous system, microalgae are obtained on a regular basis, and then followed by the incorporation of nutrients into the media for greater development. Semi-continuous cultivation methods can take place in either open ponds or closed bioreactors. Industrial quality chemicals can be produced through large-scale production of microalgae in open ponds/raceway ponds. When required, either freshwater or seawater can be used. In pilot photobioreactors, inoculum for large-scale open ponds can be created.

6 Photobioreactors

Microalgal biomass could be produced economically in open ponds. However, the efficiency of biomass is affected by the environment, sun intensity, nutritional composition, and water type. Other drawbacks of open ponds are limited cell light availability, evaporation water loss, the need for large areas of land and significant amounts of water, and low biomass yield [67]. However, comparing biomass productivity in open and closed systems might not be appropriate because it changes depending on geographical location.

6.1 Tubular Photobioreactors

For outdoor large-scale microalgal cultivation, PBR tubing is a great choice. Pumps in tubular PBR are used to cycle the culture media, which increases the concentration of dissolved oxygen (Richmond et al. 1993, [55], 87, 92). However, compared to other types of PBRs, tubular PBRs perform poorly when deployed at extremely large scales. Photoinhibition and difficulties in maintaining reactor temperature when grown outdoors are two more disadvantages. The bioreactor is built from acrylic glass tubes of varied lengths. Scale of operation and microalgal species decide the bioreactor size and dimensions.

There are a number of recorded tubular photobioreactor designs. Common layouts include: A column of bubbles and an airlift: The primary method of construction consists of a transparent vertical cylinder that lets light shine through. The carbon dioxide is supplied by bubbling gas through the bottom of the reactor. The horizontal tubular bioreactor is a transparent cylinder that is horizontal in design and has gas transmission systems. A plastic tube is wound in a spiral around an annular bioreactor.

6.2 Helical Photobioreactor

A helical photobioreactor consists of a degassing unit and a series of lighted, twisted tubes of relatively small diameter. The tube containing the culture is pumped to the degassing apparatus. As a result of enhanced mass transfer from the gas phase to the liquid phase, these structures increase the amount of CO_2 available to the culture. However, this type of system is easily contaminated by fouling. In order to maximize the yield of microalgal biomass, researchers have proposed several different configurations for helical photobioreactors. A warm round fluorescent lamp shines from the outside of an acrylic glass box with a UV-resistant translucent PVC tubing design inside [62].

Another approach for growing *Nannochloropsis* sp. was proposed by Briassoulis et al. [17]; this one involved a high ratio of culture volume to surface area and an optimized light path. Another design by Travieso et al. [86] consists of a clear PVC tube that is 60 m long and has a diameter of 1.6 cm at its center and a cylindrical tube that has a base area of 0.25 m^2. The growth medium was infused with 4% CO_2 at a rate of 1 vvm.

6.3 Vertical Photobioreactor

PBRs in the form of vertical columns are compact, economical, and easy to manage in sterile settings. They are also great for cultivating microalgae in large quantities. When compared to a tubular PBR, the reactor best suited for mass culture of microalgae [54], biomass output in airlift and bubble column PBRs was comparable. Shear over the microalgal culture imposed by the reactor architecture and the difficulties in scaling up illumination are some of the drawbacks of a vertical photobioreactor. Other issues include the limited surface area accessible for light and the requirement for complicated components during construction.

Polyethylene bag/glass tube vertical photobioreactors are the most common type of airlift and bubble column reactors. These kinds of constructions are see-through enough to let light travel through them. Proper mixing of sufficient quantities of CO_2 and efficient O_2 disposal is ensured by sparging air from the bottom of the reactor [18]. The acrylic glass long tubular reactor proposed by Torzillo et al. [85] is positioned horizontally and equipped with a diaphragm pump for efficient culture stirring and gas transmission within the tube. Using sunlight and exhaust gas, Chae et al. [19] created a device with vertical and horizontal tubes to cultivate microalgae.

6.4 Flat Panel Bioreactor

The illumination surface area of a flat panel photobioreactor is large. Solar light can reach the cells in the flat panel photobioreactor since it is built of transparent glass [92]. These reactors also have a lower dissolved oxygen deposition compared to horizontal tubular photobioreactors. Guan et al. [30] designed a membrane-based flat panel bioreactor with a thick coating and an extra support layer. Another concept for growing *Chlorella sorokiniana* was proposed by Li et al. [44], and both *Scenedesmus obliquus* and *Chlorella sorokiniana* use an airlift component composed of tubes and plates in addition to a flat panel made of clear polycarbonate (PC) divided into nine parallel flow chambers. To provide adequate mixing of the culture, the airlift consists of a riser and a down comer connected to a gas liquid splitter. A perforated sparger directs the flow of air to provide uniform turbulence. Carbon dioxide (CO_2) and air together can be used as a carbon source. A different method, proposed by Bergmann et al. [15], involves the addition of a membrane to allow for the passage of clean air and CO_2 for the cultivation of suitable microalgae. Constant dispersal occurs for the created foam and exhaust air [15].

7 Parameters Influencing the Production of Biohydrogen

7.1 Light Intensity

Like photosynthesis, bio-photolysis and the production of biohydrogen are highly dependent on the available light. Both the light reaction and the dark reaction (light-independent reaction) are necessary for photosynthesis to occur. When water is broken down in the presence of light, ATP and NADPH are produced. Then, ATP and NADPH will both play roles in the dark process to produce endogenous substrates. Since water does not undergo photolysis in the dark, the process proceeds without interruption. Therefore, the dark response is possible even when light is present. During the light phase of the photosynthesis cycle, microalgae engage in indirect bio-photolysis to increase their starch and lipid production. This results in an over-abundance of naturally occurring substrates. Increased biohydrogen synthesis is due to the exclusive production of endogenous substrates during the light cycle. PSII consumes endogenous substrates created in the dark cycle in an anaerobic condition. To reach anaerobic conditions rapidly the cells of microalgae are briefly exposed to elevated intensity light [32].

Photoinhibition is caused by high levels of light, resulting in reduced oxygen production. Photoinhibition at high light intensities has been shown to cause destruction of PSII components, even though PSI tends to be stable in producing electrons from water breakdown [24]. As a result of the elevated amounts of sulfur and oxygen usage in the culture medium, the enzymes become rapidly active, breaking down the microalgal proteins and starch reserves, resulting in rapid decomposition [2, 34]. The optimal hydrogen production is often determined by the quantity of light available during the anaerobic process. In the absence of oxygen, *Chlamydomonas reinhardtii* may use solar energy to split water into hydrogen and oxygen. In a sulfur-free environment, *Chlamydomonas reinhardtii* was exposed to light intensities between 60 and 300 E/m2/s by Kim et al. [37]. Hydrogen production was found to increase until light intensities reached 200 E/m2/s, at which point production slowed. The maximum hydrogen yield was 225 mL Hydrogen/L when exposed to 200 E/m2/s of light. In addition, the time needed to initiate hydrogen synthesis decreased from 66 to 22 h when the light intensity was raised [37]. Alternating photoperiods were tested to see how they affected the growth of *Chlamydomonas reinhardtii* in a sulfur-poor, light-driven environment. Light and dark were cycled in 2-, 3-, and 4-h photoperiod cycles. The equivalent hydrogen percentages were 47, 86, and 87% throughout these photoperiods. This demonstrated that light is required for the synthesis of hydrogen [96].

7.2 Density of Micro Algae

The amount of light that reaches microalgal cells is determined by their density. It has a crucial role in the growth of microalgae and the production of hydrogen [34]. The amount of hydrogen that can be extracted from microalgae depends on their cell density, which can be determined by cultivation methods. Cell density is also affected by other factors such as the concentration of carbon dioxide, the abundance of organic or inorganic components, and the capacity of light to penetrate the environment. Autotrophic, mixotrophic, and heterotrophic conditions are all suitable for cultivating microalgae [59]. Biomass production by the microalgae *Chlorella sorokiniana* was studied by Sajadian et al. in three distinct growth settings, with glucose supplementation in the heterotrophic and mixotrophic environments. Mixotrophic culture, with a concentration of 3.55 g/L, has produced the most biomass by dry weight, while photoautotrophic culture and heterotrophic cultures have contributed 0.68 g/L and 0.73 g/L, respectively [75].

In an autotrophic culture, microalgae use a wide variety of inorganic chemicals and light as sources of energy to produce biomass. Autotrophic microalgae often produce low biomass densities due to the inverse connection between cell concentration and light absorption. Light could only reach a depth of a few micrometers into the autotrophic culture, restricting cell growth, as stated by Jareonsin and Pumas. Heterotrophic microalgae, which do not require light to thrive, can enhance their cell density and biomass production by using organic carbon as their only carbon and energy source [32]. Some microalgae, known as mixotrophs, can grow by consuming both organic and inorganic carbon. Mixotrophic microalgae can boost cell density by CO_2 fixation and organic carbon digestion [45].

Having a high enough concentration of cells is essential. Because total cell respirations are insufficient to absorb all dissolved oxygen, anaerobic conditions cannot be produced in cultures when microalgae concentrations are too low [24]. Microalgal respiration was found to be accelerated by high cell density, leading to increased oxygen intake and the creation of an anaerobic environment [33]. However, the rapid transition to anaerobic conditions due to severe self-shading may occur if the microalgal culture has a high cell concentration. This constrains starch production, which in turn reduces hydrogen yield. Maintaining an active development stage while maintaining microalgae as a preculture necessitates replacing a portion of the culture with new media when the cell concentration approaches 1.9 107 cells per mL [34].

7.3 Effect of Substrate on Biohydrogen Production

Substrate addition is universally necessary to boost cell growth and hydrogen production by microalgae. Development modes of microalgae, such as heterotrophy, mixotrophy, and autotrophy, dictate the substrates required for a culture. For instance, chlorella can use both internal and external chemical sources for hydrogen production through heterotrophic metabolism [16, 34]. Microalgae cannot develop without exogenous carbon sources, and using them can increase the amount of hydrogen they produce. Starch is the most important endogenous substance because it provides electrons for hydrogen creation during the first stage of photosynthesis. Hydrogen can be produced by microalgae from a variety of external carbon sources, including glucose, sucrose, fructose, acetate, and organic wastewater. For biohydrogen synthesis, the initial glucose content is crucial.

The ideal concentration of glucose for efficient biohydrogen synthesis is 5 g/L, as reported by El-Dalatony et al. [24]. A study found that the optimal concentration of glucose was 30 mM, as this cut the waiting time for hydrogen production by about 10 h [1]. In addition, Sengmee et al. investigated whether *Chlorella* sp. could use raw glycerol as an external carbon source for hydrogen production in anaerobic conditions. With a glycerol concentration of 16 g/L, it was found that *Chlorella* sp. produced the maximum hydrogen, with 10.31 mL/L [77]. *Chlorella vulgaris* MSU 01 strain was able to produce 220 mL/L of hydrogen when fed a substrate consisting of 4 g/L maize stalk [4]. In addition, the use of glucose, sucrose, galactose, and fructose at concentrations of 2000 ppm was investigated for their potential roles in *Anabaena* sp. methane and hydrogen production. Fructose and sucrose were shown to be the best substrates for *Anabaena* sp. for the production of methane and hydrogen, with yields of 1.6 mmol and 1 mmol, respectively [20].

7.4 Effect of pH on the Production of Biohydrogen

The pH of the medium is the fundamental physical factor that influences the process of metabolism that produces biohydrogen from microalgae, as a little variation in pH influences both the metabolic processes to generate hydrogen and the growth of the microalgae. Furthermore, pH modulates enzyme activity and prevents endogenous substrate degradation [2]. The ideal pH depends on the species of microalgae being grown. Carbonic acid is produced from carbon dioxide and water during photosynthesis, resulting in a decrease in the original pH. Eventually, the pH increases because the microalgae take in the dissolved carbon dioxide [8]. Hydrogenase, via the pyruvate-Fd-oxidoreductase pathway, produces hydrogen at acidic pH. Hydrogen is not produced, but rather ATP and acetate due to the inhibition of [Fe–Fe]-hydrogenase [2].

While lower pH levels slow down the hydrogen creation process, higher pH levels between 5.0 and 7.0 speed up the production rate. Javed et al. [32] report that, depending on the species, microalgae can grow in a pH range from 6.0 to 9.0. *Chlorella* sp., for instance, produces the most hydrogen at a pH of 8.6, while it does so in a range of pH values from 7.0 to 9.0. Hydrogen generation in *Chlamydomonas reinhardtii* was greatest at a pH of 7.7 in sulfur-deficient circumstances [38]. A pH value less than 6 shifts the medium into an acidic environment, putting pressure on microalgae and finally causing hydrogen generation to cease. For example, the buildup of acidogenic metabolites during dark fermentation reduces pH. The acidogenic metabolites can then permeate the cell of hydrogen-producing microorganisms, causing a growth disparity, raising ionic strength, and breaking the cell [93]. As a result, hydrogen-producing microorganisms' activity and generation of hydrogen are inhibited. The very basic medium, on the other hand, with a pH greater than 9, produces stress during the metabolic activities of microalgae. Photosynthesis will be hampered at these pH levels because inorganic carbon occurs as carbonate (CO_3^{2-}), which microalgae cannot ingest. As a result, microalgal development is slowed, and the conditions for the production of hydrogen may be unfavorable [33].

7.5 *Effect of Temperature on Biohydrogen Production*

The different species of microalgae have different optimum temperature ranges for biomass growth and hydrogen production. Extreme changes in temperature can slow down or halt microalgal activities [12], hence finding the sweet spot for microalgal growth is essential. Low temperatures hinder photosynthesis by reducing carbon, while high temperatures slow down cell respiration. High temperatures also hinder photosynthesis by rendering photosynthetic proteins inactive. Because of a decline in the activity of the enzyme ribulose-1,5-bisphosphate (Rubisco), the rate of photosynthesis and, by extension, growth, slows. Because enzymes are such proteins, they become inactive when heated too much. Furthermore, the disruption of active sites at high temperatures renders them inactive. Ananthi et al. [5] classified temperatures into four distinct ranges: ambient (15–30 °C), mesophilic (30–39 °C), thermophilic (50–64 °C), and hyperthermophilic (more than 64 °C). According to [51], the ideal temperatures for growing microalgae, cyanobacteria, and PNS bacteria are 20–30 °C, 25–55 °C, and 25–35 °C, respectively. Hydrogen can be produced by microalgae like *Chaetoceros* spp. and *Anacystis nidulans* at temperatures as high as 40 °C. The different species of microalgae have different optimum temperature ranges for biomass growth and hydrogen production. Extreme changes in temperature can slow down or halt microalgal activities [12], hence finding the sweet spot for microalgae growth is essential. Low temperatures hinder photosynthesis by reducing carbon, while high temperatures slow down cell respiration. High temperatures also hinder photosynthesis by rendering photosynthetic proteins inactive. Because of a decline in the activity of the enzyme ribulose-1,5-bisphosphate (Rubisco), the rate of photosynthesis and, by extension, growth, slows.

7.6 *Effect of Sulfur, Nitrogen, Phosphorus and Magnesium*

When it comes to the components of the medium, the outcomes indicated by the Sulfur (S) deficiency are only achieved when a photosynthetic growth stage has occurred and supplied all the necessary nutrients. Alongside this method, which foresees a period of distinct growth followed by a period of deprivation fraught with a variety of operational obstacles, researchers have recently examined a modality employing low levels of S to mimic a famine scenario [86, 103]. A growth in starch reserves in the period before S deprivation can be crucial in maintaining output, thus it is important to pay careful thought to the carbon supply as well. Mixotrophic environments, which include both inorganic and organic carbon sources, appear to be preferred [26, 76].

Hydrogen production by *Chlamydomonas reinhardtii* is enhanced by nitrogen (N) deprivation, a less studied circumstance, albeit slowly and with lower yields, suggesting a different mechanism in the establishment process [65]. Metabolism of *Scenedesmus obliquus* is disrupted by nitrogen deprivation, leading to the accumulation of lipids for use as an energy source, which is counterproductive to the organism's goal of producing hydrogen. Since sulfur (S) deprivation has been shown to be ineffectual due to the availability of sulfates in seawater, the impact of phosphorous (P) insufficiency was investigated in marine *Chlorella* spp. Even though a low-density culture was required to lessen the impact associated with cellular P reservoirs, P deprivation was able to induce anaerobic metabolism with prolonged hydrogen photoproduction [64]. Magnesium (Mg) deficiency was researched along with the depletion of other, less common elements. Hydrogen production increased over time in *Chlamydomonas reinhardtii* cells when Mg was depleted compared to S depletion. The relative insignificance of Mg to cellular activities compared to S [10] may account for this reaction.

8 Metabolic Limitations and Bottlenecks

The utilization of water as the primary substrate and the sun's energy as the source of power is a distinguishing aspect of direct photolysis. Because both of its reacting substances (water and sun light) are abundant, direct photolysis is an interesting generation of biohydrogen method. Nonetheless, it faces significant hurdles in practical implementations. A major difficulty seems to be the irregular output of hydrogen. In addition to hydrogen, photosynthetic microorganisms produce molecular oxygen, which is a strong inhibitor of the [Fe]-hydrogenase enzyme [35].

According to Laurinavichene et al. [43], the photochemical conversion capability of naturally photosynthesized hydrogen is only 1.5%. A 10% conversion rate is achievable, however, if oxygen is eliminated from the system as soon as it is produced. Research into creating oxygen-sensitive algae, disabling oxygen-hydrogen cycles, and altering the link between photosynthesis and respiration have all been explored as potential solutions to the problem of excess atmospheric oxygen. In addition to reducing metabolic activity, sulfate supplementation has been demonstrated to reduce oxygen evolution [90].

To counteract this, biohydrogen systems were developed to physically or temporally partition the photometabolisms of hydrogen and oxygen. At two consecutive times in the spatial system, the oxygen and hydrogen metabolisms occur independently of one another [13]. Initially, open microalgal ponds were used to photosynthesize ambient CO_2 into carbohydrates and oxygen. After the carbohydrates were added to the solar-powered dark fermenters, acetic acid and then hydrogen were produced. The algae were revived after being recycled back to the photosynthesis ponds of stage one. A temporal separation technique was investigated by shifting the relative timing of oxygen and hydrogen metabolisms in the green algae *Chlamydomonas reinhardtii* [50]. The initial phase was devoted to the production of photosynthetic oxygen as well as carbohydrate buildup. In the second phase, sulfur shortage (indirect photolysis) reduced photosynthesis and so the production of oxygen. This aided the alga's hydrogenase activity, resulting in the production of hydrogen. Maintaining consistent hydrogen generation is a major difficulty for indirect photolysis via sulfur deprivation. Light-based hydrogen generation would be halted in a few days because the oxygen created would block hydrogenases. It was discovered that under sunlight, hydrogen was created consistently for just around 60 h before tapering off [105]. The algae need to be restored by feeding endogenous substrate through regular photosynthesis after having sulfur depleted for around 4 days [30].

The impacts of algal–bacterial co-culture on green algae's photosynthesis-based hydrogen production were investigated by Ban et al. [9]. Co-culturing the single-celled green alga *Chlamydomonas reinhardtii* with the bacterium *Pseudomonas* sp. strain D revealed a syntrophic interaction between the two organisms. After all the oxygen in the headspace and substrate had been successfully absorbed by microbial respiration, a strict anaerobic environment was formed, ideal for algal hydrogenesis. Because it decreased chlorophyll degradation, enhanced starch accumulation, and preserved protein structure, the syntrophic interaction between algae and bacteria appeared to play a substantial role in boosting algal hydrogen generation. The syntrophic connections between *Chlamydomonas reinhardtii* and *Pseudomonas* sp. strain D were also seen in the green algae *Chlorella* and *Scenedesmus*.

In photocatabolism, an internal substrate is used to generate biohydrogen. Since the processes involved in photocatabolism and NADPH-PQOR activities are not well understood, their determination presents a significant challenge [80]. In photocatabolic microorganisms, the NADPH-PQOR photosystem has complicated syntrophic interactions with electron transport. Hydrogen from anaerobically grown plants and DCMU-inhibited chloroplasts have been linked in a small number of published research, although there is substantial circumstantial evidence for this. Research suggests that only the initial stages of hydrogen generation can be tracked with any degree of accuracy [25]. This proved that photocatalytic hydrogen production is not a steady-state process [105]. This could put a damper on NADPH-PQOR activity and the ability to transport electrons. With such a high rate of hydrogen synthesis, however, further research is needed to fully explore the possibilities of biohydrogen from photocatabolism.

Dark fermentation is attractive as a method for producing biohydrogen since it requires no sun light and very little space. Rapid hydrogen synthesis, minimal energy input, simplicity of implementation, and environmental benefits all lend credence to its growing popularity. It has been shown that a variety of organic substrates and carbohydrate-rich wastewaters can be used for the inexpensive synthesis of hydrogen and energy recovery [39, 40]. In contrast to the pure hydrogen produced by photosynthetic metabolism, the gaseous result of dark fermentation is a combination of hydrogen (approximately 70%), CO_2, CH_4, H_2S, ammonia, and vapor. Hydrogen purification issues must be fixed before its practical applications can be realized [80]. Commercial hydrogen produced from fossil fuels is still uncompetitive in terms of cost, efficiency of conversion, and process dependability compared to hydrogen created by electrolysis using renewable electricity.

9 Economic Evaluation

The affordability of the technology for real-world applications is one of the main challenges facing the development of biohydrogen today. There is a dearth of literature offering thorough economic assessments of biohydrogen production. Regarding the expenses of the biohydrogen plant, operation, and materials, more thorough work needs to be done on optimizing hydrogen yield. Complete economic studies should also consider the lighting pattern, the weather, the amount of land used, the construction materials used, the mixing of biomass, the operation, stability, and maintenance of the facility [80]. Sustainable biohydrogen production expenses should, as a rule, be equivalent with or more advantageous than market pricing. The industrial cost objective for hydrogen generation in a market with competition was set at 0.30 USD per kilogram of hydrogen gas generated in a report by the International Energy Agency. The energy in gasoline, which costs 2.5 USD/GJ, was used in thermochemical methods to estimate this with relation to commercial hydrogen.

An indirect photolysis plant cost estimate was developed [13], with the open microalgal ponds of the first stage accounting for 140 ha and the photoreactors of the second stage accounting for 14 ha. The plant's goal was to generate 1200 TJ/yr with 90% efficiency, and the total capital expenditure budget was USD 43 million. Production costs were estimated at $12 million per year, or $10 per GJ of energy (or $1.30 per kilogram, assuming a calorific value of 130 MJ/kg). Akkerman et al. [3] found that, assuming a 25% annual capital charge, capital expenditures accounted for close to 90% of total costs. The open ponds were expected to cost 6 USD/m2, but the photoreactors accounted for the bulk of the initial investment and ongoing maintenance expenditures (100 USD/m2). It was mentioned that there was a significant omission in the cost connected with handling hydrogen gas. The cost of producing this product is substantially higher than that of producing gasoline (2.50 USD/GJ or 0.33 USD/kg vs. 10 USD/GJ or 1.30 USD/kg). One strategy for fostering microalgal growth is to recycle metabolic wastes from photoreactors, such as organic acids, while another is to reduce the chemical cost of nutritional supplies. Additional photoreactor effluent treatment and chemical nutrient imports could account for 80–85% of the entire cost of a commercial biohydrogen [79]. The efficiency and cost-effectiveness of a scaled-up pilot modular photoreactor field operation were analyzed [80]. Cost inventory collected from six-month field operations indicated that most expenses (84%) were related to the purchase of manufacturing components and chemical nutrients. Compared to the often-reported rates of 20–100 USD/m2 [104], the estimated cost of creation was 0.75 USD/m2. The analysis gave a comparison of the costs of materials, labor, water use, land lease, energy, and other expenses for an industrial photoreactor, despite its seemingly oversimplified design. It was also suggested that the metabolic nutrients and reactor construction materials be recycled and reused to considerably cut down on operational costs.

A cost analysis of bio-photolysis at a commercial scale was performed in a single-stage horizontal tubular microalgal photoreactor [89]. The tubular photoreactor was praised for its superior features, such as its efficient water-spray cooling mechanism and internal gas exchange, among others. The price of the photoreactor was calculated at around 50 USD/m2 based on a conversion efficiency of 10% using sunlight. Gas transportation expenses were ignored, and a low 17% annual capital charge was estimated. The materials needed to construct the photoreactor make up the bulk of the fixed capital expenses that make up the estimated 80% of the total investment. The cost of manufacturing hydrogen in a two-phase biohydrogen from organic wastes was calculated to be 1.95 USD/kg (or 15 USD/GJ according to an energy content of 130 MJ/kg) [89].

Using the full algal photosynthetic potential, a commercially viable daily hydrogen output of 20 g per square meter was calculated [49]. Based on an achievable photosynthetic yield of 50%, the price of biohydrogen was estimated to be 2.80 USD/kg (or 21.5 USD/GJ). The researchers concluded that hydrogen's cost might be comparable to gasoline at this level since one kilogram of hydrogen has the same energy content as one gallon of gasoline. However, recent studies have shown that only around 10% of an alga's photosynthetic energy potential can really be converted into biohydrogen.

Costs associated with producing biohydrogen are estimated to be between $1.30 and $2.60 per kilogram (or $10–20 per gigajoule) by Akkerman et al. [3]. Biohydrogen as an alternative fuel appears to have a realistic economic goal of such a price, as stated by Benemann [14]. It is important to note that most of the estimates are optimistic because they were based on positive assumptions. The primary goal of this discussion was to highlight the primary costs associated with producing photosynthetic biohydrogen. At the current technological level, the cost of manufacturing biohydrogen from microalgae is undeniably higher than that of conventional fuels [80]. Since demonstration or pilot testing has demonstrated that DOE Algae Biomass Program/BETO technologies cannot be expanded outside of the lab, they are falling out of favor, as reported by the National Algae Association. Many engineering and technical hurdles appear to need to be cleared before we can break through the economic constraints. However, the findings show that creating cost-effective systems and increasing the efficiency of photochemical conversion are the two most important areas for future research.

10 Conclusions

Microalgae may fix many times more carbon dioxide per unit area than trees or agricultural crops. Since microalgal fuels outperform current fossil fuels, there is a strong justification for growing microalgae to produce biofuels and other beneficial compounds for the market. Recent developments in research and development have raised hopes for the future of biohydrogen produced from microalgae as an alternative biofuel. In addition to technological advancement, the future of microalgal biohydrogen is dependent on elements including economic viability, improvements in hydrogen energy systems, and public acceptance. Costs associated with production, transportation, and storage represent a substantial technological barrier to the realization of Biohydrogen applications. Additional studies are needed to improve essential operational parameters, look for appropriate reactor systems, and screen for highly capable hydrogen-producing microalgae from a variety of environments, among other things.

References

1. Ahmed SF, Mofijur M, Nahrin M, Chowdhury SN, Nuzhat S, Alherek M, Rafa N, Ong HC, Nghiem LD, Mahlia TM (2022) Biohydrogen production from wastewater-based microalgae: progresses and challenges. Int J Hydrogen Energy 47(88):37321–37342
2. Ahmed SF, Rafa N, Mofijur M, Badruddin IA, Inayat A, Ali MS, Farrok O, Yunus Khan TM (2021) Biohydrogen production from biomass sources: metabolic pathways and economic analysis. Front Energy Res 10(9):753878
3. Akkerman I, Janssen M, Rocha J, Wijffels RH (2002) Photobiological hydrogen production: photochemical efficiency and bioreactor design. Int J Hydrogen Energy 27(11–12):1195–1208
4. Amutha KB, Murugesan AG (2011) Biological hydrogen production by the algal biomass Chlorella vulgaris MSU 01 strain isolated from pond sediment. Biores Technol 102(1):194–199
5. Ananthi V, Ramesh U, Balaji P, Kumar P, Govarthanan M, Arun A (2022) A review on the impact of various factors on biohydrogen production. Int J Hydrogen Energy
6. Antal TK, Krendeleva TE, Laurinavichene TV, Makarova VV, Ghirardi ML, Rubin AB, Tsygankov AA, Seibert M (2003) The dependence of algal H_2 production on Photosystem II and O_2 consumption activities in sulfur-deprived Chlamydomonas reinhardtii cells. Biochimica et Biophysica Acta (BBA)-Bioenergetics 1607(2–3):153–160
7. Anwar M, Lou S, Chen L, Li H, Hu Z (2019) Recent advancement and strategy on biohydrogen production from photosynthetic microalgae. Biores Technol 1(292):121972
8. Bakonyi P, Kumar G, Bélafi-Bakó K, Kim SH, Koter S, Kujawski W, Nemestóthy N, Peter J, Pientka Z (2018) A review of the innovative gas separation membrane bioreactor with mechanisms for integrated production and purification of biohydrogen. Biores Technol 1(270):643–655
9. Ban S, Lin W, Wu F, Luo J (2018) Algal-bacterial cooperation improves algal photolysis-mediated hydrogen production. Biores Technol 1(251):350–357
10. Batyrova K, Gavrisheva A, Ivanova E, Liu J, Tsygankov A (2015) Sustainable hydrogen photoproduction by phosphorus-deprived marine green microalgae Chlorella sp. Int J Mol Sci 16(2):2705–2716
11. Bayro-Kaiser V, Nelson N (2017) Microalgal hydrogen production: prospects of an essential technology for a clean and sustainable energy economy. Photosynth Res 133:49–62

12. Béchet Q, Laviale M, Arsapin N, Bonnefond H, Bernard O (2017) Modeling the impact of high temperatures on microalgal viability and photosynthetic activity. Biotechnol Biofuels 10:1–1
13. Benemann JR (1997) Feasibility analysis of photobiological hydrogen production. Int J Hydrogen Energy 22(10–11):979–987
14. Benemann JR (2000) Hydrogen production by microalgae. J Appl Phycol 12:291–300
15. Bergmann P, Ripplinger P, Beyer L, Trösch W (2013) Disposable flat panel airlift photobioreactors. Chem Ing Tec 85(1–2):202–205
16. Bolatkhan K, Kossalbayev BD, Zayadan BK, Tomo T, Veziroglu TN, Allakhverdiev SI (2019) Hydrogen production from phototrophic microorganisms: reality and perspectives. Int J Hydrogen Energy 44(12):5799–5811
17. Briassoulis D, Panagakis P, Chionidis M, Tzenos D, Lalos A, Tsinos C, Berberidis K, Jacobsen A (2010) An experimental helical-tubular photobioreactor for continuous production of Nannochloropsis sp. Biores Technol 101(17):6768–6777
18. Carvalho AP, Meireles LA, Malcata FX (2006) Microalgal reactors: a review of enclosed system designs and performances. Biotechnol Prog 22(6):1490–1506
19. Chae SR, Hwang EJ, Shin HS (2006) Single cell protein production of Euglena gracilis and carbon dioxide fixation in an innovative photo-bioreactor. Biores Technol 97(2):322–329
20. Chen PC, Fan SH, Chiang CL, Lee CM (2008) Effect of growth conditions on the hydrogen production with cyanobacterium Anabaena sp. strain CH_3. Int J Hydrogen Energy 33(5):1460–1464
21. Chisti Y (2008) Biodiesel from microalgae beats bioethanol. Trends Biotechnol 26(3):126–131
22. Dawood F, Anda M, Shafiullah GM (2020) Hydrogen production for energy: an overview. Int J Hydrogen Energy 45(7):3847–3869
23. Dong K, Dong X, Jiang Q (2020) How renewable energy consumption lower global CO_2 emissions? Evidence from countries with different income levels. World Econ 43(6):1665–1698
24. El-Dalatony MM, Zheng Y, Ji MK, Li X, Salama ES (2020) Metabolic pathways for microalgal biohydrogen production: current progress and future prospectives. Biores Technol 1(318):124253
25. Florin L, Tsokoglou A, Happe T (2001) A novel type of iron hydrogenase in the green AlgaScenedesmus obliquus is linked to the photosynthetic electron transport chain. J Biol Chem 276(9):6125–6132
26. Fouchard S, Hemschemeier A, Caruana A, Pruvost J, Legrand J, Happe T, Peltier G, Cournac L (2005) Autotrophic and mixotrophic hydrogen photoproduction in sulfur-deprived Chlamydomonas cells. Appl Environ Microbiol 71(10):6199–6205
27. Gaffron H (1939) Photoreduction of CO_2 with H_2 in green plant. Nature 143:204–205
28. Georgianna DR, Mayfield SP (2012) Exploiting diversity and synthetic biology for the production of algal biofuels. Nature 488(7411):329–335
29. Ghirardi ML, Posewitz MC, Maness PC, Dubini A, Yu J, Seibert M (2007) Hydrogenases and hydrogen photoproduction in oxygenic photosynthetic organisms. Annu Rev Plant Biol 2(58):71–91
30. Ghirardi ML, Zhang L, Lee JW, Flynn T, Seibert M, Greenbaum E, Melis A (2000) Microalgae: a green source of renewable H_2. Trends Biotechnol 18(12):506–511
31. Guan J, Xiao Y, Song J, Miao J (2014) A typical flat-panel membrane bioreactor with a composite membrane for sulfur removal. Front Earth Sci 8(1):142–149
32. Jareonsin S, Pumas C (2021) Advantages of heterotrophic microalgae as a host for phytochemicals production. Front Bioeng Biotechnol 12(9):628597
33. Javed MA, Zafar AM, Hassan AA, Zaidi AA, Farooq M, El Badawy A, Lundquist T, Mohamed MM, Al-Zuhair S (2022) The role of oxygen regulation and algal growth parameters in hydrogen production via biophotolysis. J Environ Chem Eng 10(1):107003
34. Jimenez-Llanos J, Ramirez-Carmona M, Rendón-Castrillón L, Ocampo-López C (2020) Sustainable biohydrogen production by Chlorella sp. microalgae: a review. Int J Hydrogen Energy 45(15):8310–8328

35. Kapdan IK, Kargi F (2006) Bio-hydrogen production from waste materials. Enzyme Microb Technol 38(5):569–582
36. Khetkorn W, Rastogi RP, Incharoensakdi A, Lindblad P, Madamwar D, Pandey A, Larroche C (2017) Microalgal hydrogen production–a review. Biores Technol 1(243):1194–1206
37. Kim JP, Kang CD, Park TH, Kim MS, Sim SJ (2006) Enhanced hydrogen production by controlling light intensity in sulfur-deprived Chlamydomonas reinhardtii culture. Int J Hydrogen Energy 31(11):1585–1590
38. Kosourov S, Seibert M, Ghirardi ML (2003) Effects of extracellular pH on the metabolic pathways in sulfur-deprived, H2-producing Chlamydomonas reinhardtii cultures. Plant Cell Physiol 44(2):146–155
39. Kumar G, Mudhoo A, Sivagurunathan P, Nagarajan D, Ghimire A, Lay CH, Lin CY, Lee DJ, Chang JS (2016) Recent insights into the cell immobilization technology applied for dark fermentative hydrogen production. Biores Technol 1(219):725–737
40. Kumar G, Nguyen DD, Sivagurunathan P, Kobayashi T, Xu K, Chang SW (2018) Cultivation of microalgal biomass using swine manure for biohydrogen production: impact of dilution ratio and pretreatment. Biores Technol 1(260):16–22
41. Kumar R, Babu P, Vinoth V, Vignesh K, Rani S, Renganathan S, Sarma V (2023) Analyses of challenges faced during the production of amylolytic enzymes and bio-ethanol using left-over cooked-rice, and rice unfit for human consumption. Asian J Mycol 6(2):1–25. https://doi.org/10.5943/ajom/6/2/1
42. Lakaniemi AM, Hulatt CJ, Thomas DN, Tuovinen OH, Puhakka JA (2011) Biogenic hydrogen and methane production from Chlorella vulgaris and Dunaliella tertiolecta biomass. Biotechnol Biofuels 4:1–2
43. Laurinavichene TV, Kosourov SN, Ghirardi ML, Seibert M, Tsygankov AA (2008) Prolongation of H_2 photoproduction by immobilized, sulfur-limited Chlamydomonas reinhardtii cultures. J Biotechnol 134(3–4):275–277
44. Li J, Stamato M, Velliou E, Jeffryes C, Agathos SN (2015) Design and characterization of a scalable airlift flat panel photobioreactor for microalgae cultivation. J Appl Phycol 27:75–86
45. Li T, Zheng Y, Yu L, Chen S (2014) Mixotrophic cultivation of a Chlorella sorokiniana strain for enhanced biomass and lipid production. Biomass Bioenerg 1(66):204–213
46. Li Y, Horsman M, Wu N, Lan CQ, Dubois-Calero N (2008) Biofuels from microalgae. Biotechnol Prog 24(4):815–820
47. Manish S, Banerjee R (2008) Comparison of biohydrogen production processes. Int J Hydrogen Energy 33(1):279–286
48. Marone A, Ayala-Campos OR, Trably E, Carmona-Martínez AA, Moscoviz R, Latrille E, Steyer JP, Alcaraz-Gonzalez V, Bernet N (2017) Coupling dark fermentation and microbial electrolysis to enhance bio-hydrogen production from agro-industrial wastewaters and by-products in a bio-refinery framework. Int J Hydrogen Energy 42(3):1609–1621
49. Melis A, Happe T (2001) Hydrogen production. Green algae as a source of energy. Plant Physiol 127(3):740–748
50. Melis A, Zhang L, Forestier M, Ghirardi ML, Seibert M (2000) Sustained photobiological hydrogen gas production upon reversible inactivation of oxygen evolution in the green alga Chlamydomonas reinhardtii. Plant Physiol 122(1):127–136
51. Melitos G, Voulkopoulos X, Zabaniotou A (2021) Waste to sustainable biohydrogen production via photo-fermentation and biophotolysis—a systematic review. Renew Energy Environ Sustain 6:45
52. Meuser JE, Ananyev G, Wittig LE, Kosourov S, Ghirardi ML, Seibert M, Dismukes GC, Posewitz MC (2009) Phenotypic diversity of hydrogen production in chlorophycean algae reflects distinct anaerobic metabolisms. J Biotechnol 142(1):21–30
53. Meuser JE, D'Adamo S, Jinkerson RE, Mus F, Yang W, Ghirardi ML, Seibert M, Grossman AR, Posewitz MC (2012) Genetic disruption of both Chlamydomonas reinhardtii [FeFe]-hydrogenases: insight into the role of HYDA2 in H_2 production. Biochem Biophys Res Commun 417(2):704–709

54. Mirón AS, Garcıa MC, Camacho FG, Grima EM, Chisti Y (2002) Growth and biochemical characterization of microalgal biomass produced in bubble column and airlift photobioreactors: studies in fed-batch culture. Enzyme Microb Technol 31(7):1015–1023
55. Molina E, Fernández J, Acién FG, Chisti Y (2001) Tubular photobioreactor design for algal cultures. J Biotechnol 92(2):113–131
56. Mona S, Kumar SS, Kumar V, Parveen K, Saini N, Deepak B, Pugazhendhi A (2020) Green technology for sustainable biohydrogen production (waste to energy): a review. Sci Total Environ 1(728):138481
57. Morya R, Raj T, Lee Y, Pandey AK, Kumar D, Singhania RR, Singh S, Verma JP, Kim SH (2022) Recent updates in biohydrogen production strategies and life–cycle assessment for sustainable future. Biores Technol 19:128159
58. Nagarajan D, Dong CD, Chen CY, Lee DJ, Chang JS (2021) Biohydrogen production from microalgae—major bottlenecks and future research perspectives. Biotechnol J 16(5):2000124
59. Nagarajan D, Lee DJ, Kondo A, Chang JS (2017) Recent insights into biohydrogen production by microalgae–from biophotolysis to dark fermentation. Biores Technol 1(227):373–387
60. Newell R, Raimi D, Villanueva S, Prest B (2020) Global energy outlook 2020: energy transition or energy addition. Resources for the Future
61. Oron G, Shelef G, Levi A (1979) Growth of Spirulina maxima on cow-manure wastes. Biotechnol Bioeng 21(12):2169–2173
62. Oswald WJ. Micro-algae and waste-water treatment. In: Borowitzka MA, Borowitzka LJ (eds) Microalgal biotechnology
63. Pallam RB, Ramamoorthy NK, Sadras SR, Sahadevan R, University P, Sarma VV, Vengadesan VK (2023) Ein konsolidietres, selbsttragendes, kreislaufwirtschaftliches Bioraffineriesystem. (A consolidated, self-sustainable, circular economical bio-refinery model) German Patent and Trademark Office DE 20 2022 107 273
64. Papazi A, Korelidou A, Andronis E, Parasyri A, Stamatis N, Kotzabasis K (2018) Bioenergetic reprogramming plasticity under nitrogen depletion by the unicellular green alga Scenedesmus obliquus. Planta 247:679–692
65. Philipps G, Happe T, Hemschemeier A (2012) Nitrogen deprivation results in photosynthetic hydrogen production in Chlamydomonas reinhardtii. Planta 235:729–745
66. Posewitz MC, King PW, Smolinski SL, Zhang L, Seibert M, Ghirardi ML (2004) Discovery of two novel radical S-adenosylmethionine proteins required for the assembly of an active [Fe] hydrogenase. J Biol Chem 279(24):25711–25720
67. Posten C, Schaub G (2009) Microalgae and terrestrial biomass as source for fuels—a process view. J Biotechnol 142(1):64–69
68. Pulz O (2001) Photobioreactors: production systems for phototrophic microorganisms. Appl Microbiol Biotechnol 57:287–293
69. Ramamoorthy NK, Vengadesan V, Pallam RB, Sadras SR, Sahadevan R, Sarma VV (2023a) A pilot-scale sustainable biorefinery, integrating mushroom cultivation and in-situ pretreatment-cum-saccharification for ethanol production. Prep Biochem Biotechnol. https://doi.org/10.1080/10826068.2022.2162922
70. Ramamoorthy NK, Pallam RB, Sahadevan R, Sarma VV (2023b) Cellulase production from disposed COVID-19 personal protective equipment (PPE) using cyclic fed-batch strategies. Process Biochem 127:112–126. https://doi.org/10.1016/j.procbio.2023.02.013
71. Ramamoorthy NK, Pallam RB, Subash Chandrabose K, Sahadevan R, Vemuri VS (2023c) A critical process variable-regulated, parameter-balancing auxostat, performed using disposed COVID-19 personal protective equipment-based substrate mixture, yields sustained and improved endoglucanase titers. Prep Biochem Biotechnol. https://doi.org/10.1080/10826068.2023.2204479
72. Ramamoorthy NK, Vengadesan V, Pallam RB, Krishnasamy V, Sadras SR, Sahadevan R, Vemuri VS (2023d) A commercial-scale, circular-economical bio-refinery model for sustainable yields of mushrooms, cellulase-complex, bio-priming agents, bio-ethanol, and bio-fertilizer. Kavaka 59:26–39. https://doi.org/10.36460/Kavaka/59/2/2023/26-39

73. Richmond A (2003) Biological principles of mass cultivation. In: Handbook of microalgal culture: biotechnology and applied phycology, pp 125–177
74. Rumpel S, Siebel JF, Diallo M, Farès C, Reijerse EJ, Lubitz W (2015) Structural insight into the complex of ferredoxin and [FeFe] hydrogenase from Chlamydomonas reinhardtii. ChemBioChem 16(11):1663–1669
75. Sajadian SF, Morowvat MH, Ghasemi Y (2018) Investigation of autotrophic, heterotrophic, and mixotrophic modes of cultivation on lipid and biomass production in Chlorella vulgaris. Nat J Physiol Pharm Pharmacol 8(4):594–599
76. Scoma A, Durante L, Bertin L, Fava F (2014) Acclimation to hypoxia in Chlamydomonas reinhardtii: can biophotolysis be the major trigger for long-term H_2 production? New Phytol 204(4):890–900
77. Sengmee D, Cheirsilp B, Suksaroge TT, Prasertsan P (2017) Biophotolysis-based hydrogen and lipid production by oleaginous microalgae using crude glycerol as exogenous carbon source. Int J Hydrogen Energy 42(4):1970–1976
78. Seshadri CV, Thomas S (1979) Mass culture of Spirulina using low-cost nutrients. Biotech Lett 1:287–291
79. Show KY, Lee DJ, Chang JS (2011) Bioreactor and process design for biohydrogen production. Biores Technol 102(18):8524–8533
80. Show KY, Lee DJ, Tay JH, Lin CY, Chang JS (2012) Biohydrogen production: current perspectives and the way forward. Int J Hydrogen Energy 37(20):15616–15631
81. Show KY, Yan Y, Ling M, Ye G, Li T, Lee DJ (2018) Hydrogen production from algal biomass–advances, challenges and prospects. Biores Technol 1(257):290–300
82. Sivaramakrishnan R, Shanmugam S, Sekar M, Mathimani T, Incharoensakdi A, Kim SH, Parthiban A, Geo VE, Brindhadevi K, Pugazhendhi A (2021) Insights on biological hydrogen production routes and potential microorganisms for high hydrogen yield. Fuel 1(291):120136
83. Skjånes K, Rebours C, Lindblad P (2013) Potential for green microalgae to produce hydrogen, pharmaceuticals and other high value products in a combined process. Crit Rev Biotechnol 33(2):172–215
84. Srirangan K, Pyne ME, Chou CP (2011) Biochemical and genetic engineering strategies to enhance hydrogen production in photosynthetic algae and cyanobacteria. Biores Technol 102(18):8589–8604
85. Stripp ST, Goldet G, Brandmayr C, Sanganas O, Vincent KA, Haumann M, Armstrong FA, Happe T (2009) How oxygen attacks [FeFe] hydrogenases from photosynthetic organisms. Proc Natl Acad Sci 106(41):17331–17336
86. Tamburic B, Dechatiwongse P, Zemichael FW, Maitland GC, Hellgardt K (2013) Process and reactor design for biophotolytic hydrogen production. Phys Chem Chem Phys 15(26):10783–10794
87. Torzillo G, Pushparaj B, Bocci F, Balloni W, Materassi R, Florenzano G (1986) Production of Spirulina biomass in closed photobioreactors. Biomass 11(1):61–74
88. Travieso L, Hall DO, Rao KK, Benıtez F, Sánchez E, Borja R (2001) A helical tubular photobioreactor producing Spirulina in a semicontinuous mode. Int Biodeterior Biodegradation 47(3):151–155
89. Tredici MR, Zittelli GC (1998) Efficiency of sunlight utilization: tubular versus flat photobioreactors. Biotechnol Bioeng 57(2):187–197
90. Turner J, Sverdrup G, Mann MK, Maness PC, Kroposki B, Ghirardi M, Evans RJ, Blake D (2008) Renewable hydrogen production. Int J Energy Res 32(5):379–407
91. Ueno Y, Kurano N, Miyachi S (1999) Purification and characterization of hydrogenase from the marine green alga, Chlorococcum littorale. FEBS Lett 443(2):144–148
92. Ugwu CU, Aoyagi H, Uchiyama H (2008) Photobioreactors for mass cultivation of algae. Biores Technol 99(10):4021–4028
93. Usman TM, Banu JR, Gunasekaran M, Kumar G (2019) Biohydrogen production from industrial wastewater: an overview. Bioresour Technol Rep 1(7):100287
94. Vargas SR, Santos PV, Giraldi LA, Zaiat M, Calijuri MD (2018) Anaerobic phototrophic processes of hydrogen production by different strains of microalgae C hlamydomonas sp. FEMS Microbiol Lett 365(9):fny073

95. Vignais PM, Billoud B, Meyer J (2001) Classification and phylogeny of hydrogenases. FEMS Microbiol Rev 25(4):455–501
96. Vijayaraghavan K, Karthik R, Nalini SK (2009) Hydrogen production by Chlamydomonas reinhardtii under light driven sulfur deprived condition. Int J Hydrogen Energy 34(19):7964–7970
97. Vonshak A, Cohen Z, Richmond A (1985) The feasibility of mass cultivation of Porphyridium. Biomass 8(1):13–25
98. Wang H, Fan X, Zhang Y, Yang D, Guo R (2011) Sustained photo-hydrogen production by Chlorella pyrenoidosa without sulfur depletion. Biotech Lett 33:1345–1350
99. Williams CR, Bees MA (2014) Mechanistic modeling of sulfur-deprived photosynthesis and hydrogen production in suspensions of Chlamydomonas reinhardtii. Biotechnol Bioeng 111(2):320–335
100. Winkler M, Heil B, Heil B, Happe T (2002) Isolation and molecular characterization of the [Fe]-hydrogenase from the unicellular green alga Chlorella fusca. Biochimica et Biophysica Acta (BBA)-Gene Struct Expr 1576(3):330–334
101. Woon JM, Khoo KS, Akermi M, Alanazi MM, Lim JW, Chan YJ, Goh PS, Chidi BS, Lam MK, Zaini J, Bilad MR (2023) Reviewing biohydrogen production from microalgal cells through fundamental mechanisms, enzymes and factors that engendering new challenges and prospects. Fuel 15(346):128312
102. Wünschiers R, Stangier K, Senger H, Schulz R (2001) Molecular evidence for a Fe-hydrogenase in the green alga Scenedesmus obliquus. Curr Microbiol 42:353–360
103. Yagi T, Yamashita K, Okada N, Isono T, Momose D, Mineki S, Tokunaga E (2016) Hydrogen photoproduction in green algae Chlamydomonas reinhardtii sustainable over 2 weeks with the original cell culture without supply of fresh cells nor exchange of the whole culture medium. J Plant Res 129:771–779
104. Zaborsky OR, Benemann JR, Matsunaga T, Miyake J, San Pietro A (eds) (1998) Biohydrogen. Springer US, Boston, MA
105. Zhang L, Happe T, Melis A (2002) Biochemical and morphological characterization of sulfur-deprived and H_2-producing Chlamydomonas reinhardtii (green alga). Planta 214:552–561
106. Zhen GU, Zhao-An CH, Xing-Ju YU, Mei-Fang JI, Wei LI, Zhang W (2007) Subcellular localization and identification of hydrogenase isolated from the marine green alga Platymonas subcordiformis using immunoprecipitation and MALDI-TOF MS. Chin J Biotechnol 23(2):297–302

Social Accounting and Creditability of Biomass Resources Utilization in Various Applications

V. Deekshitha Sree, V. Varalakksmi, S. Sudalai, and A. Arumugam

Abstract The effective utilization of biomass resources has gained significant attention due to its potential to address sustainability challenges and reduce reliance on fossil fuels. This chapter examines the social, economic, and ecological consequences of different biomass applications, highlighting the interplay between resource utilization and societal well-being. The study also emphasizes the UN's Sustainable Development Goals (SDGs) associated with biomass utilization, emphasizing the importance of policy frameworks for social inclusion of minorities. It highlights the need for transparent and accountable reporting mechanisms to ensure credibility of biomass utilization practices, promoting informed decision-making and responsible resource management. The integration of social accounting principles into the assessment of biomass resource utilization provides a holistic understanding of its impacts, beyond traditional economic and environmental metrics. By recognizing the social dimensions and involving stakeholders, the credibility of biomass applications can be significantly enhanced.

Keywords Biomass utilization · Sustainability · Credibility · Social accounting

1 Introduction

In an era marked by growing environmental concerns and the need for sustainable resource management, the utilization of biomass resources has emerged as a pivotal solution across various applications. Social accounting, a framework that assesses the social and environmental impacts of business activities, plays a crucial role in evaluating the credibility of biomass resource utilization practices. As societies and

V. Deekshitha Sree · V. Varalakksmi · A. Arumugam (✉)
Advanced Biorefinery & Catalysis (ABC) Lab, Centre for Bioenergy, School of Chemical and Biotechnology, SASTRA Deemed University, Thirumalaisamudram, Thanjavur 613401, India
e-mail: aruchemxl@scbt.sastra.edu

S. Sudalai
Centre for Pollution Control and Environmental Engineering, School of Engineering and Technology, Pondicherry University, Puducherry 605014, India

S. Kumar et al. (eds.), *Clean Energy Transition-via-Biomass Resource Utilization*, Green Energy and Technology, https://doi.org/10.1007/978-981-97-6321-4_8

industries strive to transition toward greener alternatives, understanding the intricate relationship between biomass utilization and its broader societal implications has become essential. This exploration delves into the realm of social accounting and the assessment of credibility concerning the utilization of biomass resources across diverse applications, shedding light on its multifaceted benefits and challenges in the pursuit of a more ecologically responsible future. From energy generation to material production, examining the social dimensions and accountability aspects of biomass utilization offers a comprehensive perspective on its role in shaping sustainable pathways forward.

Furthermore, as the global demand for energy, materials, and food continues to rise, the exploration of biomass resources as a renewable and potentially carbon–neutral solution gains prominence. Biomass, obtained from organic materials like agricultural residues and forestry by-products, along with organic waste, has the ability to reduce greenhouse gas emissions and decrease dependence on finite fossil fuels. However, realizing these benefits requires a comprehensive understanding of the broader impacts of biomass utilization.

Social accounting in the context of biomass resources goes beyond mere financial metrics, encompassing the assessment of social, environmental, and ethical dimensions [10]. It involves evaluating the positive and negative effects of biomass utilization on local communities, biodiversity, air and water quality, and other ecological factors. Additionally, it considers factors such as job creation, livelihood improvement, and the equitable distribution of benefits among various stakeholders. The credibility of biomass resource utilization hinges on the transparent and accurate measurement of these social and environmental impacts. This involves methodologies to quantify carbon emissions reductions, water savings, land use changes, and other relevant indicators. Establishing credible accounting practices is essential not only for ensuring the effectiveness of biomass utilization strategies but also for fostering trust and accountability within communities, industries, and regulatory bodies.

In recent years, the sustainable development concept has gained traction globally, prompting industries and governments to re-evaluate their practices and embrace more environmentally friendly alternatives. Biomass resources, as a potentially renewable source with a lower carbon footprint, offer a compelling avenue for reducing the ecological burden associated with traditional fossil fuel-based industries. The versatility of biomass utilization spans across a range of sectors, including energy production, agriculture, forestry, and waste management. From generating biofuels to producing bio-based materials, these applications present innovative solutions to mitigate climate change and minimize the depletion of finite resources. However, the effectiveness of these solutions relies on accurate social accounting to ensure that the intended benefits are indeed realized. In the realm of energy generation, for instance, utilizing biomass for power production can contribute to reducing greenhouse gas emissions and diversifying energy sources [27]. Nevertheless, this potential benefit needs to be evaluated against potential adverse effects, such as land use competition with food crops or habitat destruction. Social accounting practices play a crucial role in determining whether the overall societal impact of such initiatives is positive or negative.

Moreover, the credibility of biomass utilization becomes increasingly vital as industries and consumers demand transparent and responsible practices. Companies striving to incorporate biomass resources into their supply chains need to demonstrate a commitment not only to economic gains but also to social and environmental stewardship. This requires comprehensive reporting mechanisms that quantify the environmental benefits, potential risks, and the strategies in place to mitigate any negative consequences. In various applications, from bioenergy production to the manufacturing of bioplastics and biochemicals, the utilization of biomass resources presents both opportunities and challenges. While it offers a potential pathway toward a more sustainable future, there are concerns about potential negative impacts, such as deforestation, land use conflicts, and unintended shifts in agricultural practices. The credibility of biomass utilization strategies relies on addressing these challenges and effectively communicating the benefits and trade-offs associated with such practices. A challenge in the field of biomass resource utilization lies in establishing standardized methodologies for social accounting that can be universally applied. Different applications may have distinct indicators to measure, and the context in which biomass is utilized can greatly influence its overall impact. Developing reliable and consistent assessment frameworks is key to ensuring that the credibility of biomass utilization remains high across various sectors and regions.

This Chapter delves into the intricate relationship between biomass, its multifaceted applications, and the role of transparent and comprehensive social accounting in shaping a more sustainable and responsible future.

2 Biomass Utilization and SDG

Biomass utilization has drawn significant attention in recent periods due to its potential as a renewable and sustainable source of energy and raw materials. Biomass is utilized across a variety of industrial sectors for a variety of purposes. There is a need to evaluate the credibility of biomass utilization in terms of its environmental contributions, economic viability, and societal acceptance [37].

Biomass sources include wood and its derivatives such as sawdust and wood pellets, which can be used directly to generate heat or processed into biofuels. Agricultural residues such as rice husks and corn stalks offer post-harvest resource recycling opportunities. Energy crops such as switchgrass and Chinese reeds are specially grown because of their high biomass content and the potential for higher energy yields. Municipal solid waste, consisting of food leftovers and garden waste, can be recovered by composting or anaerobic digestion. Algae, which grow rapidly and have an oil-rich composition, are expected to have various applications. Animal manure serves as a valuable resource for biogas production through anaerobic digestion. Aquatic plants and forest residues are other resources that can be used for biomass

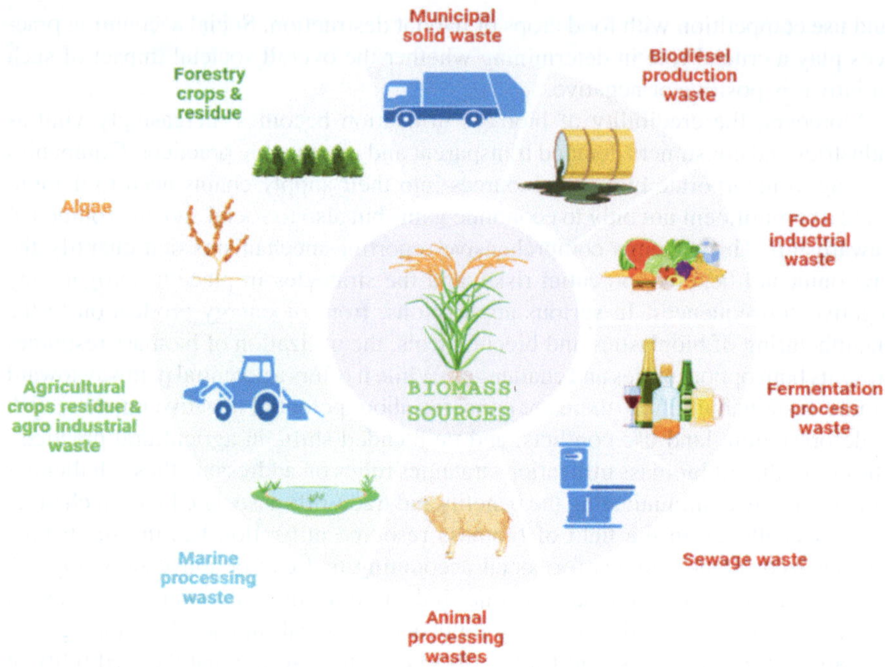

Fig. 1 Various sources of biomass utilized for different purposes [20]

production. By-products from food processing, urban green waste, crop residues, and wastewater treatment plants add to this diverse biomass source (Fig. 1) [20].

2.1 *Environmental Impacts*

Biomass utilization offers significant environmental benefits, primarily through its ability to reduce greenhouse gas emissions compared to fossil fuels. Biomass can be considered carbon–neutral when it is sourced sustainably and its growth absorbs as much carbon dioxide as is released when it is used. Utilizing biomass for energy production can help reduce net greenhouse gas emissions compared to using fossil fuels [14] Biomass utilization can divert organic waste from landfills, reducing methane emissions from decomposition. Additionally, utilizing biomass waste streams for energy, biofuels, and bioproducts contributes to a circular economy by transforming waste into valuable resources. Some biomass cultivation practices, such as cover cropping and agroforestry, can improve soil health and prevent erosion, enhancing overall land quality and ecosystem services. Biomass can serve as a renewable energy source, which reduces the dependence on fossil fuels, decreases air pollution, and mitigates climate change. Properly managed biomass utilization practices can contribute to habitat restoration and biodiversity conservation, particularly when integrated with sustainable land management strategies [12].

2.2 Economic Viability

The economic viability of biomass utilization is a crucial factor in determining its long-term feasibility. It depends on various factors, like the type of biomass feedstock, the conversion technology used, market demand, policy support, and the local economic context. The key considerations that influence the economic feasibility of biomass utilization are as follows.

2.2.1 Biomass Feedstock

The cost and availability of biomass feedstock play a crucial role. Abundant and low-cost feedstock sources, such as agricultural residues or forestry waste, can contribute to economic viability [30]. Consistency and quality of the feedstock impact the efficiency of conversion processes as well. High-quality feedstock with consistent properties often leads to better yields and lower operational costs.

2.2.2 Conversion Technology

Efficient biomass conversion technologies are crucial for maximizing energy or product output, as higher conversion efficiency typically leads to improved economic viability. The initial capital investment and ongoing operating costs linked to the chosen technology play a vital role in determining economic feasibility. Furthermore, the size of the biomass conversion facility impacts both costs and revenues, with economies of scale often making larger facilities more cost-effective.

2.2.3 Market Demand

Biomass energy prices must be competitive with other energy sources in the market. Market fluctuations can impact the economic viability of biomass-based energy production. Moreover, biomass can be used for producing not only energy but also bio-based products like biofuels, biochemicals, and bioplastics. The demand and market prices for these products influence economic viability [25].

2.2.4 Policy and Incentives

Government subsidies, tax incentives, and renewable energy mandates can significantly improve the economic viability of biomass utilization by reducing upfront costs or enhancing revenue streams. Favorable policies that prioritize renewable energy sources can create a more stable and supportive market environment for biomass energy [23].

2.2.5 Environmental Considerations

Policies that put a price on carbon emissions can make biomass utilization more economically attractive, as it can offset the costs associated with greenhouse gas emissions from fossil fuels. In cases where biomass utilization involves waste-to-energy approaches, the cost savings associated with waste disposal can enhance economic viability as well.

2.2.6 Technological Advancements

Advances in biomass conversion technologies can lead to increased efficiency, reduced costs, and improved economic viability. Innovative approaches, such as integrated biorefineries that produce multiple products from biomass, can create additional revenue streams and enhance economic feasibility.

2.2.7 Local Context and Infrastructure

The presence of existing infrastructure for biomass collection, transportation, and conversion can lower costs and improve economic viability. Local factors such as climate, distance to markets, and regional energy demands can impact the economics of biomass utilization.

2.3 Societal Acceptance and Challenges

Public perception and societal acceptance are essential components of biomass utilization. It involves the willingness of communities, stakeholders, and the general public to embrace and support the utilization of biomass for various purposes, including energy production, materials, and other applications [24]. Achieving social acceptance is essential for the successful implementation of biomass projects.

Factors Influencing Social Acceptance are

1. Perceived Environmental Benefits

2. Local Economic Opportunities
3. Community Engagement
4. Health and Air Quality
5. Cultural Considerations
6. Education and Awareness.

Biomass utilization, while offering numerous benefits, also presents several challenges to be addressed for an efficient implementation. These challenges can vary based on the specific application and regional context [22]. Some key challenges associated with biomass utilization are (Fig. 2).

1. Feedstock Availability and Sustainability challenges such as land degradation, deforestation, and competing land usage.
2. Technical and Economic Challenges such as Conversion efficiency, high costs, and scaling-up issues.
3. Environmental Impacts such as air pollution due to combustion.
4. Infrastructure and Logistics challenges such as supply chain management and infrastructure development.
5. Social and Community Considerations such as cultural sensitivity and local opposition.
6. Policy and Regulatory Issues.
7. Technological Advancements.
8. Integration with Existing Energy Systems.
9. Sustainability Certification and Standards.

Addressing these challenges requires collaboration among governments, industries, researchers, communities, and stakeholders. As the biomass utilization field

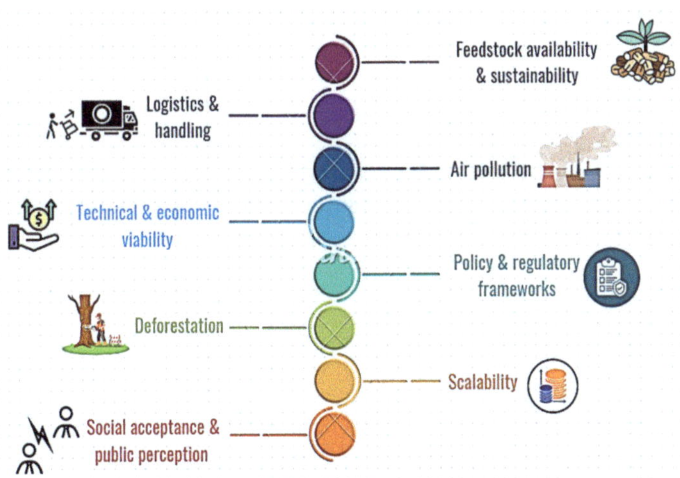

Fig. 2 Challenges associated with biomass utilization [22]

evolves, advancements in technology, policy frameworks, and stakeholder engagement are essential to overcome these challenges and to realize the full potential of biomass as a sustainable resource [31].

2.4 Credibility Assessment

Credibility assessment of biomass utilization is essential for ensuring that projects contribute positively to sustainable development and do not cause unintended negative impacts. It requires an extensive approach that considers environmental, economic, and social factors, and entails working among various stakeholders, including governments, industries, communities, and NGOs. Credibility assessment involves evaluating the environmental, economic, and social aspects of biomass projects to determine their authenticity, transparency, and overall sustainability. This assessment can be conducted through various methods, including Life Cycle Analysis (LCA), Economic Viability Assessment, Social Impact Assessment, Sustainability Certification and Standards, Transparency and Reporting, Compliance with Regulations, Independent Assessment, and Continuous Improvement [16].

The Environmental Impact Assessment (LCA) evaluates the entire lifecycle of biomass utilization, including greenhouse gas emissions, energy consumption, resource depletion, biodiversity and land use, air and water quality, and economic viability. The Economic Viability Assessment evaluates the economic benefits of biomass utilization against costs, market analysis, and policy and incentive evaluation. The Social Impact Assessment involves stakeholder engagement, assessing the positive social impacts, and evaluating equity and justice.

Sustainability Certification and Standards are essential for ensuring the sustainability of biomass projects. Certification programs and credible organizations provide third-party verification of sustainable practices [8]. Transparency and Reporting ensure that data related to biomass sourcing, conversion processes, emissions, and impacts are transparently reported and easily accessible. Compliance with Regulations ensures that biomass projects adhere to local, regional, and national regulations related to environmental protection, land use, and waste management. Independent Assessment and Continuous Improvement encourage biomass projects to adopt adaptive management practices.

2.5 SDGs Associated with Biomass Utilization

The Sustainable Development Goals (SDGs), often known as the 2030 Agenda for Sustainable Development, are a group of 17 global objectives that were defined by the United Nations in 2015. The SDGs seek to steer nations and stakeholders toward a more sustainable and equitable world by the year 2030 by addressing a wide variety of social, economic, and environmental concerns [21]. Utilizing biomass helps to achieve several SDGs, including SDG 1, SDG 2, SDG 7 SDG 8, SDG 12, SDG 13, and SDG 15 (Fig. 3).

SDG 1—No Poverty and SDG 8—Decent Work and Economic Growth

Biomass utilization can have significant socio-economic impacts, particularly in rural areas. SDG 1 aims to end poverty in all its forms, and SDG 8 promotes inclusive economic growth and decent work. Biomass projects, if designed with inclusivity and local empowerment in mind, can create job opportunities, generate income, and improve livelihoods in communities where these resources are abundant.

SDG 2—Zero Hunger

While there may be potential competition between biomass utilization for energy and biomass utilization for food, responsible biomass practices can still contribute to SDG 2. For example, biomass waste can be used to generate energy, while the nutritious components can be returned to the soil as organic fertilizers, enhancing agricultural productivity and reducing food waste.

Fig. 3 Sustainable developmental goals for biomass utilization [21]

SDG 7—Affordable and Clean Energy

Biomass utilization, particularly in the form of bioenergy, aligns closely with SDG 7. This objective is aimed at ensuring that everyone has access to modern, sustainable, economical energy. Several types of energy, including biofuels, biogas, and biomass-based electricity, can be produced from biomass resources. By using these resources, nations may increase the variety of their energy sources, lessen their reliance on fossil fuels, and help to create a more sustainable energy future.

SDG 12—Responsible Consumption and Production

SDG 12 emphasizes sustainable patterns of consumption and production to reduce waste and environmental impact. Biomass utilization can contribute to responsible production by using organic waste materials to generate energy or produce bio-based products. This approach reduces waste generation and promotes a circular economy, aligning with the principles of sustainable resource management and reducing the strain on finite resources.

SDG 13—Climate Action

Biomass utilization has the potential to mitigate climate change, which is the core focus of SDG 13. By replacing fossil fuels with biomass-derived energy sources, carbon emissions can be reduced, contributing to the global effort to limit temperature increases and combat climate change. Additionally, certain biomass utilization practices, such as afforestation and reforestation, contribute to carbon sequestration and ecosystem restoration.

SDG 15—Life on Land

Biomass utilization should be aligned with SDG 15, which focuses on protecting, restoring, and promoting sustainable use of terrestrial ecosystems. Appropriate land use planning and management are essential to avoid negative impacts on land resources, biodiversity, and local communities [26].

3 Bioenergy Prospects

Contemporary bioenergy stands as the largest global contributor to renewable energy, constituting 55% of the renewable energy share and encompassing over 6% of the world's energy provision. In alignment with the Net Zero Emissions by 2050 (NZE) Scenario, there emerges a swift surge in bioenergy adoption, anticipated to supplant fossil fuels by 2030. The utilization of contemporary bioenergy has observed an average annual growth of around 3% from 2010 to 2022, maintaining an ascending trajectory. However, intensified endeavors are imperative to expedite the deployment of modern bioenergy, aligning with the NZE Scenario's trajectory, necessitating an annual expansion of 8% between 2022 and 2030. This trajectory is to be navigated while vigilantly ensuring that bioenergy production remains devoid of adverse social and environmental implications. Sustainable bioenergy offers a significant avenue for generating employment and income in rural communities [28]. It alleviates the disproportionate burden on women, who frequently bear the responsibility of fuel collection. This transition brings about health advantages by curbing air pollution and facilitating proper waste management. Additionally, it plays a role in diminishing methane emissions stemming from waste decomposition. It is paramount to further advance efforts aimed at phasing out the conventional utilization of biomass, considering that its absolute usage has demonstrated limited change since 2016.

3.1 Access to Clean Fuel

The issue of access to clean fuel has emerged as a pivotal concern for sustainable development, environmental preservation, and equitable progress. Clean fuel, characterized by its minimal environmental impact and reduced greenhouse gas emissions, plays a transformative role in mitigating the challenges posed by climate change, air pollution, and energy security [39]. Biomass derived fuel hence gained a lot of importance as an alternative to conventional fuel. Several kinds of biofuels were developed from a variety of sources. A recent estimate of 2019 estimated 159 billion liters of biofuel production that year. Furthermore, an annual growth rate increase of 12% was observed during the period of 2000–2019.

3.1.1 Bioethanol

Bioethanol, also known as ethyl alcohol (C_2H_5OH), is generated through biological processes. It is one of the biofuels with the most traction. Instead of lead, bioethanol is mostly added to petrol (petrol) as an octane booster. It is derived from biomass, primarily from crops like corn, sugarcane, and other plant materials. The production of bioethanol involves a multi-step process that converts biomass into a viable and environmentally friendly fuel source.

Basic steps involved in Bioethanol production

1. Saccharification
2. Fermentation
3. Distillation
4. Dehydration
5. Denaturation
6. Blending.

Complex sugars in the biomass are broken down into simpler sugars like glucose. These sugars are then mixed with yeast or microorganisms in fermentation tanks, producing ethanol along with carbon dioxide as by-products. To separate the ethanol from water and impurities, the mixture is distilled. Dehydration is then performed to remove the remaining water, resulting in a higher concentration of ethanol. If necessary, ethanol produced for fuel usage may be denatured to prevent its use for consumption.

However, the concern that dedicating land and crops to biofuel production could potentially divert resources from the production of food, leading to higher food prices and potential shortages, particularly in regions with food insecurity is a critical factor to consider. Second and subsequent generations of bioethanol have been developed to address concerns about competition with food supplies. Unlike first-generation, these newer generations utilize non-food raw materials. Second-generation bioethanol is primarily derived from lignocellulosic biomass. Additionally, industrial by-products like whey or crude glycerol can serve as feedstock. Such biomass is often locally available and economically viable.

Lignocellulose, a key component of these non-edible sources, is considered to be renewable and sustainable. It's present in various plant materials and its availability depends on climate conditions [5]. The process of converting lignocellulose into sugars is more complex compared to starch. Several varieties of plant biomass have been explored for biofuel production. This includes dedicated energy crops like herbaceous plants and perennial grasses that grow on low-quality soil. Agricultural wastes are also potential sources of lignocellulosic biomass (e.g., Rice husk, bagasse, corn cob, etc.). Forest biomass, as well as waste from parks and gardens, have also been explored. Industrial by-products like brewer's spent grains and distilleries' spent grains, along with municipal solid wastes have been considered as well. Overall, these diverse sources offer promising alternatives that mitigate the concerns associated with biofuels competing against food supplies [29].

3.1.2 Biodiesel

An alternative fuel in resemblance to conventional diesel is Biodiesel. It can be generated from many sources like vegetable oil, oil seed, waste cooking oil, animal oil/fats, tallow, and many other agricultural sources, even from algae.

Basic steps in Biodiesel production involve

1. Feedstock Selection
2. Preprocessing
3. Transesterification
4. Separation and Washing
5. Drying.

Transesterification is a process used to convert oil to Biodiesel. The triglycerides in the feedstock react with an alcohol, often methanol or ethanol, facilitated by catalysts (homogeneous or heterogeneous). This catalyzed chemical transformation yields biodiesel (methyl or ethyl esters) and glycerol. After the reaction, separation and washing are conducted to isolate the biodiesel from glycerol and cleanse it of residual impurities.

The Basic and main advantage of biodiesel is that it is considered to be "carbon–neutral." In actuality, this is not entirely correct because CO_2 is emitted during the fertilizer's manufacturing, which is necessary to fertilize the areas where oil crops are cultivated. The esterification process followed by the downstream processes are other sources of pollution related to the manufacture of biodiesel [1]. Fertilizer production is not the sole source of pollution. All of these activities need an energy input, which can come from fuel or electricity, both of which typically emit greenhouse gasses into the atmosphere. Since biodiesel is absolutely non-toxic and quickly biodegrades, spills pose significantly less of a threat than those with fossil diesel. In 2021, the market for biodiesel was projected to be worth USD 32.09 billion. From 2022 to 2030, it is expected to grow at a CAGR of 10.0%.

3.1.3 Biogas

Biogas emerges as a renewable gaseous energy source, derived from diverse sources like agro waste, municipal waste, sewage, green waste, wastewater, food waste, etc. The generation of biogas involves anaerobic digestion within anaerobic digesters, biodigesters, or bioreactors, facilitated by anaerobic organisms or methanogens. The gas primarily comprises methane along with carbon dioxide (CO_2), with potential traces of hydrogen sulfide (H_2S) with moisture, and siloxanes. Methane and hydrogen within the gas can be ignited or combined with oxygen, resulting in energy release. This energy-rich nature enables biogas utilization as a fuel, applicable in fuel cells, heating purposes like cooking, and gas engines for electricity and heat generation (Fig. 4) [42].

In order to encourage domestic production and consumption of solid and gaseous biogas through 2026, India expanded its Biomass Programme in 2022. In 2019, there were 62.3 billion m^3 of biogas produced worldwide, with a 1.43 EJ equivalent energy content. From 2000 to 2019, the industry witnessed an average annual growth rate of 9%.

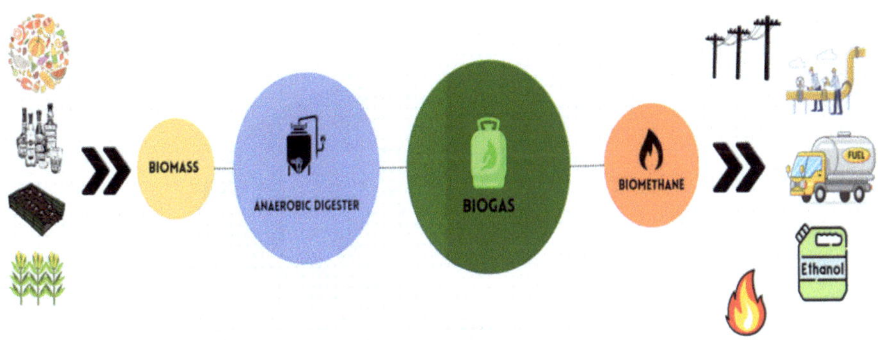

Fig. 4 Process flow for biomethane production from biomass [42]

3.2 Access to Electricity

Electricity generated from biomass uses the energy derived from organic materials like wood and agricultural waste. It is a renewable source with several benefits, including being carbon–neutral and lowering greenhouse gas emissions, recycling garbage to reduce environmental problems, and creating local jobs to stimulate economies. Biomass electricity stands out as a viable option for a more environmentally friendly future as we work to produce cleaner energy (Fig. 5) [9]. Around 637 TWh of power, or 2% of the total electricity generated worldwide, comes from modern bioenergy methods. Globally, 655 TWh of energy were produced from biomass in 2019. Solid biomass sources account for 68% of the total amount of biopower produced, while municipal and industrial waste accounted for 17% and biogas accounted for 13% [3].

Biocoal, also known as torrefied biomass, is a type of solid fuel produced by heating biomass in the absence of oxygen in a process known as torrefaction. Torrefaction takes place at temperatures between 200 and 300 °C (392 and 572 °F). The biomass goes through physical and chemical changes throughout this process, producing a more hydrophobic and energy-dense end product. The traditional torrefaction method, however, demands extending the processing time since energy and heat are often transferred through conduction and convection.

In recent studies, Wang et al., Tan et al., and Satpathy et al. have introduced a microwave-induced torrefaction technique for biomass processing. This innovative method offers a significant reduction in processing time compared to traditional methods. The rapidity of microwave-induced torrefaction is attributed to the distinct conversion of electromagnetic energy into thermal energy through interactions among materials possessing dielectric properties. Consequently, this phenomenon facilitates exceptional heat transfer, even heat distribution, and accelerated heating

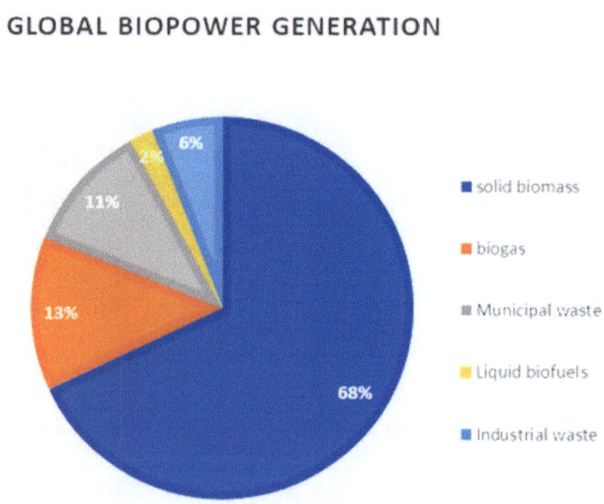

Fig. 5 Chart representing global biopower generation utilizing various biomass sources [3]

rates, resulting in expedited torrefaction when compared to traditional/conventional procedures. Cheng et al., have recently documented an alternative method comprising rapid pyrolysis followed by atmospheric distillation for the generation of biocoal from lignocellulosic biomass sources like rice husk, wheat straw, etc. This procedure involves subjecting the biomass to rapid pyrolysis at 500 °C under anaerobic conditions to generate both biochar and bio-oil. Subsequently, the bio-oil undergoes distillation at 240 °C under reasonable atmospheric conditions to separate liquid chemicals and produce biocoal. The resulting biocoal exhibits calorific values ranging from 25.4 to 27.7 MJ/kg, equivalent to those of commercial coal. This similarity in energy content underscores the potential of this approach to efficiently convert biomass into renewable fuels. In conclusion, these innovative methods offer promising avenues for transforming biomass into valuable renewable resources, contributing to the development of sustainable energy solutions [40].

3.3 Access to Energy

Energy in the form of heat or electricity through biomass such as wood, agricultural residues, and animal waste can be termed Bioenergy. Utilizing biomass as a renewable energy source, when managed sustainably, offers the advantage of locally sourced energy, reducing reliance on imported fossil fuels, while also aiding in the responsible disposal of organic waste and mitigating environmental pollution [36]. Furthermore, biomass energy serves as a crucial means of providing access to energy in rural and remote regions where alternative energy sources may be lacking (Fig. 6). A few non-conventional bioenergy sources are discussed here. These are used for several purposes like electrode material, catalyst, energy storage, etc.

3.3.1 Biochar

Biochar is a kind of charcoal-like substance that is produced through the pyrolysis (heating in the absence of oxygen) of organic materials, such as agricultural waste, wood chips, or other biomass. [41]This process converts the organic matter into a stable form of carbon that can be used for various purposes.

Biochar production through gasification involves converting biomass into a combination of gasses (syngas), liquids (bio-oil), and solid residues (biochar) using controlled heating in a low-oxygen environment. Gasification can be a more complex process compared to traditional pyrolysis, as it requires precise control over the reaction conditions to achieve the desired outcomes. Biochar production through hydrothermal technology involves subjecting wet biomass to high temperatures and pressures in the presence of water, which results in the formation of hydrochar, a biochar-like material. This method, known as hydrothermal carbonization (HTC), offers a unique approach to biochar production, as the presence of water during the process can influence the properties of the resulting hydrochar [4].

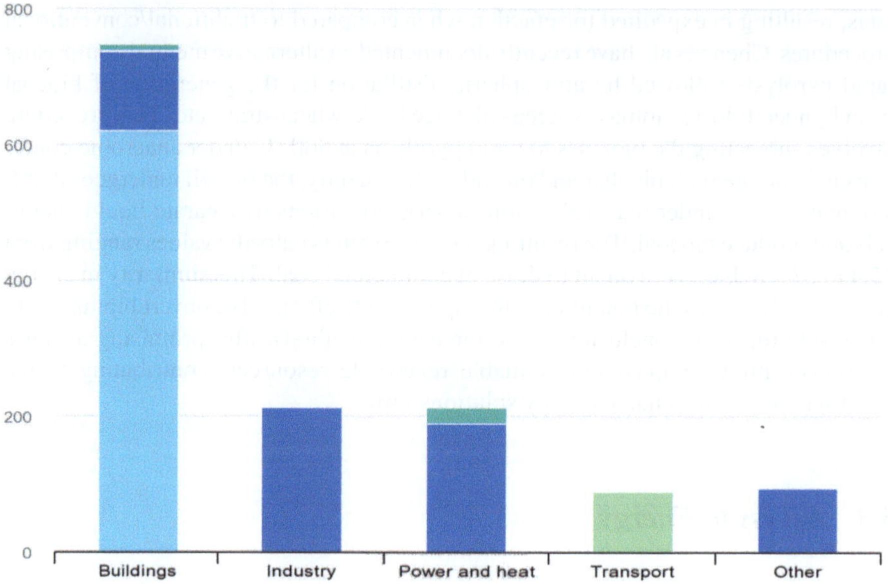

Fig. 6 Global bioenergy consumption by different types of sources and sectors [36]

Biochar is known for its potential to improve soil fertility and carbon sequestration. When added to soil, it can enhance soil structure, water retention, nutrient availability, and microbial activity. These properties can contribute to increased crop yields, especially in degraded or nutrient-poor soils. Additionally, biochar has the ability to sequester carbon for long periods, which can help mitigate climate change by removing carbon dioxide from the atmosphere and storing it in a stable form. Aside from its agricultural and environmental applications, biochar can also be used as an electrode material in various applications due to its unique properties. Its high carbon content and porous structure make it a suitable candidate for energy storage and conversion systems, such as supercapacitors and batteries [15]. Biochar electrodes can store and release electrical energy efficiently, offering a sustainable and cost-effective alternative to traditional electrode materials. This application capitalizes on biochar's capacity to enhance conductivity and provide a platform for electrochemical reactions, contributing to the development of more environmentally friendly energy storage technologies. Biochar can serve as a catalyst for hydrogen production through a process known as thermochemical or catalytic steam reforming. In this context, biochar acts as a catalyst support material that enhances the efficiency of hydrogen gas generation from various feedstocks, such as biomass, waste, or even fossil fuels.

3.3.2 Advanced Carbon Materials

Advanced carbon nanomaterials produced from biomass involve the transformation of natural organic materials, such as agricultural waste or plant-derived substances, into highly engineered carbon-based materials with unique properties [18]. These materials have applications in electronics, energy storage, catalysis, medicine, and environmental remediation. Examples include carbon nanotubes (CNTs), graphene (a single layer of carbon atoms arranged in a two-dimensional honeycomb lattice), carbon quantum dots (CQDs), activated carbon (a highly porous carbon material with a large surface area), biomass-derived carbon nanospheres, biochar-derived nanomaterials, and nanocellulose. These materials offer sustainable and eco-friendly applications, utilizing renewable biomass resources while providing enhanced properties due to their nano-scale nature. Researchers are continuously exploring innovative methods for producing and functionalizing these materials to unlock their full potential in various technological fields.

3.3.3 Biohydrogen

Biohydrogen, often referred to as hydrogen gas (H_2) produced through biological processes, is a clean and renewable energy carrier. It is generated by certain microorganisms under specific conditions through a process called "biological hydrogen production" or "biohydrogen production". In this process, certain types of bacteria, algae, or archaea use organic materials, such as sugars or organic waste, as substrates. Through a series of biochemical reactions, these microorganisms produce hydrogen gas as a metabolic byproduct. Biohydrogen serves as a clean transportation fuel, powering vehicles via combustion or fuel cells. These fuel cells also generate electricity for various uses, including backup power systems. In industrial processes, biohydrogen replaces conventionally produced hydrogen, curbing carbon emissions [6]. It facilitates hydrogenation reactions in food, pharmaceutical, and chemical industries. Additionally, biohydrogen aids agriculture, soil remediation, and bioremediation, fostering plant growth and pollutant breakdown. With potential in hydrogen storage and fueling stations, biohydrogen presents a sustainable pathway toward greener energy solutions, despite ongoing challenges in efficiency and cost-effectiveness (Fig. 7).

4 Biochemical Prospects

Biomass utilization through biochemical processes offers a diverse range of prospects that contribute to sustainable development while necessitating transparent social accounting practices. These prospects involve the conversion of organic matter into valuable bio-based products, including biofuels, bioplastics, biochemicals, and other

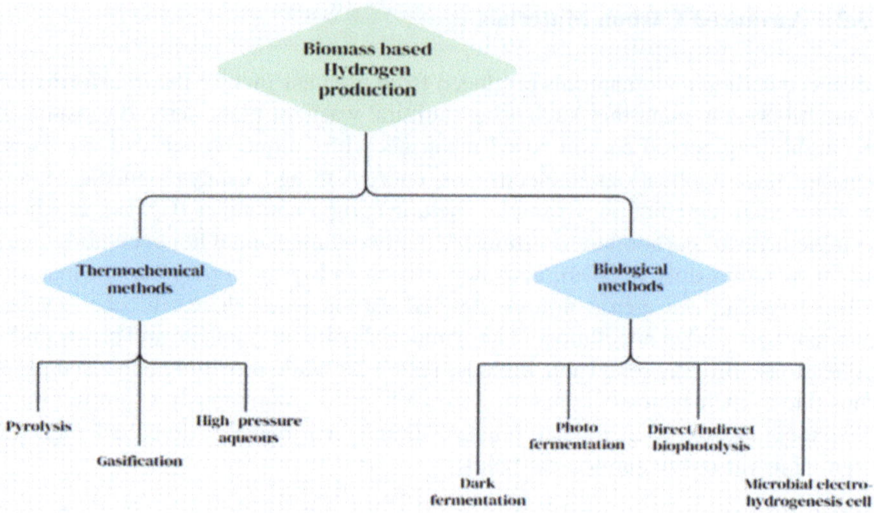

Fig. 7 Methods of biomass-based production [6]

bioproducts. This section elaborates on the biochemical prospects of biomass utilization, emphasizing their contributions to social accounting and credibility across various applications.

4.1 *Biofuels and Energy Generation*

Biofuels, such as bioethanol, biodiesel, and biogas, are generated from biomass feedstocks through biochemical pathways like fermentation, enzymatic hydrolysis, and thermochemical processes. These biofuels can serve as alternative or supplementary energy sources, thereby mitigating greenhouse gas emissions and enhancing energy security. From a social accounting perspective, biofuels can create local employment opportunities, particularly in rural areas where biomass feedstock production and processing occur. This aligns with SDG 8 (Decent Work and Economic Growth) and supports inclusive development. Transparent social accounting practices would involve quantifying the number of jobs created, assessing the socio-economic impact on local communities, and reporting on the equitable distribution of benefits [7].

4.2 *Bioplastics and Sustainable Materials*

Biomass-derived materials, such as bioplastics, hold the potential to decrease the environmental impact associated with conventional plastics. Bioplastics are produced from renewable sources, often through biochemical processes that convert biomass-derived sugars into polymer precursors. By replacing fossil fuel-based plastics, bioplastics contribute to SDG 12 (Responsible Consumption and Production) and SDG 13 (Climate Action) by reducing plastic waste and decreasing reliance on non-renewable resources. Social accounting in this context involves evaluating the reduction in plastic waste, measuring the carbon footprint of bioplastics production, and ensuring that bioplastics adhere to sustainability standards throughout their life-cycle. Transparent reporting of these metrics enhances the credibility of bioplastics projects and their environmental benefits [38].

4.3 Biochemicals and Green Chemistry

Biomass resources can be harnessed to produce a wide spectrum of high-value biochemicals utilized in industries such as pharmaceuticals, cosmetics, and agriculture. Biochemical production pathways often employ green chemistry principles, leading to reduced reliance on environmentally harmful chemicals and processes. The utilization of biomass for the production of pharmaceutical intermediates, bioactive compounds, and sustainable agricultural inputs aligns with SDG 3 (Good Health and Well-being) and SDG 9 (Industry, Innovation, and Infrastructure). Social accounting in this context entails assessing the environmental benefits of bio-based products, quantifying the reduction in hazardous chemical usage, and ensuring that the production processes follow sustainable practices. Transparent documentation of these aspects fosters credibility and accountability in the biochemical industry [2].

Few examples of Bio-based chemicals and Pharmaceuticals are,

1. Platform Chemicals

   Biomass can serve as a source of platform chemicals like glucose, xylose, and lignin. These chemicals can be used as building blocks for a wide range of downstream chemical processes, leading to the production of various valuable products.
2. Biobutanol

   Microbial fermentation of biomass-derived sugars can yield biobutanol, which has applications as a biofuel and a chemical precursor.
3. Organic Acids

   Biomass can be converted into organic acids such as citric acid, succinic acid, and lactic acid. These acids are utilized in the food, pharmaceutical, and chemical industries.
4. Bioactive Compounds

   Biomass can yield bioactive compounds like polyphenols, flavonoids, and alkaloids. These compounds have potential applications in pharmaceuticals and nutraceuticals, contributing to health and wellness industries.
5. Pharmaceutical Intermediates

Biomass-derived compounds can serve as intermediates in the synthesis of pharmaceuticals, reducing the reliance on fossil-derived starting materials.

4.4 Waste-To-Energy and Circular Economy

Biochemical conversion of biomass waste streams, such as agricultural residues and organic waste, into energy represents a sustainable waste-to-energy solution. Processes like anaerobic digestion and microbial fermentation transform organic matter into biogas, which can be used for electricity and heat generation. This approach contributes to SDG 7 (Affordable and Clean Energy) and addresses waste management challenges, supporting SDG 12 (Responsible Consumption and Production). From a social accounting perspective, the diversion of organic waste from landfills reduces environmental hazards, and the generation of renewable energy contributes to local economies [19] Transparent social accounting practices encompass tracking the amount of waste diverted, assessing the impact on nearby communities' well-being, and ensuring equitable distribution of economic benefits. Few important examples are

1. Enzyme Production
   Biomass can be a source of raw material for enzyme production. Enzymes have diverse applications, including food processing, textile manufacturing, and biofuel production.
2. Bioethanol Production
   Microorganisms like yeast can ferment biomass-derived sugars (e.g., from corn, sugarcane, or cellulosic materials) to produce bioethanol, a renewable alternative to gasoline.
3. Microbial Conversion
   Certain microorganisms can convert complex biomass structures into simpler molecules, which can then be used to synthesize valuable chemicals.
4. Biorefineries
   Biorefineries use various types of biomass feedstocks (e.g., agricultural residues, forestry waste, algae) to produce multiple products, including fuels, chemicals, materials, and energy. Fractionation method is employed to separate biomass into its various components (cellulose, hemicellulose, lignin, etc.), each of which can be utilized for different purposes.
5. Waste Valorization
   Biomass waste from agriculture and food processing, such as crop residues, fruit peels, and husks, can be transformed into valuable bio-based chemicals through biochemical processes. Organic waste from households and municipalities can be converted into bio-based chemicals and biofuels, reducing landfill burden and promoting circular economy principles.
6. Nutrient Recycling
   Biomass can be converted into organic fertilizers like compost, contributing to soil health, enhancing nutrient availability, and promoting sustainable agriculture.

7. Sustainable Textiles

   Biomass-derived cellulose fibers can be used to produce sustainable textiles. These fibers are biodegradable and have a lower environmental impact compared to synthetic textiles.
8. Renewable Energy Co-production

   During biochemical biomass processing, anaerobic digestion of organic materials can produce biogas, primarily composed of methane. This biogas can be used as a renewable energy source.

4.5 Algae and Aquatic Biomass

Algae and other aquatic biomass hold potential as sources of biofuels, food supplements, and biochemicals. Algae cultivation can occur in non-arable land or even in wastewater environments, reducing competition for limited resources. Products derived from aquatic biomass contribute to SDG 2 (Zero Hunger) and SDG 14 (Life Below Water) by potentially providing sustainable protein sources and promoting aquatic ecosystem health. In terms of social accounting, it is essential to evaluate the impact of algae cultivation on local aquatic ecosystems, quantify the potential for job creation in algae-related activities, and ensure that resource extraction does not disrupt ecosystem services. Transparent reporting of these aspects ensures the credibility of algae-based projects and their contributions to sustainability [34].

The biochemical prospects of biomass utilization offer a transformative pathway toward sustainable development while necessitating robust social accounting practices. These prospects span across sectors including energy, materials, chemicals, and waste management. To ensure the credibility of biomass projects and their positive impacts, comprehensive social accounting mechanisms must be in place to assess the social, economic, and environmental implications. By promoting inclusivity, accountability, and stakeholder engagement, biomass utilization can play a pivotal role in achieving the Sustainable Development Goals, advancing the circular economy, and fostering a more sustainable and equitable future.

5 Policy Framework for Social Inclusion

The paramount objective of any organization or project should be to contribute to the betterment of society. To achieve this, it is imperative that all stakeholders, regardless of their social, political, cultural, or economic backgrounds, have the opportunity to partake in and reap the benefits of advancements. A policy framework becomes essential in order to devise and execute measures aimed at tackling systemic inequalities and barriers, thereby fostering social inclusion for individuals from diverse backgrounds. A comprehensive policy framework should duly address several critical factors, few of which include.

5.1 Education and Training

Providing opportunities for individuals from all backgrounds to acquire the necessary knowledge and skills to actively engage in the endeavors of the organization or project. A comprehensive perspective on education is of utmost importance. Well-rounded early childhood care and education programs have been shown to enhance children's well-being, effectively prepare them for primary school, and significantly improve their likelihood of success throughout their educational journey. This impact is particularly pronounced among the most disadvantaged and vulnerable children. The literacy level of adults, particularly mothers, has a substantial bearing on the educational participation of their children, with a notable effect on the attendance of their daughters in particular [35].

The integration of inclusivity within broader developmental objectives serves as a catalyst for the transformation of education systems, poverty alleviation, and the realization of the Millennium Development Goals. An inclusive educational framework extends its benefits to all learners, without bias or discrimination toward any individual or group. This approach is rooted in democratic principles, tolerance, and a genuine appreciation for diversity.

5.2 Employment Opportunities

The formulation of a robust policy framework for social inclusion in employment necessitates the integration of various key elements to ensure equal opportunities, diversity, and equitable participation in the workforce. The imperative of social inclusion within employment is far-reaching, encompassing diverse benefits. A culture of inclusion enriches perspectives and ignites innovation by melding varied backgrounds and experiences. This environment enhances employee satisfaction and performance, fueled by a sense of belonging and psychological safety. It also attracts a wide talent pool, nurturing unique skills and viewpoints that elevate problem-solving. Inclusive workplaces tend to experience lower turnover rates, curbing recruitment and training costs [13].

Moreover, social inclusion broadens market reach, appealing to a diverse customer base. Ethically and legally, it aligns with equal opportunity mandates and enhances an organization's reputation. Economically, it bolsters growth by integrating a broader workforce. Socially, it fosters unity and empathy. Thus, social inclusion in employment transcends a moral obligation, fuel innovation, engagement, and prosperity across organizational and societal landscapes.

5.3 Women Empowerment

The empowerment of women plays a pivotal role in the landscape of global development. Its significance is closely intertwined with economic progress, as it effectively mitigates inequalities between genders, thereby fostering an environment conducive to developmental advancement. Notably, the encouragement of women's participation in the public sector aligns with the targets set by the Millennium Development Goals for 2020. These goals are aimed at promoting gender equality and women's empowerment within developing nations. It is important to emphasize that women's involvement in livelihood pursuits or public sector endeavors does not detract from their primary roles as homemakers.

In various low-income and developing economies, the establishment of seaweed farming has emerged as a robust strategy for empowering women in coastal areas. This approach not only holds the potential to elevate the socio-economic status of women but also contributes to broader developmental objectives [33].

5.4 Economic Setting

Developing economies encounter challenges in adopting renewable energy sources due to economic barriers related to the setup and upkeep of such systems, despite their significant biomass production potential. To bolster the expansion of renewable energy technology in these economies, fiscal policy support emerges as pivotal. This support serves to stimulate the augmentation of capacity, the advancement of technology, and the engagement of private enterprises in promoting these technologies. Essential to this approach is the allocation of public funds toward establishing substantial energy generation facilities, including expansive solar parks, wind farms, waste-to-energy plants, and hydroelectric projects. Furthermore, the reduction of domestic and import taxes coupled with the implementation of subsidies emerges as a strategy to incentivize private sector entities to establish smaller-scale energy installations at both residential and commercial levels.

Monetary policies can play a pivotal role in fostering the uptake of renewable technology within developing economies, particularly by mitigating challenges associated with private investment financing. Introducing accommodating financial frameworks, endorsed by central banking authorities, could encompass elements such as reduced interest rates, extended grace periods, and elongated repayment schedules. Moreover, the establishment of dedicated funds tailored for non-governmental organizations and micro-finance institutions can serve as a catalyst for encouraging the integration of indigenous renewable technology within rural, isolated, and environmentally sensitive areas. Despite its promising potential, bioenergy has not received significant investments. It is imperative to implement effective policies to tap into the vast capabilities of bioenergy, thereby creating pathways toward a more environmentally friendly future [17].

5.5 Health and Safety Standards

Robust regulations must be in place to protect the workforce involved in bioprocess industries from potential occupational hazards, fostering a secure and conducive working environment for all. Incorporating social inclusion into health and safety standards is crucial to fostering workplaces that prioritize the welfare of all employees. This approach acknowledges the diverse needs and circumstances of individuals, ensuring that health and safety measures are accessible and effective for everyone.

Socially inclusive health and safety standards encompass considerations such as gender, age, ability, ethnicity, and socio-economic background. This entails creating physically accessible environments, offering clear and culturally sensitive communication, and accommodating various learning styles and languages. By tailoring safety protocols to a wide range of needs, organizations demonstrate their commitment to fairness and equity. Inclusive health and safety standards lead to a more engaged and supportive workforce. When employees perceive that their unique requirements are recognized, they are more likely to actively engage in safety practices. Moreover, these standards cultivate a sense of belonging and psychological well-being, fostering positive morale and a harmonious work environment.

From a risk management standpoint, integrating social inclusion minimizes potential hazards and prevents accidents. By identifying and addressing risks that might disproportionately affect specific groups, organizations exhibit proactive safety measures. This not only safeguards employees' physical and mental health but also bolsters productivity and organizational cohesion. By infusing social inclusion into health and safety standards, organizations fulfill legal obligations, uphold ethical responsibilities, enhance their corporate image, and strengthen community ties. Ultimately, this integration creates workplaces that prioritize the safety and well-being of every individual, reinforcing a culture of respect and inclusivity [11].

6 Conclusion

While biomass energy offers numerous advantages, it's equally essential to thoroughly examine its drawbacks. This comprehensive evaluation is crucial to effectively address the challenges associated with biomass energy and ensure its beneficial availability to society. Several disadvantages are directly linked to the utilization of biomass as a fuel source. Moreover, there are additional secondary costs connected to the generation and application of biomass energy. It's important to acknowledge that both these direct and indirect implications hold significance in understanding the practical implementation of biomass energy. Considering the potential downsides alongside the benefits allows for a well-informed and balanced perspective. This approach is instrumental in navigating the complexities of biomass energy and facilitating its responsible incorporation for the benefit of society at large. As we navigate the dynamic landscape of biomass resource utilization, it becomes evident that the synergy between social accounting and credibility is indispensable. By integrating social perspectives, fostering transparent methodologies, and valuing diverse voices, we can forge a path toward sustainable biomass utilization that not only bolsters our energy and material needs but also nurtures resilient communities and safeguards the integrity of our ecosystems. The journey toward a greener future requires a steadfast commitment to both responsible resource management and a society that reaps the benefits equitably and conscientiously. The intersection of social accounting and the credibility of biomass resources' utilization is pivotal in shaping a more sustainable future. By rigorously evaluating the social, environmental, and ethical dimensions of biomass applications, we can harness their potential benefits while avoiding unintended consequences [32]. As research and innovation continue to advance, a comprehensive understanding of the multifaceted nature of biomass utilization will serve as a compass guiding us toward a greener and more responsible path forward.

References

1. Ahmed R, Huddersman K (2022) Review of biodiesel production by the esterification of wastewater containing fats oils and grease (FOGs). J Ind Eng Chem 110:1–14. https://www.sciencedirect.com/science/article/pii/S1226086X22001071
2. Ahmed SF et al (2022) Green approaches in synthesising nanomaterials for environmental nanobioremediation: technological advancements, applications, benefits and challenges. Environ Res 204:111967. https://www.sciencedirect.com/science/article/pii/S0013935121012627
3. Akinbami OM, Oke SR, Bodunrin MO (2021) The state of renewable energy development in South Africa: an overview. Alexandria Eng J 60(6):5077–5093. https://www.sciencedirect.com/science/article/pii/S1110016821002295
4. Amalina F et al (2022) Biochar production techniques utilizing biomass waste-derived materials and environmental applications—a review. J Hazard Mater Adv 7:100134. https://www.sciencedirect.com/science/article/pii/S2772416622000900
5. Anwar Z, Gulfraz M, Irshad M (2014) Agro-industrial lignocellulosic biomass a key to unlock the future bio-energy: a brief review. J Radiat Res Appl Sci 7(2):163–173. https://www.sciencedirect.com/science/article/pii/S1687850714000119
6. Argun H, Kargi F (2011) Bio-hydrogen production by different operational modes of dark and photo-fermentation: an overview. Int J Hydrogen Energy 36(13):7443–7459. https://www.sciencedirect.com/science/article/pii/S0360319911007063
7. Ben-Iwo J, Manovic V, Longhurst P (2016) Biomass resources and biofuels potential for the production of transportation fuels in Nigeria. Renew Sustain Energy Rev 63:172–192. https://www.sciencedirect.com/science/article/pii/S1364032116301605
8. Bishop G, Styles D, Lens PNL (2021) Environmental performance comparison of bioplastics and petrochemical plastics: a review of life cycle assessment (LCA) methodological decisions. Resour Conserv Recycl 168:105451. https://www.sciencedirect.com/science/article/pii/S0921344921000586
9. Bisht AS, Thakur NS (2019) Small scale biomass gasification plants for electricity generation in India: resources, installation, technical aspects, sustainability criteria & policy. Renew Energy Focus 28:112–126. https://www.sciencedirect.com/science/article/pii/S1755008418303211
10. Campbell-Johnston K, Vermeulen WJV, Reike D, Brullot S (2020) The circular economy and cascading: towards a framework. Resour Conserv Recycl X 7:100038. https://www.sciencedirect.com/science/article/pii/S2590289X20300098

11. Demski C et al (2015) Public values for energy system change. Glob Environ Change 34:59–69. https://www.sciencedirect.com/science/article/pii/S0959378015000953
12. Fernández-Getino AP, Alonso-Prados JL, Santín-Montanyá MI (2018) Challenges and prospects in connectivity analysis in agricultural systems: actions to implement policies on land management and carbon storage at EU level. Land Use Policy 71:146–159. https://www.sciencedirect.com/science/article/pii/S0264837716310365
13. Gielen D et al (2019) The role of renewable energy in the global energy transformation. Energy Strategy Rev 24:38–50. https://www.sciencedirect.com/science/article/pii/S2211467X19300082
14. Gustavsson L, Börjesson P, Johansson B, Svenningsson P (1995) Reducing CO_2 emissions by substituting biomass for fossil fuels. Energy 20(11):1097–1113. https://www.sciencedirect.com/science/article/pii/036054429500065O
15. Husain Z et al (2022) Nano-sized mesoporous biochar derived from biomass pyrolysis as electrochemical energy storage supercapacitor. Mater Sci Energy Technol 5:99–109. https://www.sciencedirect.com/science/article/pii/S2589299121000719
16. Jaiswal KK et al (2022) Renewable and sustainable clean energy development and impact on social, economic, and environmental health. Energy Nexus 7:100118. https://www.sciencedirect.com/science/article/pii/S2772427122000687
17. Kabir M, Kabir Z, Sultana N (2022) "Chapter 14—Climate change, sustainability, and renewable energy in developing economies. In: Khan I (ed) Renewable energy and sustainability. Elsevier, pp 377–415. https://www.sciencedirect.com/science/article/pii/B9780323886680000012
18. Malode SJ et al (2023) Biomass-derived carbon nanomaterials for sensor applications. J Pharm Biomed Anal 222:115102. https://www.sciencedirect.com/science/article/pii/S0731708522005234
19. Manikandan S et al (2023) Critical review of biochemical pathways to transformation of waste and biomass into bioenergy. Bioresour Technol 372:128679. https://www.sciencedirect.com/science/article/pii/S0960852423001050
20. Mehmood MA et al (2017) Biomass production for bioenergy using marginal lands. Sustain Prod Consumption 9:3–21. https://www.sciencedirect.com/science/article/pii/S2352550916300197
21. Moallemi EA et al (2020) Achieving the sustainable development goals requires transdisciplinary innovation at the local scale. One Earth 3(3):300–313. https://www.sciencedirect.com/science/article/pii/S2590332220304152
22. Munawar MA et al (2021) Challenges and opportunities in biomass ash management and its utilization in novel applications. Renew Sustain Energy Rev 150:111451. https://www.sciencedirect.com/science/article/pii/S1364032121007346
23. Newell RG, Pizer WA, Raimi D (2019) U.S. Federal Government Subsidies for clean energy: design choices and implications. Energy Econ 80:831–41. https://www.sciencedirect.com/science/article/pii/S0140988319300751
24. Paletto A et al (2019) Assessment of environmental impact of biomass power plants to increase the social acceptance of renewable energy technologies. Heliyon 5(7):e02070. https://www.sciencedirect.com/science/article/pii/S2405844019357305
25. Popp J et al (2021) Bioeconomy: biomass and biomass-based energy supply and demand. New Biotechnol 60:76–84. https://www.sciencedirect.com/science/article/pii/S1871678420301825
26. Priyanka S, Karthick Raja Namasivayam S, Arvind Bharani RS, John A (2023) Biocompatible green technology principles for the fabrication of food packaging material with noteworthy mechanical and antimicrobial properties—a sustainable developmental goal towards the effective, safe food preservation strategy. Chemosphere 336:139240. https://www.sciencedirect.com/science/article/pii/S0045653523015072
27. Rahman MN, Wahid MA (2021) Renewable-based zero-carbon fuels for the use of power generation: a case study in Malaysia supported by updated developments worldwide. Energy Rep 7:1986–2020. https://www.sciencedirect.com/science/article/pii/S2352484721002195

28. Rather RA et al (2022) Bioenergy: a foundation to environmental sustainability in a changing global climate scenario. J King Saud Univ Sci 34(1):101734. https://www.sciencedirect.com/science/article/pii/S1018364721003967
29. Robak K, Balcerek M (2018) Review of second generation bioethanol production from residual biomass. Food Technol Biotechnol 56(2):174–187
30. Saleem M (2022) Possibility of utilizing agriculture biomass as a renewable and sustainable future energy source. Heliyon 8(2):e08905. https://www.sciencedirect.com/science/article/pii/S2405844022001931
31. Scarlat N, Dallemand J-F, Monforti-Ferrario F, Nita V (2015) The role of biomass and bioenergy in a future bioeconomy: policies and facts. Environ Dev 15:3–34. https://www.sciencedirect.com/science/article/pii/S2211464515000305
32. Stringer LC et al (2020) Adaptation and development pathways for different types of farmers. Environ Sci Policy 104:174–189. https://www.sciencedirect.com/science/article/pii/S1462901119305209
33. Sultana F et al (2023) Seaweed farming for food and nutritional security, climate change mitigation and adaptation, and women empowerment: a review. Aquac Fish 8(5):463–480. https://www.sciencedirect.com/science/article/pii/S2468550X22001460
34. Ullah K et al (2014) Algal biomass as a global source of transport fuels: overview and development perspectives. Progr Nat Sci Mater Int 24(4):329–339. https://www.sciencedirect.com/science/article/pii/S1002007114000860
35. Urmetzer S, Lask J, Vargas-Carpintero R, Pyka A (2020) Learning to change: transformative knowledge for building a sustainable bioeconomy. Ecol Econ 167:106435. https://www.sciencedirect.com/science/article/pii/S0921800919301685
36. Varjani S et al (2022) Sustainable management of municipal solid waste through waste-to-energy technologies. Bioresour Technol 355:127247. https://www.sciencedirect.com/science/article/pii/S0960852422005764
37. Vijay V et al (2022) Sustainable utilization of biomass resources for decentralized energy generation and climate change mitigation: a regional case study in India. Environ Res 212:113257. https://www.sciencedirect.com/science/article/pii/S0013935122005849
38. Walker TR (2021) (Micro)Plastics and the UN sustainable development goals. Curr Opin Green Sustain Chem 30:100497. https://www.sciencedirect.com/science/article/pii/S2452223621000535
39. Wang F et al (2021) Technologies and perspectives for achieving carbon neutrality. Innovation 2(4):100180. https://www.sciencedirect.com/science/article/pii/S2666675821001053
40. Wang MJ et al (2012) Microwave-induced torrefaction of rice husk and sugarcane residues. Energy 37(1):177–184. https://www.sciencedirect.com/science/article/pii/S0360544211007845
41. Yaashikaa PR, Senthil Kumar P, Varjani S, Saravanan A (2020) A critical review on the biochar production techniques, characterization, stability and applications for circular bioeconomy. Biotechnol Rep 28:e00570. https://www.sciencedirect.com/science/article/pii/S2215017X20300023
42. Yadav K, Sircar A (2022) Fundamentals and developments of compressed biogas in city gas distribution network in India: a review. Pet Res 7(3):409–418. https://www.sciencedirect.com/science/article/pii/S2096249521000958

Economic, Social, and Organizational Challenges in Biorefineries

Heitor B. S. Bento, Íkaro Tessaro, Janaina Martins Theodoro, Willian de Souza Matias Reis, Gabrielle Policarpo, Cristiano E. Rodrigues Reis, and Ana Karine Furtado de Carvalho

Abstract This chapter investigates the economic, social, and organizational challenges faced by biorefineries, with the aim of promoting sustainability. Biorefineries play a crucial role in the circular economy by transforming biomass and waste materials into value-added products such as fuels, chemicals, and materials, ultimately reducing dependency on fossil resources. Some of the economic challenges related to initial investment, feedstock availability, market dynamics, and policy frameworks that impact the viability of biorefineries are presented herein. The social challenges explored in this chapter encompass public perception, environmental concerns, employment opportunities, land use conflicts, and the role of biorefineries in achieving global sustainability goals. Additionally, the organizational challenges addressed include integration with existing industries, supply chain management, intellectual property, workforce training, and risk management strategies. To provide practical insights, case studies of successful biorefinery implementations and lessons learned from struggling projects are also introduced and presented. The chapter highlights the importance of addressing these challenges and harnessing the potential of emerging technologies to ensure the success of biorefineries in fostering a sustainable future.

Keywords Biorefineries · Sustainability · Circular economy · Economic · social · and organizational challenges · Emerging technologies

H. B. S. Bento
School of Pharmaceutical Sciences, State University of Sao Paulo, Araraquara, Sao Paulo, Brazil

Í. Tessaro · J. M. Theodoro · W. de Souza Matias Reis · G. Policarpo · A. K. F. de Carvalho (✉)
Engineering School of Lorena, University of Sao Paulo, Lorena, Sao Paulo, Brazil
e-mail: anacarvalho@usp.br

C. E. R. Reis
EARTH University, Guácimo, Limón, Costa Rica

S. Kumar et al. (eds.), *Clean Energy Transition-via-Biomass Resource Utilization*, Green Energy and Technology, https://doi.org/10.1007/978-981-97-6321-4_9

1 Introduction

The exponential world population growth, projected to reach 10 billion people by 2057, is associated with several concerns about environmental issues, food security, global warming, waste generation, and natural resource reduction. Biorefineries have emerged as a promising approach to mitigate the global challenges of sustainable development, energy security, and better natural resource usage. These multifaceted facilities integrate many processes and technologies to convert renewable biomass into valuable products such as biofuels, biochemicals, and bio-based materials. However, as the world transitions towards a bioeconomy, there are significant economic, social, and organizational challenges that must be addressed to ensure the successful large-scale implementation of biorefineries [84].

Biorefineries, as promising hubs for sustainable production and processing of biomass-derived resources, face a range of social challenges that demand careful consideration. One primary concern is the potential competition between biorefineries and food production, as both rely on agricultural resources. Balancing the demands of bioenergy and bio-based products with the need to ensure food security becomes crucial. Additionally, biorefineries often emerge in rural areas, and their establishment may impact local communities and livelihoods. Issues like land use conflicts, water availability, and increased traffic can strain the social fabric [139].

The implementation of biorefineries, while offering promising prospects for sustainable development, comes with its fair share of economic and organizational challenges. One significant hurdle lies in the initial capital investment required to establish and operate these sophisticated facilities. Biorefineries often demand substantial funding for state-of-the-art equipment and research and development to optimize processes. Securing adequate financial support and attracting investors can be daunting, particularly when traditional fossil fuel industries still dominate the market [7].

Organizational challenges also loom large, as biorefineries typically involve complex supply chains and intricate coordination among various stakeholders. Efficient collection, transportation, and processing of diverse biomass feedstocks are crucial for smooth operations, but these processes often necessitate the cooperation of multiple entities, ranging from farmers and suppliers to technology providers and end-users. Establishing strong partnerships and collaborations becomes critical to addressing bottlenecks and ensuring a steady supply of feedstocks. Furthermore, the regulatory landscape poses another obstacle. Biorefineries must navigate through a maze of environmental, safety, and health regulations, which can vary widely across regions and countries. Compliance with these stringent guidelines can be costly and time-consuming, necessitating meticulous planning and legal expertise [7, 84, 139].

Addressing these economic and organizational challenges requires a comprehensive and interdisciplinary approach. Governments, private sectors, and research institutions need to collaborate closely to provide financial incentives, streamline regulations, and foster innovation. By investing in research and development and promoting sustainable practices, the potential of biorefineries to revolutionize the energy and chemical industries can be fully realized, ensuring a more sustainable and greener future [7, 84, 139].

2 Economic Challenges

Biorefineries play a crucial role in the transition towards a more sustainable, resource-based economy [41]. However, the implementation of these units faces significant economic challenges such as high upfront investment costs, the need for adequate infrastructure, the availability and cost of biomass, market dynamics and uncertainties, government policies and regulations, and the need for technological advancements and scale [107].

In light of these challenges, the objective is to explore innovative strategies and solutions to overcome them and drive the success of biorefineries. By identifying effective approaches to address the economic challenges, it is possible to maximize financial efficiency, ensure a sustainable biomass supply, understand the dynamics of constantly changing markets, influence favorable policies, and promote technological advancements to scale operations. Through these strategies, it is possible to enhance the viability of biorefineries and move towards a more sustainable, resource-based economy [36].

2.1 Initial Investment and Infrastructure Costs

When addressing the topic of initial investment costs and necessary infrastructure to establish a biorefinery, it is crucial to conduct a comprehensive analysis of the elements contributing to these costs and identify strategies aimed at reducing them and improving financial efficiency [17]. A thorough examination of the key components of investment costs, such as equipment, facilities, technology, and labor, will allow for a deeper understanding of the challenges faced in this area [14]. Furthermore, exploring opportunities arising from public–private partnerships becomes relevant, as such collaborations can facilitate resource sharing and consequently contribute to cost reduction [158].

Furthermore, the pursuit of financing and government incentives stands as a crucial strategy to alleviate the initial financial burden and stimulate investments in the biorefinery sector [151]. It is essential to examine concrete examples of funding programs and incentives already implemented in different regions, aiming to analyze the impact of these initiatives on the economic viability of biorefineries [55]. By presenting these strategies and exploring their potential advantages and challenges, it will be possible to provide readers with a comprehensive view of the options available to reduce investment costs and enhance financial efficiency in the implementation of biorefineries [161].

The initial investment costs and necessary infrastructure to establish a biorefinery can vary widely, typically ranging from millions to billions of dollars, depending on the plant size, technology, location, and project specifics [154]. Some of the key components of these costs include initial investment, equipment procurement, facility construction, utilization of advanced technologies, auxiliary infrastructure, research and development, obtaining licenses and permits, and hiring skilled labor [117]. It is important to note that these figures are general indications, and actual costs can vary significantly. Conducting detailed feasibility studies and consulting industry experts are recommended to accurately assess the investment costs and infrastructure requirements in a specific biorefinery project [61].

2.2 *Feedstock Availability, Cost, and Competition*

When exploring the topic of biomass availability as a feedstock, associated costs, and competition, it is crucial to adopt an efficient approach based on reliable data [160]. For instance, the availability of biomass as a feedstock in biorefineries is a crucial aspect to consider in the transition towards a more sustainable, resource-based economy. An in-depth analysis of the current availability of biomass and its sources in different regions is essential to understand the viability of these renewable energy sources. Challenges such as geographical restrictions, seasonality, and accessibility can impact biomass availability, necessitating the exploration of strategies to ensure a sustainable supply [127].

Diversifying biomass sources is a promising approach to address these challenges. In addition to traditional biomass sources such as agricultural and forestry residues, it is important to consider the potential of energy crops, algae, and other cellulosic materials. The cultivation of dedicated crops specifically for the production of biofuels and renewable chemicals is also an area of interest. However, a careful examination of the environmental, social, and economic implications of these strategies is necessary, ensuring that there is no competition with food production and other essential uses of biomass [6, 86].

In addition to availability, the costs associated with biomass play an important role in the economic viability of biorefineries. Analyzing the costs of biomass acquisition and processing, including harvesting, transportation, storage, and pretreatment, is crucial in assessing the financial sustainability of these ventures [102]. Comparing biomass costs with other energy sources and raw materials such as fossil fuels and petroleum-derived chemicals provides a comprehensive understanding of the challenges faced [89, 136].

Strategies to reduce biomass costs should be explored, taking into consideration efficiency and technological innovation [156]. The development of more efficient harvesting technologies, optimization of transportation logistics, and utilization of agricultural waste and by-products are areas of research and development that can contribute to cost reduction [80, 130]. Furthermore, incentive policies and government support can play a vital role in the economic feasibility of biorefineries [73].

Competition for biomass between biorefineries and other industries such as food production, animal feed, and construction materials is a challenge that requires balance and cooperation among sectors [43]. The discussion on the importance of partnerships between the agricultural, industrial, and energy sectors is crucial to ensure a sustainable supply of biomass, avoiding conflicts and ensuring availability for all necessary uses [34, 143].

The implementation of sustainability initiatives, such as certifications and traceability standards, also plays a crucial role in promoting responsible biomass utilization [148]. These initiatives ensure that the biomass used in biorefineries comes from sustainable sources, minimizing environmental and social impacts [46, 138].

2.3 Market Dynamics and Uncertainties

The analysis of market dynamics is crucial to understanding the economic viability of biorefineries. It involves carefully identifying key elements that impact the sector, such as supply, demand, government policies, consumer preferences, and technological advancements. These interrelated factors shape the competitive landscape and financial sustainability of biorefineries [125].

Moreover, it is essential to assess current and future market trends, including the projected growth in demand for biofuels, renewable chemicals, and other biomass-derived products. Understanding these trends is crucial for making informed strategic decisions, such as investments, product diversification, and identifying market opportunities [4, 97].

It is equally important to explore the potential of emerging markets and export opportunities, considering regulatory requirements and international quality standards. Seizing these opportunities drives the growth of biorefineries, enabling entry into new markets and forming strategic partnerships [70].

For a solid market analysis, reliable data, market studies, and up-to-date information on policies and regulations are crucial. Understanding the factors shaping the biomass and end-products market allows biorefineries to anticipate changes, identify market niches, and develop effective strategies for long-term success [20, 106].

According to Martinez-Valencia et al. [91], strategies to mitigate market uncertainties are crucial for the viability of biorefineries. Some effective approaches include:

- **Long-term contracts:** Establishing long-lasting contracts with biomass suppliers and customers provides stability in raw material flows and sales. This reduces exposure to price and demand fluctuations, providing financial predictability [91].
- **Diversification of end products:** Identifying new applications and markets for biorefinery products helps mitigate market risks. Diversifying the product

line allows for greater adaptability to demand fluctuations and price variations, reducing dependence on a single product [22, 91].

- **Strategic partnerships:** Establishing partnerships with technology suppliers, distributors, and logistics companies strengthens the competitive position of the biorefinery. These partnerships allow for resource sharing, knowledge exchange, and experience sharing, mitigating risks and collectively addressing market challenges [91, 129].

2.4 Policy and Regulatory Frameworks

Analyzing the role of government policies and regulations is of utmost importance in assessing the viability of biorefineries in the transition towards a more sustainable, resource-based economy. In this context, it is essential to explore and understand the various aspects related to this topic [138].

First and foremost, government policies play a crucial role in creating an enabling environment for the development and success of biorefineries. The existence of clear and stable policies encompassing financial incentives, subsidies, production targets, and sustainability standards is essential. These policies provide predictability and legal certainty for entrepreneurs and investors in the sector, stimulating investments and contributing to the economic viability of biorefineries [39, 54].

Furthermore, government policies have a direct influence on the competitiveness and profitability of biorefineries. Through subsidies and financial incentives, it is possible to reduce investment and operational costs, making biorefineries more economically attractive. Production targets and sustainability standards also impact market demands, driving the adoption of more efficient and sustainable technologies in biorefineries [40, 116].

To promote favorable government policies for biorefineries, it is essential for the sector to engage with policymakers. This can be done through participation in lobbying groups, where relevant data and information can be provided to inform discussions [90]. Additionally, raising awareness campaigns about the economic and environmental benefits of biorefineries are crucial to sensitize policymakers and society at large. Seeking partnerships with governments and related institutions can also strengthen the sector's influence in shaping policies that support the growth and sustainability of biorefineries [21].

When exploring this topic, it is recommended to use case studies that demonstrate the impact of government policies on biorefineries, highlighting successful examples and the benefits generated [113]. It is also important to present effective strategies to influence the formulation of favorable policies, showcasing how active industry participation and partnerships can play a significant role in this process [25]. This in-depth analysis will allow for a comprehensive understanding of the role of government policies and regulations in the viability and growth of biorefineries, thus driving the transition towards a more sustainable, resource-based economy [78].

2.5 *Prospects*

In summary, the analysis of government policies and regulations is essential to ensure the viability of biorefineries. The presence of clear, stable, and comprehensive policies, along with financial incentives, subsidies, production targets, and sustainability standards, provides a favorable environment for the development and success of these units [106]. These policies directly influence the competitiveness and profitability of biorefineries by reducing investment and operational costs, stimulating the adoption of sustainable technologies, and shaping market demands [74].

To promote favorable policies, it is crucial for the biorefinery sector to actively engage with policymakers through lobbying groups, providing relevant data and information, and raising awareness about the economic and environmental benefits of biorefineries [95]. Additionally, seeking strategic partnerships with governments and related institutions strengthens the sector's influence in shaping policies that support the growth and sustainability of biorefineries [54].

Overcoming the economic challenges faced in the implementation of sustainable biorefineries requires an integrated approach, considering financial, political, technological, and market aspects [11]. Collaboration and international partnerships are key factors in driving the economic viability of biorefineries, promoting the development of more efficient technologies, diversifying end products, exploring new markets, and strengthening collaboration between the public and private sectors [79]. With continuous efforts in these areas, biorefineries have the potential to play a crucial role in the transition towards a sustainable economy, reducing dependence on non-renewable resources, and contributing to a more sustainable future [138, 144].

3 Social Challenges

In addition to the economic challenges, biorefineries confront a series of considerable social challenges during their establishment and operation. These challenges encompass reconciling the demand for raw materials with food and energy security, potential land-use conflicts, and public acceptance by impacted communities. Addressing these challenges demands a meticulous approach that contemplates the social repercussions and pursues sustainable solutions to foster the balanced growth of biorefineries (Fig. 1).

3.1 *Public Perception and Acceptance of Biorefineries*

Although biorefineries are considered a crucial part of advancing toward a sustainable bioeconomy in the future, the implementation of these facilities still faces social challenges related to public perception and acceptance by local communities, especially

Fig. 1 Biorefinery: social impacts

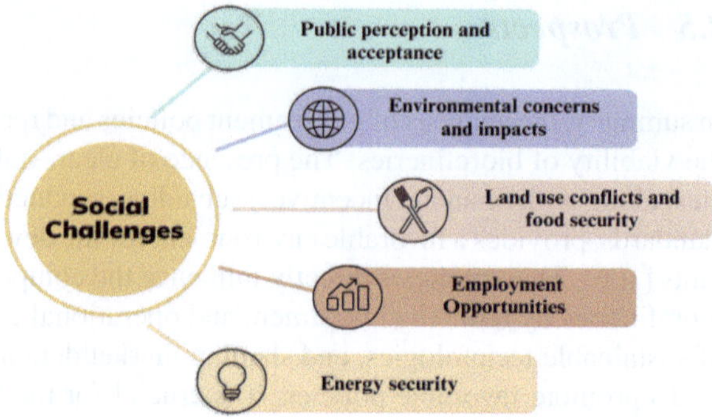

in areas where agriculture is the main economic activity [13, 75]. Acceptance by the local community is an essential factor to consider for the development and success of a biorefinery project, as a lack of understanding and accurate information about these facilities can result in resistance from local society [43, 100]. A receptive community can provide incentives that help offset costs, reducing implementation delays and indirectly lowering project expenses. Conversely, a less receptive community can cause permit and execution delays for a biorefinery due to protests and blockades, which can increase project costs and, in extreme cases, require forced relocation [75, 82].

Local society often expresses concerns regarding projects in their region due to their strong connection to the community and the greater influence they experience from the potential consequences of these projects. They often believe that negative aesthetics or visually displeasing aspects also represent harmful environmental impacts [82, 95]. Furthermore, the perception of specific risks and uncertainties plays an important role in assessing social acceptability. If a project, technology, or product is considered high-risk or uncertain, it is likely to be rejected by society [95]. Therefore, to ensure the growth of biorefineries, it is essential to address community concerns and seek their acceptance, as a lack of social support can create significant obstacles to successful implementation [58]. The participation of stakeholders and citizen associations is crucial as it promotes transparency, mutual understanding, and highlights the benefits of these solutions for energy and environmental challenges in a sustainable manner [23, 75].

3.2 *Environmental Concerns and Impacts*

The concept of biorefineries has attracted considerable social interest, especially in the context of the global sustainable development agenda. This approach is widely recognized for its positive technical, economic, and social impact on the environment, as it employs non-fossil resources as the basis for its operations. However, it is important to consider that this view may not hold true in all cases, depending on the systemic approach adopted [16, 72].

The critical issue of sustainability plays a key role in the acceptance of biorefineries by industry, political sectors, and society at large. Therefore, a proactive and sustainable approach to managing the environmental impacts of biorefineries is crucial to ensure their long-term viability and maximize their benefits for transitioning to a cleaner and more sustainable economy. This involves responsible use of biomass, implementation of clean technologies, proper waste management, emission reduction, and compliance with applicable environmental regulations [108].

The establishment and operation of these facilities can have various effects on the surrounding environment, and these issues are of great importance to communities near biorefineries. These communities have legitimate concerns about the air quality they breathe, the availability and quality of water, and the preservation of local ecosystems. Therefore, to identify and assess the environmental impacts and ensure the sustainability of these biorefineries, it is essential to conduct a comprehensive analysis through a sustainability framework. This will ensure the implementation of a safe and community-acceptable project, establishing biorefineries as environmentally responsible and socially accepted solutions [138, 150].

Among the various existing approaches, Life Cycle Assessment (LCA) is widely used in studies aiming to assess environmental impact [56, 110]. This tool analyzes each individual stage of the life cycle, from raw material acquisition to the final product, taking into account environmental, social, and economic impacts through a quantitative assessment. In this way, it is possible to identify critical points and understand the true impact of the product throughout the entire production chain, resulting in a comprehensive evaluation of process sustainability [28, 72, 145].

3.3 *Land-use Conflicts and Food Security*

In addition to the social challenges related to environmental concerns, biorefineries also face the challenge of land use conflicts. The need to ensure a constant and reliable supply of feedstock to compete cost-effectively with petroleum refineries implies a considerable demand for land. This can lead to conflicts, especially in areas where land availability is limited and there is already competition for agricultural resources [47, 51, 108].

One of the critical aspects associated with land use conflicts is food security due to the expansion of biorefineries, which can lead to the conversion of productive agricultural land into areas dedicated to the production of feedstocks for biofuels or chemicals. This can reduce the availability of land for food production and supply, biodiversity, and result in increased food prices, affecting local and regional food security, especially in developing countries [8, 159]. Biorefineries must be carefully planned and operated in a way that mitigates any negative impact on food security, as this is a fundamental issue, and it is essential to ensure that biorefinery activities do not compromise this basic right of the population. Therefore, to address such challenges, it is crucial to adopt sustainable approaches, such as the use of feedstocks that do not directly compete with food crops [51].

Biorefineries can be classified into first-generation, second-generation, and third-generation biorefineries. First-generation feedstocks are those that can impact food security as they are crops used for human consumption, such as edible oil, sugar, or starch, which can trigger competition for water and land, driving up the costs of food crops [51, 98]. Second-generation feedstocks are residues from agro-industry, agriculture, forestry, households, and other industries, providing advantages to biorefineries due to their low cost [93]. Third-generation feedstocks are based on microbial biomass, which have oils used in the production of biofuels like biodiesel, however, they still face significant challenges to achieve commercially viable production [66]. Furthermore, promoting sustainable agriculture and encouraging land-use practices that optimize land use can also play an important role in reconciling land-use conflicts. This can include the implementation of efficient cultivation techniques, the use of agroecological practices, and the promotion of agricultural diversification, ensuring the coexistence of biomass production and food production [118, 159].

In summary, land-use conflicts and food security are important challenges related to biorefinery projects, as it is crucial to ensure that their activities do not compromise the resources of the local community. Therefore, balanced solutions need to be found that consider biomass production and food security, involving an integrated and sustainable approach that encompasses economic, energy, social, and environmental aspects.

3.4 Employment Opportunities and Labor Force Development

The pursuit of sustainability has driven and motivated society towards an economy grounded in biological foundations. In this perspective, the use of low-cost renewable resources for the development of bioproducts stimulates the transition towards advancing the bioeconomy [37, 49, 67]. As mentioned earlier, biorefineries have the capacity to utilize various types of biomass from forestry, agriculture, aquaculture, as well as waste from industries and households. Among these sources, those commonly used today are related to the agricultural sector [56]. The cultivation of these feedstocks requires the involvement of labor for activities such as planting, harvesting, and resource management, and as a result, companies are increasingly recognizing market opportunities in this field [56]. This scenario can bring significant benefits to local farmers by offering an additional source of income and allowing for the diversification of their agricultural activities. Overall, biorefineries can contribute to the generation of new employment opportunities in both rural regions, where biomass is cultivated and harvested, and urban areas, where processing and production of end products occur, creating jobs in a variety of areas including agriculture, chemistry, engineering, logistics, and entrepreneurship [156].

From a socio-economic perspective, the implementation of a biorefinery presents significant benefits for the development and success of the bioeconomy. These benefits are associated with the possibility of supporting new industrial investments, driving regional growth and competitiveness, reducing dependence on fossil fuels, stimulating development in rural and local areas, job creation, energy production, as well as promoting the economy by avoiding landfill costs and reducing greenhouse gas emissions [37, 104, 156]. In this regard, the implementation of a local biorefinery can contribute to achieving the Sustainable Development Goals (SDGs) by boosting local employment and representing a means of activating and promoting socio-economic growth [138].

Therefore, it is expected that biorefineries will play a crucial role in improving the competitiveness and prosperity of countries, meeting the growing demand for a wide range of bio-based products and energy in an economically, socially, and environmentally sustainable manner [35, 56]. To leverage the employment opportunities generated by biorefineries, it is crucial to invest in workforce training and the development of new skills [56]. It is essential for companies to establish partnerships with educational institutions and training programs to ensure that workers have the necessary skills to meet the demands of the industry. In this way, biorefineries not only create jobs but also drive professional growth and contribute to the sustainable development of communities.

3.5 Energy Security and the Role of Biorefineries in Achieving Global Sustainability Goals

In a context of growing environmental awareness and the need for concrete actions to address energy and environmental challenges, biorefineries represent a promising solution in the pursuit of energy security and global sustainability. By utilizing renewable feedstocks, biorefineries offer a sustainable alternative to fossil fuels, enabling the production of biofuels and other renewable products [53].

Biorefineries play a crucial role in the transition towards a cleaner and more sustainable energy matrix, with one of their main advantages being the ability to utilize biomass to produce biofuels that partially or completely replace highly polluting fossil fuels. These biofuels offer a renewable and low-carbon emission alternative [30, 51].

Energy security in the context of biorefineries is achieved through maximizing energy self-sufficiency. By locally producing biofuels and other renewable products, biorefineries reduce the need for fossil fuel imports, thus increasing the energy autonomy of a region. This is particularly relevant for countries heavily dependent on energy imports. The inclusion of an energy security indicator is already considered a step forward in incorporating sustainability concerns related to the bioeconomy [108].

Furthermore, biorefineries play an important role in the production of renewable chemicals, replacing fossil fuel-derived counterparts such as plastics, solvents, and fertilizers with biomass-derived products. This substitution reduces dependence on non-renewable resources and mitigates the environmental impacts associated with their extraction and production.

Therefore, the substitution of fossil fuels with renewable sources such as biomass and organic waste emerges as a highly attractive alternative, significantly driving the advancement of biorefinery systems. This approach plays a fundamental role in ensuring global energy security by reducing dependence on fossil fuels and mitigating greenhouse gas emissions. Moreover, the "green" energy derived from these renewable sources proves promising both for environmental preservation and for achieving global sustainability goals [53]. Thus, bio-refineries contribute to the attainment of the Sustainable Development Goals (SDGs) outlined in the 2030 agenda, significantly encompassing social issues, as illustrated in Fig. 2.

Fig. 2 Sustainable development goals related to biorefineries

4 Process-related Challenges

4.1 Integration of Biorefinery Processes with Existing Industries

Brazil is a major player in global biomass production, with plentiful resources from crops like sugarcane, corn, soybean, rice, and cotton. The country's production of biochemicals and biofuels from biomass offers environmental and economic advantages, providing an alternative to traditional chemicals and fuels. These materials have various applications and are garnering significant attention and research in Brazil's biorefinery sector [26].

The production of biochemicals and biofuels from biomass presents distinctive environmental and economic advantages, positioning them as viable substitutes for traditional chemicals and fuels. Additionally, biofuels and biochemicals, such as organic acids, amino acids, enzymes, and biopolymers, are valuable products with diverse applications and a promising market outlook. However, continued technological progress is needed to facilitate large-scale production and commercialization of these products [24, 33].

In recent decades, significant progress has been made in biomass processing for sugar recovery, fermentation, and product separation and purification, not only at the laboratory scale but also at pilot, demonstration, and industrial scales. This progress has facilitated the development of biofuels and biochemicals' production from biomass in Brazil [85, 119]. In addition to biofuels and biochemicals, nanocellulose and products derived from kraft pulping and biorefinery plants have gained considerable attention in the Brazilian biorefinery sector. These materials have diverse applications and represent an area of growing interest and research within the broader biorefinery [60, 119].

Sustainable development is among the most important issues for both the global research and political agenda today. This is due to serious challenges facing society, including climate change, resource depletion, and environmental degradation. Implementing the bioeconomy will promote the use of renewable biological resources for food, materials, and energy production while balancing factors such as waste production and/or greenhouse gas emission [65].

The bioindustry utilizes a range of bio-based raw materials, including animal fats, vegetable oils (e.g., rapeseed, palm, soybean oil), sugar, starch crops (e.g., maize, wheat, sugar beet), lignocellulosic and algal biomass (e.g., wood, straw, sugarcane bagasse, corn stover, seaweed), and waste-derived feedstocks (e.g., brewer's spent grain, organic waste from municipal solid waste, residues from the paper and pulp industry). Additionally, intermediate compounds and coproducts like glycerol and ethanol are considered platform chemicals by these companies [99].

Biorefineries have gained attention as a technology for mitigating climate change and promoting sustainable development. As a result, the sustainability assessment of biorefineries has become a critical issue [57, 135]. Biorefinery platforms focus on converting various forms of biomass into renewable energy sources and essential specialty chemicals, such as organic or aromatic acids and vegetable oils. Consequently, biorefinery is seen as a promising biotechnological process that could potentially replace petroleum refineries, enabling the production of biofuels and biochemicals derived from renewable bioresources [9, 69].

Integrated biorefineries are facilities that combine material and energy production using biomass or biomass-derived feedstocks. Following the principle of cascade use, various value-added products and energy are produced from biogenic feedstocks through specific processes. The process in integrated biorefineries involves two main steps: primary refining, where biomass is pretreated and separated, and secondary refining, where the separated materials are converted into various products using physical, mechanical, thermochemical, and biotechnological processes [88].

Integrated biorefineries play a vital role in maximizing feedstock utilization and minimizing waste in biorefinery operations. By integrating processes and utilizing all material streams, these biorefineries enhance feedstock utilization and reduce costs compared to single-product production. The energy generated within the biorefinery can be used to meet energy demands, leading to reduced operational costs. This integrated approach provides a more sustainable and efficient utilization of biomass resources, maximizing their value and minimizing waste generation [3, 134].

A successful biorefinery is characterized by a sustainable business plan that establishes clear value chains, connecting raw materials, intermediate products, and end-products. An example of a successful biorefinery is one based on sugarcane utilization, where sugar production and first- and second-generation biofuel production are integrated [29]. The surplus steam and water from the process are utilized in the second-generation unit, making it economically competitive [85]. Flexibility allows for the selection of processes, optimal operational conditions, and adaptation to specific feedstock characteristics. This flexibility enables the production of a wide range of marketable products. Overall, integrated biorefineries aim to maximize the utilization of biomass resources, reduce waste, and create sustainable value chains for multiple products [109].

Integrating biorefineries and promoting resource efficiency and environmental sustainability can drive the adoption of circular economy principles and industrial symbiosis. These innovative business models, such as eco-industrial parks, go beyond traditional supply chain relations. They enable manufacturers and service providers to collaborate by exchanging waste or residues as valuable resources. If both parties benefit from these exchanges, a symbiotic partnership is formed. For example, industrial symbiosis can optimize energy efficiency by cascading waste heat, utilizing biomass residues as renewable fuel alternatives, and developing shared utilities among co-located businesses to reduce overall energy consumption [77, 147].

Some industrial sectors already have well-established and developed integrated biorefineries, such as carbohydrate-based, lipid-based, lignocellulosic-based, and waste- and residue-based biorefineries [155].

Carbohydrate-based biorefineries utilize sugars (e.g., glucose, galactose, xylose) and starch as feedstocks. Sugars are derived from crops like sugarcane, sugar beet, and sorghum, while starch is found in crops such as wheat and corn. Starch is typically converted into sugars for subsequent processing, including fermentation or the production of valuable chemicals [126].

Lipid-based biorefineries utilize oils and fats, such as soybean, palm, rapeseed, and sunflower oils, as raw materials. These oils can be used in biodiesel production through transesterification or as feedstocks for the production of hydrocarbons through processes like hydrodeoxygenation and decarboxylation. However, the use of edible oils in biorefineries faces challenges due to competition with the food industry, leading to the exploration of non-edible crops as alternative sources [12].

Lignocellulosic biomass, sourced from industries like pulp and paper mills and sugar mills, contains significant amounts of lignin. However, a large portion of this lignin is currently discarded as waste, resulting in underutilization and increased production costs. Biorefineries primarily focus on utilizing the cellulose and hemicellulose components of biomass, leaving lignin as a residue [38]. Lignin, as a major component of biomass, holds significant potential for generating chemicals and materials due to its unique characteristics. Isolating and recovering lignin during pretreatment enables its use in producing high-purity lignin and converting it into low-molecular-weight aromatics. These products have diverse applications in industries such as cosmetics, plastics, pharmaceuticals, and fuel [2].

The sugar platform of biorefineries focuses on converting C6 and C5 sugars derived from lignocellulosic biomass into a variety of bioproducts. Alongside bioalcohols like butanol and organic acids, fermentation or conversion processes can yield diols, lactic acid, muconic acid, xylitol, furans, and levulinic acid. Several companies have ventured into the production of these bioproducts, which serve as important base or intermediate chemicals for the generation of various commodities [26, 27, 38].

Waste- and residue-based biorefineries utilize biomass from various sources like municipal solid waste, manure, and fruit/vegetable industry residues. However, the high moisture and heterogeneous nature of food waste present challenges for effective conversion. Algal treatment systems offer an economical solution for wastewater treatment and simultaneously produce valuable bio-based products such as lipids, Omega 3 fatty acids, single-cell proteins, and carotenoids. The extraction and valorization of waste have gained significant attention, and the concept of biorefinery plays a crucial role in transitioning to a circular economy by maximizing biomass utilization for marketable products [50].

Therefore, the integration of biorefinery processes with existing industries is a key strategy for building a circular and sustainable economy. This approach aims to maximize natural resource efficiency, reduce waste, and mitigate the environmental impact of industrial production. The mentioned examples demonstrate the potential of such collaboration, driving the bioeconomy and accelerating the transition towards a more sustainable and environmentally conscious society.

4.2 Optimization of Reaction Conditions and Catalysts

The optimization of reaction conditions and catalysts is crucial for improving the efficiency, selectivity, and overall performance of biorefinery processes. By fine-tuning these factors, biorefineries can enhance product yields, reduce energy consumption, minimize waste generation, and improve economic viability [114]. Through systematic experimental studies and process optimization techniques, optimal operating conditions can be identified to maximize desired product formation while minimizing undesired side reactions [1].

Optimizing reaction conditions and supply chain efficiency, including transportation and storage, can improve the environmental and economic performance of conversion processes in biorefineries. The cost of producing bio-based products is often higher compared to fossil-based products due to the significant investment required for capital expenditures (CAPEX). To reduce CAPEX and operating expenses (OPEX) of biorefineries, process optimization, the development of new conversion strategies, and minimizing utility requirements are essential [64].

Nowadays, advancements have been made in catalyst design and development to enhance their performance in biorefinery processes. Catalyst optimization techniques include the modification of catalyst structures, surface functionalization, and the incorporation of promoter materials [157]. These strategies aim to improve catalyst activity, selectivity, stability, and resistance to deactivation. Additionally, the use of novel catalyst materials, such as metal–organic frameworks, zeolites, and mesoporous materials, has shown promise in biorefinery applications due to their unique properties and tunability [1].

Catalysts play a crucial role in biorefinery processes, driving the conversion of feedstock into valuable products. They accelerate reaction rates, improve selectivity, and enhance overall process efficiency. Catalyst selection and optimization involve factors like activity, stability, selectivity, and cost-effectiveness. Biorefineries utilize different types of catalysts, including heterogeneous catalysts (such as metal catalysts supported on solid materials) and enzymatic catalysts (such as enzymes or whole cells). These catalysts enable efficient and sustainable transformation of biomass resources into high-value products [15, 71].

The lignin biorefinery has a variety of catalysts applied in reactions for the depolymerization of its structure since lignin has a complex and diverse chemical structure [60]. The depolymerization-valorization processes of lignin, including kraft, hydrolysis, and organosolv, have been extensively researched. Hydrogenolysis involves moderate temperatures, high hydrogen pressure, and metal catalysts like Pt, Ni, Pd, Cr, and Cu supported on various materials. This process aims to break C–O and C–C bonds in lignin, particularly the β-O-4 ether bond, to produce smaller fragments or monomeric compounds such as phenolics, aromatics, and alkanes [81, 141].

Recent advancements in green chemistry and engineering have focused on the development of recyclable catalysts, nontoxic metal catalysts, and biocatalysts. These innovations have led to improved efficiency, waste reduction, recyclability, cost reduction, and safer processes. Inorganic and polymer-supported catalysts, as well as immobilized and nanocatalysts, have been employed, contributing to the principles of green chemistry and engineering [152].

Biocatalysis has emerged as a dynamic research field, offering increased selectivity, activity, and sustainability. It enables the catalysis of reactions under mild operating conditions, resulting in reduced energy consumption and environmental impact. The use of enzymes as biocatalysts allows for precise control over reaction selectivity and provides access to a wide range of chemistries. Biocatalytic processes also have the potential for high productivity and can be integrated into sustainable and efficient biorefinery systems [96, 120]. Computer modeling and directed evolution techniques have facilitated the design and development of efficient enzymes as biocatalysts for chemical reactions [115].

In the vegetable oil biorefinery, the catalysts used for transesterification in biodiesel production are classified as alkali, acid, or enzyme. Alkali catalysts, such as sodium hydroxide and potassium hydroxide, are more effective and allow high conversion levels in short reaction times. However, alkali-catalyzed transesterification has drawbacks, including energy intensity, glycerol recovery challenges, catalyst removal, and wastewater treatment. Acid catalysts require higher temperature and pressure but can handle feedstocks with high free fatty acid (FFA) content and are less sensitive to water [141].

Enzymatic catalysts, specifically lipases, offer effective transesterification of triglycerides in both aqueous and non-aqueous systems, providing solutions to the challenges mentioned earlier. Lipases allow for easy removal of glycerol without complex processes and complete conversion of free fatty acids in waste oils and fats to alkyl esters [10, 153]. However, the production cost of lipase catalysts is generally higher compared to alkaline catalysts, to overcome this challenge, lipase immobilization techniques have been used, which allow the recovery of the immobilized biocatalyst, allowing its use in repeated reactions [31, 121].

4.3 Waste Minimization and Resource Efficiency

Secondary biomass feedstocks are derived from the processing of primary biomass feedstocks, resulting in substantial physical or chemical breakdown. Examples include sawdust from sawmills, black liquor from paper making, and cheese whey from cheesemaking. Waste biomass from the agro-food supply chain, such as agricultural residues and food manufacturing waste, provides renewable materials that don't compete with the food and feed industries for land. These organic residues offer opportunities for reuse and valorization in the production of sustainable bio-derived products [111, 140].

Waste valorization is a promising concept that promotes circular economy principles by reusing, recycling, and converting waste materials into valuable products or energy sources. Biorefineries, utilizing biological processes, are at the forefront of waste valorization. Rather than simply disposing of pollutants, biorefineries transform them into valuable compounds or chemical building blocks, reducing the need for finite natural resources [131].

Integrated biorefineries using residual biomasses offer diverse product possibilities and contribute to the circular economy while reducing environmental impacts. However, the heterogeneous nature of residual biomass presents challenges for efficient and profitable processes. Fractionation techniques play a crucial role in breaking down biomass components while minimizing the loss of valuable compounds [131].

The implementation of these biorefineries requires careful consideration of factors such as waste origin, quantity, recyclability, and appropriate management approaches. Local and national authorities need to have knowledge of waste biorefinery issues and identify suitable waste valorization methods. Factors like biorefinery type, construction, location, desired products, and feedstock pretreatment also need to be considered [112].

These biorefineries provide economic benefits through energy recovery, the production of value-added products, land savings, job creation, and cost savings from reduced landfilling. They also offer significant environmental benefits, including the reduction of greenhouse gas emissions and the conservation of natural resources. Waste biorefineries contribute to a more sustainable and efficient use of waste materials while promoting environmental stewardship and economic growth [132].

The bioconversion or valorization of lignocellulosic residues and industrial wastes into biofuels, biochemicals, and biomaterials through biorefinery processes using green technologies is a promising strategy for achieving economic and sustainable goals [124]. Due to the complexity of the lignocellulosic structure, it is necessary to process this material for its use, involving different biomass processing unit operations, like size reduction, pretreatment, enzymatic hydrolysis, and fermentation. In the biorefining approach, it is crucial to consider both waste-stream management and waste-treatment technologies. In the ethanol production process, for example, three primary waste streams are generated: pretreatment liquor, undigested residue from enzymatic hydrolysis, and vinasse from ethanol fermentation [92].

Another residue with high added value is vegetable oil residues, which have a great capacity to be used as raw materials in biomanufacturing. Waste vegetable oil, which is no longer suitable for consumption, presents a significant disposal challenge due to its increasing volume alongside the global consumption of cooking oil. However, this waste oil can be effectively converted into products like biodiesel, fatty acids and glycerol, emollient esters, and lubricants [44].

4.4 Energy Consumption and Greenhouse Gas Emissions

Climate change is a critical global environmental issue driven primarily by the combustion of fossil fuels. To combat climate change, a union of countries is implementing strategies that include reducing energy demand, improving energy efficiency, and seeking to develop technologies for producing energy from renewable matrices [9].

Carbon neutrality involves reducing emissions by 60–80% and offsetting or sequestering the remaining 20–40% through sustainable development initiatives. Bioenergy, derived from biomass, is seen as a promising solution for achieving carbon neutrality. Carbon neutrality aims to achieve a balance between emitted and absorbed carbon, emphasizing emission reduction and renewable optimization [149]. Aiming at reducing carbon dioxide (CO_2) emissions, the concept of carbon neutrality has been developed and encouraged in biorefineries.

Biorefineries are advanced facilities that efficiently convert biomass into a wide range of valuable bioproducts, with the aim of increasing economic potential and achieving carbon neutrality within a circular bioeconomy framework. These complex facilities optimize and maximize the efficiency of biomass resources by reducing waste, conserving raw materials, and establishing a symbiotic relationship between products and feedstock, preserving social, economic, and environmental benefits [29, 140].

Bioenergy from biorefineries as a renewable energy source is clean and aids in achieving grid stability while providing means to distance the economy from fossil fuel-based energy sources. Furthermore, unconventional energy sources, such as bioenergy, have the potential to outperform fossil fuels. In this context, biomass biofuels play a crucial role in reducing greenhouse gas emissions because the carbon in biomass originates from the air. With an efficient rate of biomass processing, carbon can be recycled, potentially achieving a carbon balance, and thus zero CO_2 emissions can be achieved, reducing the greenhouse effect associated with burning fossil fuels [29].

Lignocellulosic biomass is increasingly used for biofuel production, but its growing demand may lead to higher costs. To ensure sustainability, low-grade organic materials like forest residues and agro-industrial waste can be used as cost-effective alternatives. Aquatic biomass, such as microalgae and macroalgae, also offer advantages like efficient sunlight conversion, rapid growth, and CO_2 sequestration, without competing with agriculture. Utilizing nutrient-rich wastewater for algal cultivation improves sustainability and reduces pollution [141].

Industrial biowaste can be sustainably converted into bioproducts such as bioenergy, biopolymers, biochemicals, bioplastics, and biofertilizers for application in the industrial, commercial, agricultural, and energy sectors. Furthermore, the global market value of biowaste technologies for bioenergy is around US$25 billion and is projected to reach US$40 billion by 2023 [37].

Therefore, to address the challenges of energy consumption and greenhouse gas emissions, biorefineries are a suitable strategy for energy efficiency by utilizing renewable energy sources, implementing carbon capture and storage technologies, and developing advanced conversion processes. The combination of these approaches makes it possible to achieve a more sustainable bioeconomy sector, contributing to the mitigation of climate change and the transition to a low-carbon economy.

4.5 Advances in Separation and Purification Technologies

Residual biomass offers opportunities for integrated processes to obtain multiple valuable products within a circular economy concept [77, 155]. However, the heterogeneous structure of biomass presents challenges in designing efficient unit operations. Fractionation processes play a crucial role in deconstructing biomass, minimizing the loss of valuable compounds. Integrated biorefinery concepts have explored sequential systems to improve feedstock structures and yields. It is necessary to overcome challenges such as reducing the consumption of chemical products, minimizing the production of waste at each fractionation stage, recycling effluents, and improving operating conditions [134].

A biorefinery is a facility that converts biomass into a range of products, including biofuels and biochemicals. Similar to a petroleum refinery, there are upstream processes involved in growing, harvesting, and extracting biomass for the biorefinery. The raw material undergoes primary processing in the biorefinery, which requires subsequent downstream recovery processes to produce products that meet specifications. Unconverted materials or byproducts can be recycled or utilized in other ways [133]. Downstream recovery and separation processes are crucial in biorefineries for the purification and recovery of desired products while managing by-products and unconverted raw materials. Incomplete conversion of raw materials during biomass processing often leads to the formation of by-products or coproducts, necessitating downstream processing to obtain high-quality products. Unconverted raw materials can be recycled for further processing, and by-products can be utilized to add value to the biorefinery [142].

Separation technologies are crucial in integrated biorefineries, accounting for a significant portion of production costs. Efficient separation strategies are essential for successful bioprocessing, including feedstock preparation, product recovery and purification, impurity removal, and water management. Physical and chemical differences between species determine separation methods, considering factors such as size, shape, solubility, polarity, and molecular interactions. Purity requirements influence the design of separation schemes, impacting operation scale, energy consumption, and costs [32]. Integration of reactions with product separations enhances overall process performance. Economic considerations, including purity requirements, co-product utilization, waste disposal, energy and substrate use, capital and operating costs, and system footprint, drive the choice of separation trains [162]. The extraction process is crucial for recovering bioactive compounds without losing their beneficial properties. Techniques such as conventional solid–liquid extraction, sonication-assisted extraction, microwave-assisted extraction, supercritical fluid extraction, and ohmic heating extraction are used to recover bioactive compounds. After extraction, purification is necessary to remove interferences and impurities. Purification techniques include crystallization, distillation, membrane filtration (microfiltration, ultrafiltration, reverse osmosis), and chromatography (partition, change, adsorption, ion exchange, size exclusion, or gel filtration). These techniques ensure the production of consistent and high-quality bioactive compound products [137, 162].

5 Challenges and Lessons Learned from Deployments

Large-scale biorefineries still face challenges to be economically competitive compared to fossil fuels [45]. However, it is crucial to identify high-performance biorefinery concepts in terms of technical feasibility, positive environmental impact, and social acceptance [59]. These elements are essential to drive the adoption and success of biorefineries as a sustainable alternative to fossil fuels [94]. Viability assessment plays a crucial role in the decision to invest in biorefineries, as it is necessary to understand the feasibility and potential returns on investment [76].

Considering factors such as the biomass market, technology costs, market demand, and energy policies is essential for a comprehensive economic evaluation [160]. The guidance that integrates dynamic biomass market models and techno-economic models can help provide a more comprehensive insight into the success aspects of large-scale biorefinery deployment [63]. By considering the dynamics of the biomass market, such as feedstock availability and prices, along with the technological and economic aspects of biorefinery operation, it is possible to assess the risks and opportunities associated with the venture [128]. Such frameworks can aid in identifying suitable geographical regions for biorefinery deployment, taking into account biomass availability, existing infrastructure, market demand, and favorable government policies [52]. Additionally, by considering market dynamics, the framework can help understand how changes in the biomass market and prices can impact the overall long-term viability of the biorefinery [160].

Noteworthy cases are widespread in different parts of the globe. In Lappeenranta, Finland, UPM Biofuels is leading the way with the world's first commercial-scale wood-based biorefinery, capable of producing 160 million liters per year of advanced biofuels and a range of other products including plastics and adhesives [146]. Similarly, in Amsterdam, Nedstar's Herengracht Biorefinery specializes in industrial ethanol production from a variety of sources such as beet, cane, wheat, grain, and grape alcohol, with organic ethyl alcohol and extra neutral alcohol as other products [103].

In the United States, the Hugoton Biorefinery by Abengoa Bioenergy stands as a prominent example of cellulosic ethanol production, utilizing wheat straw and corn residues to produce an impressive 842 million gallons per year [105]. Moving to Brazil, GranBio's Alagoas Biorefinery uses a process of hemicellulose extraction from steam-treated biomass to produce 2G ethanol, along with cellulosic sugar, acetic acid, and derivatives [48]. Other parts of Europe also host a variety of biorefinery initiatives, such as the Ghent Biorefinery in Belgium by Tereos, producing 730 million liters/year of bioethanol from plant material fermentation, and Novozymes' Kalundborg Biorefinery in Denmark that focuses on bioenergy from residual raw materials. Additionally, the Rotterdam Biorefinery in the Netherlands by SkyNRG produces 100,000 tons of aviation biodiesel annually, and Beta Renewables in Italy specializes in 2G ethanol from sugarcane biomass [105]. These examples collectively demonstrate the global reach and diversity of biorefinery applications, emphasizing the adaptability and innovative utilization of various feedstocks and technologies.

Analyzing this information, one can correlate strategic location choices of biorefineries, considering feedstock availability, existing infrastructure, and market potential. Furthermore, observation of the main products and other products generated by these biorefineries provides insight into the diversity of biomass applications, including the production of biofuels, renewable chemicals, sustainable materials, and more. This information highlights the wide range of opportunities and advancements achieved in the successful implementation of biorefineries in different regions. Through the study of the experiences of these industries, valuable lessons can be extracted about the strategies, technologies, and approaches that have contributed to their success.

Economic viability is a fundamental factor in biofuel production, affecting many plants' ability to compete with fossil fuels. This is due to a combination of high production costs, varying raw material prices, and unpredictable market conditions. Additionally, government policies can have significant effects on plant viability. Changes or instability in these policies, such as reduced subsidies, modified production targets, or inconsistent regulatory support, can create challenges for plants.

One important consideration is regulatory uncertainty. Stricter environmental restrictions and unclear sustainability standards can create confusion and uncertainty for investors. Fluctuations in the prices of raw materials, along with competition from more cost-effective fossil fuels, also contribute to plant closures. Furthermore, the lack of proper investments and financing can negatively affect the financial sustainability of plants, as uncertainty regarding returns can lead to closure.

A case in point is Range Fuels, a North American company that aimed to produce ethanol from woody biomass but ceased operations in 2011 due to managerial and market challenges [122]. Another example is Neste, a Finnish company that planned to build a biorefinery in Singapore but canceled the project due to financial difficulties, although the plan is still under consideration [123].

The Crescentino biorefinery, designed to produce advanced biofuels from agricultural residues, provides another illustrative example. While initially successful, the company operating the biorefinery, Beta Renewables, faced financial issues and stopped its operations in 2015. The reasons for the shutdown included economic factors, lack of financial viability, difficulties in achieving competitive prices, regulatory issues, and lack of government incentives. The facility's experience offers valuable insights for future strategies and policies to support the sustainability of second-generation bioethanol plants. However, a new acquisition offers hope for the facility's revitalization [18].

5.1 Innovative Process Designs and Strategies to Overcome Challenges

Advanced biorefineries are considered crucial for implementing a sustainable economy based on biological resources. To overcome challenges in this context, innovative process designs and strategies are needed. Complete valorization of biomass can drive industrial adoption of biorefineries, enabling efficient utilization of all its components. Enzyme customization can play a crucial role in optimizing process performance, reducing loads and associated costs. Additionally, the application of Big Data Analytics can offer better decision-making in biorefineries, providing valuable insights into operational performance and process efficiency. Lastly, it is important to highlight that interdisciplinary collaboration is essential for the development of sustainable biorefineries, involving experts in biotechnology, process engineering, materials science, and other fields, to address challenges and capitalize on opportunities offered by this innovative approach [36].

Interdisciplinary collaboration is fundamental to the development of biorefineries, which involve complex processes of converting biomass into useful products [2]. By bringing together experts from fields such as biology, chemical engineering, materials science, and agronomy, unique perspectives and specialized knowledge can be obtained to address technical, economic, and environmental challenges [19]. This collaboration facilitates a comprehensive understanding of challenges in biorefineries and enables the identification of more effective solutions, such as process integration, the development of advanced technologies, and the use of new materials. The exchange of ideas between different disciplines stimulates innovation and drives more efficient and sustainable approaches [13].

Interdisciplinary collaboration is not limited to academic experts but also involves the participation of companies, research institutions, governments, and other stakeholders [5]. These partnerships promote knowledge transfer, resource sharing, and joint implementation of projects, accelerating the development and implementation of innovative technologies in biorefineries [83]. In summary, interdisciplinary collaboration is essential for addressing the complex challenges of biorefineries, driving innovation, improving process efficiency, and promoting a sustainable economy based on biological resources [101].

In a global context, it is common to find studies that highlight the importance of successful partnerships in the field of biorefineries [87]. These partnerships involve collaboration among companies, research institutions, governments, and local communities with the aim of developing and implementing biomass conversion technologies into value-added products such as biofuels, renewable chemicals, and biodegradable materials [79]. These partnerships seek to optimize the sustainable use of renewable resources, promote innovation and efficiency in biorefinery processes, and drive the transition towards a greener, bio-based economy [68].

The São Paulo Research Foundation (FAPESP) and the Biotechnology and Biological Sciences Research Council (BBSRC) jointly promote collaborative research proposals in biorefineries for the production of advanced biofuels. With the aim of fostering collaboration between groups in Brazil and the United Kingdom, they encourage interdisciplinary projects related to the biological processes of integrated biorefinery approaches [42].

A great example of international collaborative actions is the Task 42 program, which is an initiative within the International Energy Agency (IEA) focused on the development and implementation of efficient and sustainable biorefineries. The objective is to explore the potential of biorefineries as an approach for the production of biofuels, chemicals, materials, and energy from biomass. Through the Global Biorefineries Atlas Portal, researchers, companies, and other interested parties can explore examples of biorefineries in different countries, gain insights into the technologies and processes employed, and identify potential collaboration opportunities. This platform serves as a valuable tool for the dissemination of knowledge, exchange of experiences, and promotion of international partnerships in the field of biorefineries. It contributes to the advancement of research, development, and implementation of innovative and sustainable solutions for the utilization of biomass as a renewable resource [62].

References

1. Agharezaei P et al (2023) Designing catalysts via evolutionary-based optimization techniques. Comput Mater Sci 216:111833
2. Ahmad E, Pant KKL (2018) Lignin conversion: a key to the concept of lignocellulosic biomass-based integrated biorefinery. In: Waste biorefinery. Elsevier, pp 409–444
3. Alexandri M et al (2020). Biorefineries in Germany. In: Waste biorefinery. Elsevier, pp 601–629
4. An H, Wilhelm WE, Searcy SW (2011) Biofuel and petroleum-based fuel supply chain research: a literature review. Biomass Bioenergy 35(9):3763–3774
5. Annan-Diab F, Molinari C (2017) Interdisciplinarity: practical approach to advancing education for sustainability and for the Sustainable Development Goals. Int J Manage Educ 15(2):73–83
6. Antar M et al (2021) Biomass for a sustainable bioeconomy: an overview of world biomass production and utilization. Renew Sustain Energy Rev 139:110691
7. Arias A, Feijoo G, Moreira MT (2023) Biorefineries as a driver for sustainability: key aspects, actual development and future prospects. J Clean Prod 137925
8. Aristizábal-Marulanda V, Cardona Alzate CA (2019) Methods for designing and assessing biorefineries. Biofuels Bioprod Biorefin 13(3):789–808
9. Arora R, Sharma NK, Kumar S (2018) Valorization of by-products following the biorefinery concept: commercial aspects of by-products of lignocellulosic biomass. In: Advances in sugarcane biorefinery. Elsevier, pp 163–178
10. Attique SA et al (2023) Enzyme immobilization approaches. In: Biocatalyst immobilization. Academic Press, pp 37–54

11. Awasthi MK et al (2022) Agricultural waste biorefinery development towards circular bioeconomy. Renew Sustain Energy Rev 158:112122
12. Badgujar KC, Bhanage BM (2018) Dedicated and waste feedstocks for biorefinery: an approach to develop a sustainable society. In: Waste biorefinery. Elsevier, pp 3–38
13. Bauer F et al (2017) Technological innovation systems for biorefineries: a review of the literature. Biofuels Bioprod Biorefin 11(3):534–548
14. Becerik-Gerber B et al (2012) Application areas and data requirements for BIM-enabled facilities management. J Constr Eng Manag 138(3):431–442
15. Beckham GT et al (2016) Opportunities and challenges in biological lignin valorization. Curr Opin Biotechnol 42:40–53
16. Bernstad Saraiva A (2017) System boundary setting in life cycle assessment of biorefineries: a review. Int J Environ Sci Technol 14(2):435–452
17. Biddy MJ et al (2020) Integrated strategies to enable lower-cost biofuels. National Renewable Energy Lab
18. News B (2021) World's first commercial second-generation bioethanol facility shuts down. Biofuels News 14(10):12–15
19. Birch K, Levidow L, Papaioannou T (2010) Sustainable capital? The neoliberalization of nature and knowledge in the European "knowledge-based bio-economy." Sustainability 2(9):2898–2918
20. Black MJ et al (2016) Developing database criteria for the assessment of biomass supply chains for biorefinery development. Chem Eng Res Des 107:253–262
21. Boom-Cárcamo E, Peñabaena-Niebles R (2022) Analysis of the development of industrial symbiosis in emerging and frontier market countries: barriers and drivers. Sustainability 14(7):4223
22. Borras SM Jr et al (2016) The rise of flex crops and commodities: Implications for research. J Peasant Stud 43(1):93–115
23. Butler C et al (2015) Public values for energy futures: framing, indeterminacy and policy making. Energy Policy 87:665–672
24. Byun J, Han J (2020) Sustainable development of biorefineries: integrated assessment method for co-production pathways. Energy Environ Sci 13(8):2233–2242
25. Cacari-Stone L et al (2014) The promise of community-based participatory research for health equity: a conceptual model for bridging evidence with policy. Am J Public Health 104(9):1615–1623
26. Chandel AK et al (2021) Brazilian biorefineries from second biomass: critical insights from industry and future perspectives. Biofuels Bioprod Biorefin 15(4):1190–1208
27. Chandel AK et al (2022) Lignocellulose biorefinery: technical challenges, perspectives on industrialization, and solutions. In: Production of top 12 biochemicals selected by USDOE from renewable resources. Elsevier, pp 1–39
28. Chaturvedi T et al (2020) Developing process designs for biorefineries—definitions, categories, and unit operations. Energies 13(6):1493
29. Chen H (2015) Lignocellulose biorefinery product engineering. Lignocellul Biorefinery Eng 11:125
30. Chew KW et al (2017) Microalgae biorefinery: high value products perspectives. Biores Technol 229:53–62
31. Costa-Silva TA et al (2022) Highly effective Candida rugosa lipase immobilization on renewable carriers: integrated drying and immobilization process to improve enzyme performance. Chem Eng Res Des 183:41–55
32. Datta S, Lin YJ, Snyder SW (2014) Current and emerging separations technologies in biorefining. In: Advances in biorefineries. Woodhead Publishing, pp 112–151
33. De Almeida CD, Queirós O (2022) Special issue "Novel developments in the bioproduction of biochemicals and biomaterials". Appl Sci 12(20):10631
34. De Besi M, McCormick K (2015) Towards a bioeconomy in Europe: national, regional and industrial strategies. Sustainability 7(8):10461–10478

35. De Jong E, Jungmeier G (2015) Biorefinery concepts in comparison to petrochemical refineries. In: Industrial biorefineries & white biotechnology. Elsevier, pp 3–33
36. Dragone G et al (2020) Innovation and strategic orientations for the development of advanced biorefineries. Biores Technol 302:122847
37. Duan Y et al (2022) Sustainable biorefinery approaches towards circular economy for conversion of biowaste to value added materials and future perspectives. Fuel 325:124846
38. Duque A et al (2021) Advanced bioethanol production: from novel raw materials to integrated biorefineries. Processes 9(2):206
39. Ebadian M et al (2020) Biofuels policies that have encouraged their production and use: an international perspective. Energy Policy 147:111906
40. Ebrahimi S et al (2022) Renewable jet fuel supply chain network design: application of direct monetary incentives. Appl Energy 310:118569
41. Falcone PM et al (2021) Towards a bioenergy transition in Italy? Exploring regional stakeholder perspectives towards the Gela and Porto Marghera biorefineries. Energy Res Soc Sci 80:102238
42. FAPESP (2023). Joint FAPESP-BBSRC call for collaborative proposals in integrated biorefinery approaches for the manufacture of advanced biofuels [Online]. https://fapesp.br/9759/joint-fapesp-bbsrc-call-for-collaborative-proposals-in-integrated-biorefinery-approaches-for-the-manufacture-of-advanced-biofuels. Accessed 15 May 2023
43. Fava F et al (2015) Biowaste biorefinery in Europe: opportunities and research & development needs. New Biotechnol 32(1):100–108
44. Ferreira JA, Agnihotri S, Taherzadeh MJ (2019) Waste biorefinery. In: Sustainable resource recovery and zero waste approaches. Elsevier, pp 35–52
45. Forsberg C, Dale B (2020) Replacing liquid fossil fuels with liquid biofuels from large-scale nuclear biorefineries. In: Proceedings of applied energy symposium MIT A+B. Cambridge, MA
46. Gavrilescu M (2014) Biomass potential for sustainable environment, biorefinery products and energy. In: Sustainable energy in the built environment-steps towards nZEB: proceedings of the conference for sustainable energy (CSE). Springer International Publishing, pp 169–194
47. Ghatak HR et al (2011) Biorefineries from the perspective of sustainability: feedstocks, products, and processes. Renew Sustain Energy Rev 15(8):4042–4052
48. GranBio (n.d.) GreenPower—Tecnologia. Retrieved 15 May 2023
49. Guimarães CR et al (2023) Opportunities and challenges for the production of fuels and chemicals: materials and processes for biorefineries. Biotechnol Microb Enzymes 551–620
50. Gupta K, Kumar R, Bordoloi N (2022) Sustainable bio-based technologies for waste and wastewater treatment integrated with value-added products. In: Environmental sustainability and industries. Elsevier, pp 147–158
51. Gutiérrez CDB, Serna DLR, Alzate CAC (2017) A comprehensive review on the implementation of the biorefinery concept in biodiesel production plants. Biofuel Res J 4(3):691–703
52. Hajibabai L, Ouyang Y (2013) Integrated planning of supply chain networks and multimodal transportation infrastructure expansion: model development and application to the biofuel industry. Comput-Aided Civil Infrastruct Eng 28(4):247–259
53. Haribabu K et al (2021) Thematic issue: bioenergy and biorefinery approaches for environmental sustainability. Biomass Convers Biorefinery 1–1
54. Hellsmark H et al (2016) Innovation system strengths and weaknesses in progressing sustainable technology: the case of Swedish biorefinery development. J Clean Prod 131:702–715
55. Hina M et al (2022) Drivers and barriers of circular economy business models: where we are now, and where we are heading. J Clean Prod 333:130049
56. Hingsamer M, Jungmeier G (2019) Biorefineries. In: The role of bioenergy in the bioeconomy. Academic Press, pp 179–222
57. Holm-Nielsen JB, Ehimem EA (2014) Biorefinery plant design, engineering and process optimisation. In: Advances in biorefineries. Woodhead Publishing, pp 89–111
58. Hosseini A et al (2018) Social acceptance of renewable energy in developing countries: challenges and opportunities. Distrib Gener Altern Energy J 33(1):31–48

59. Hrbek J (2022) Technical, economic and environmental (TEE) assessment of integrated biorefineries
60. Hu J, Zhang Q, Lee DJ (2018) Kraft lignin biorefinery: a perspective. Biores Technol 247:1181–1183
61. Hytonen VE (2011) Methodology for identifying promising retrofit integrated forest biorefinery strategies-design decision making under uncertainty. Doctoral dissertation, École Polytechnique de Montréal
62. International Ennergy Agency (2023) Available at https://www.iea.org/reports/world-energy-outlook-2023
63. Iakovou E, Bochtis D, Vlachos D (2021) Supply chain management for sustainable food networks. Wiley & Sons Inc.
64. Iaquaniello G, Palo E, Salladini A (2022) Renewable sources to biorefineries, biomass conversion, and membrane technology. In: Membrane engineering in the circular economy. Elsevier, pp 177–196
65. Igbokwe VC et al (2022) Biochemical biorefinery: a low-cost and non-waste concept for promoting sustainable circular bioeconomy. J Environ Manage 305:114333
66. Jacob-Lopes E et al (2019) Bioactive food compounds from microalgae: an innovative framework on industrial biorefineries. Curr Opin Food Sci 25:1–7
67. Jain A et al (2022) Bioenergy and bio-products from bio-waste and its associated modern circular economy: current research trends, challenges, and future outlooks. Fuel 307:121859
68. Janssen M, Chambost V, Stuart PR (2008) Successful partnerships for the forest biorefinery. Ind Biotechnol 4(4):352–362
69. Junghare M et al (2023) Biorefineries: current scenario, feedstocks, challenges, and future perspectives. In: Nanotechnology for biorefinery. Elsevier, pp 1–25
70. Junginger HM et al (2019) The future of biomass and bioenergy deployment and trade: a synthesis of 15 years IEA Bioenergy Task 40 on sustainable bioenergy trade. Biofuels Bioprod Biorefin 13(2):247–266
71. Karp SG et al (2021) Lignocellulosic biorefinery for value-added products: the emerging bioeconomy. In: Biomass, biofuels, biochemicals. Elsevier, pp 291–321
72. Katakojwala R, Mohan SV (2021) A critical view on the environmental sustainability of biorefinery systems. Curr Opin Green Sustain Chem 27:100392
73. Khoshnevisan B et al (2021) A critical review on livestock manure biorefinery technologies: sustainability, challenges, and future perspectives. Renew Sustain Energy Rev 135:110033
74. Kim Y, Lee J, Ahn J (2019) Innovation towards sustainable technologies: a socio-technical perspective on accelerating transition to aviation biofuel. Technol Forecast Soc Chang 145:317–329
75. Kokkinos K et al (2018) Fuzzy cognitive map-based modeling of social acceptance to overcome uncertainties in establishing waste biorefinery facilities. Front Energy Res 6:112
76. Kumar D et al (2022) Techno-economic analysis for evaluating biorefinery strategies. In: Biomass utilization: conversion strategies. Springer International Publishing, pp 229–245
77. La Rosa AD, Ramakrishna S (2021) Industrial symbiosis for circular economy: a possible scenario in Norway. In: An introduction to circular economy, pp 95–106
78. Ladu L et al (2020) The role of the policy mix in the transition toward a circular forest bioeconomy. Forest Policy Econ 110:101937
79. Lange L et al (2021) Developing a sustainable and circular bio-based economy in EU: by partnering across sectors, upscaling and using new knowledge faster, and for the benefit of climate, environment & biodiversity, and people & business. Front Bioeng Biotechnol, p 1456
80. Lautala PT et al (2015) Opportunities and challenges in the design and analysis of biomass supply chains. Environ Manage 56:1397–1415
81. Lazaridis PA et al (2018) Catalytic fast pyrolysis of kraft lignin with conventional, mesoporous and nanosized ZSM-5 zeolite for the production of alkyl-phenols and aromatics. Front Chem 6:295
82. Lee GE, Loveridge S, Joshi S (2017) Local acceptance and heterogeneous externalities of biorefineries. Energy Econ 67:328–336

83. Lee WJ, Mwebaza R (2020) The role of the Climate Technology Centre and Network as a climate technology and innovation matchmaker for developing countries. Sustainability 12(19):7956
84. Leong HY, Chang CK, Khoo KS, Chew KW, Chia SR, Lim JW, Show PL et al (2021) Waste biorefinery towards a sustainable circular bioeconomy: a solution to global issues. Biotechnol Biofuels 14(1):1–15
85. Lewis AJ, Borole AP (2019) Bio-electro-refinery: conversion, sustainability, and policy. In: Microbial electrochemical technology. Elsevier, pp 1059–1085
86. Limayem A, Ricke SC (2012) Lignocellulosic biomass for bioethanol production: current perspectives, potential issues and future prospects. Prog Energy Combust Sci 38(4):449–467
87. Lin CSK et al (2013) Food waste as a valuable resource for the production of chemicals, materials and fuels: current situation and global perspective. Energy Environ Sci 6(2):426–464
88. Lin R et al (2021) A perspective on the efficacy of green gas production via integration of technologies in novel cascading circular bio-systems. Renew Sustain Energy Rev 150:111427
89. Mahapatra S et al (2021) Biofuels and their sources of production: a review on cleaner sustainable alternative against conventional fuel, in the framework of the food and energy nexus. Energy Nexus 4:100036
90. Mariath AB, Martins APB (2020) Ultra-processed products industry operating as an interest group. Revista de Saúde Pública 54
91. Martinez-Valencia L et al (2021) Supply chain configuration of sustainable aviation fuel: review, challenges, and pathways for including environmental and social benefits. Renew Sustain Energy Rev 152:111680
92. Mathew AK et al (2018) Lignocellulosic biorefinery wastes, or resources? In: Waste biorefinery. Elsevier, pp 267–297
93. Mathioudakis V et al (2017) The water footprint of second-generation bioenergy: a comparison of biomass feedstocks and conversion techniques. J Clean Prod 148:571–582
94. McCormick K, Kautto N (2013) The bioeconomy in Europe: an overview. Sustainability 5(6):2589–2608
95. McGuire JB et al (2017) Social acceptability of establishing forest-based biorefineries in Maine, United States. Biomass Bioenerg 105:155–163
96. Mendes AA et al (2023) Recent advances and future prospects for biolubricant base stocks production using lipases as environmentally friendly catalysts: a mini-review. World J Microbiol Biotechnol
97. Menon V, Rao M (2012) Trends in bioconversion of lignocellulose: biofuels, platform chemicals & biorefinery concept. Prog Energy Combust Sci 38(4):522–550
98. Moncada J et al (2016) Design strategies for sustainable biorefineries. Biochem Eng J 116:122–134
99. Moreno AD et al (2020) Overview of bio-based industries. In: Biobased products and industries. Elsevier, pp 1–40
100. Morone P, Imbert E (2020) Food waste and social acceptance of a circular bioeconomy: the role of stakeholders. Curr Opin Green Sustain Chem 23:55–60
101. Morone P et al (2019) Food waste: challenges and opportunities for enhancing the emerging bio-economy. J Clean Prod 221:10–16
102. Musa M et al (2019) Factors affecting microalgae production for biofuels and the potentials of chemometric methods in assessing and optimizing productivity. Cells 8(8):851
103. Nedstar (n.d.) Wheat alcohol. Retrieved May 15, 2023, from https://www.nedstar.com/products/raw-materials/sugar-beet-alcohol
104. Nizami AS et al (2017) Waste biorefineries: enabling circular economies in developing countries. Biores Technol 241:1101–1117
105. NovaCana (2023) Abengoa bioenergy fecha usina nos EUA e diminui exportação de etanol. Retrieved from https://www.novacana.com/noticias/abengoa-bioenergy-09012013
106. Pan SY et al (2015) Strategies on implementation of waste-to-energy (WTE) supply chain for circular economy system: a review. J Clean Prod 108:409–421

107. Panoutsou C et al (2021) Advanced biofuels to decarbonise European transport by 2030: markets, challenges, and policies that impact their successful market uptake. Energ Strat Rev 34:100633
108. Parada MP, Osseweijer P, Duque JAP (2017) Sustainable biorefineries, an analysis of practices for incorporating sustainability in biorefinery design. Ind Crops Prod 106:105–123
109. Patel A, Shah AR (2021) Integrated lignocellulosic biorefinery: gateway for production of second generation ethanol and value added products. J Bioresour Bioprod 6(2):108–128
110. Pérez ATE et al (2017) Key challenges and requirements for sustainable and industrialized biorefinery supply chain design and management: a bibliographic analysis. Renew Sustain Energy Rev 69:350–359
111. Pfaltzgraff LA, Clark JH (2014) Green chemistry, biorefineries and second generation strategies for re-use of waste: an overview. In: Advances in biorefineries, pp 3–33
112. Philp J, Winickoff DE (2018) Realising the circular bioeconomy
113. Pietzsch N et al (2022) Benefits, challenges and critical factors of success for Zero Waste: a systematic literature review. Waste Manage 67:324–353
114. Pinto GP et al (2022) Exploiting enzyme evolution for computational protein design. Trends Biochem Sci 47(5):375–389
115. Pinto RLS et al (2022) An overview on the production of synthetic fuels from biogas. Bioresour Technol Rep 18:101104
116. Pio DT, Tarelho LAC, Pinto PCR (2020) Gasification-based biorefinery integration in the pulp and paper industry: a critical review. Renew Sustain Energy Rev 133:110210
117. Pohl J, Rosselot N (2020) Acquisition-and ownership-related policies to safeguard essential security interests–current and emerging trends, observed designs, and policy practice in 62 economies (Table of Contents). Observed Des Policy Pract 62
118. Popp J et al (2014) The effect of bioenergy expansion: food, energy, and environment. Renew Sustain Energy Rev 32:559–578
119. Qiao J et al (2022) Integrated biorefinery approaches for the industrialization of cellulosic ethanol fuel. Biores Technol 360:127516
120. Ramos MDN et al (2022) Enzymatic catalysis as a tool in biofuels production in Brazil: current status and perspectives. Energy Sustain Dev 68:103–119
121. Reis WSM, Ferreira RDM, Pereira EB (2020) Immobilization of commercial lipase onto different supports: characterization and application in esterification reaction. Int J Eng Res Sci 6:15–24
122. Reuters (2011) Range fuels shuts down pioneering cellulosic ethanol plant. Retrieved from https://www.reuters.com/article/idUS154341571120110114
123. Reuters (2023) Neste to start production in Singapore as renewables fuels expansion takes off. Retrieved from https://www.reuters.com/business/energy/neste-start-production-singapore-renewables-fuels-expansion-weeks-official-2023-04-05
124. Romani A et al (2018) Valorization of wastes from agrofood and pulp and paper industries within the biorefinery concept: Southwestern Europe scenario. In: Waste biorefinery. Elsevier, pp 487–504
125. Roy BB, Tu Q (2022) A review of system dynamics modeling for the sustainability assessment of biorefineries. J Ind Ecol 26(4):1450–1459
126. Rozina, Ahmad M, Zafar M (2022) Biomass as sustainable material for bioethanol production. In: Handbook of smart materials, technologies, and devices: applications of industry 4.0. Springer International Publishing, pp 1–24
127. Saini R et al (2020) Lignocellulosic biomass-based biorefinery: an insight into commercialization and economic standout. Curr Sustain Renew Energy Rep 7:122–136
128. Sajid Z (2021) A dynamic risk assessment model to assess the impact of the coronavirus (COVID-19) on the sustainability of the biomass supply chain: a case study of a US biofuel industry. Renew Sustain Energy Rev 151:111574
129. Salvador R et al (2021) Key aspects for designing business models for a circular bioeconomy. J Clean Prod 278:124341

130. Sambrana A, Sørensen CAG, Kristensen EF (2008) Optimized harvest and logistics for biomass supply chain. In: Proceedings of European biomass conference and exhibition, Valencia, Spain
131. Santos EV et al (2023) The electro-refinery in organics: a new arising concept for valorization of wastes. Curr Opin Electrochem 101267
132. Saral JS, Ajmal RS, Ranganathan P (2022) Bioeconomy of hydrocarbon biorefinery processes. In: Hydrocarbon biorefinery. Elsevier, pp 355–385
133. Sarup B (2022) Overview of downstream product recovery technology in biorefineries. In: AZ of biorefinery. Elsevier, pp 509–521
134. Scapini T et al (2023) Bioenergy and beyond: Biorefinery process sources, research, and advances. In: Green approach to alternative fuel for a sustainable future. Elsevier, pp 335–347
135. Schebek L, Mrani O (2014) Environmental and sustainability assessment of biorefineries. In: Advances in biorefineries. Woodhead Publishing, pp 67–88
136. Serrano-Ruiz JC, Luque R, Sepúlveda-Escribano A (2011) Transformations of biomass-derived platform molecules: from high added-value chemicals to fuels via aqueous-phase processing. Chem Soc Rev 40(11):5266–5281
137. Sinha R et al (2022) Advances in downstream processing techniques. In: Current developments in biotechnology and bioengineering. Elsevier, pp 353–371
138. Solarte-Toro JC, Alzate CAC (2021) Biorefineries as the base for accomplishing the sustainable development goals (SDGs) and the transition to bioeconomy: technical aspects, challenges and perspectives. Biores Technol 340:125626
139. Solarte-Toro JC, Laghezza M, Fiore S, Berruti F, Moustakas K, Alzate CAC (2022) Review of the impact of socio-economic conditions on the development and implementation of biorefineries. Fuel 328:125169
140. Speight JG (2022) The biorefinery. In: Biomass processes and chemicals. Elsevier, pp 257–292
141. Taran OP, Gromov NV, Parmon VN (2018) Catalytic processes and catalyst development in biorefining. In: Sustainable catalysis for biorefineries
142. Thongchul N et al (2022) Perspectives, challenges and future directions. In: AZ of biorefinery. Elsevier, pp 739–750
143. Thrän D et al (2020) Governance of sustainability in the German biogas sector—adaptive management of the Renewable Energy Act between agriculture and the energy sector. Energy Sustainab Soc 10:1–18
144. Tye YY et al (2011) Second-generation bioethanol as a sustainable energy source in Malaysia transportation sector: status, potential and future prospects. Renew Sustain Energy Rev 15(9):4521–4536
145. Ubando AT, Felix CB, Chen WH (2020) Biorefineries in circular bioeconomy: a comprehensive review. Biores Technol 299:122585
146. UPM Biofuels (n.d.) Retrieved May 15, 2023, from https://www.upmbiofuels.com/about-upm-biofuels/
147. Vahidzadeh R, Bertanza G (2022) Industrial symbiosis and eco-industrial parks. In: Environmental sustainability and industries. Elsevier, pp 405–431
148. Van Dam J et al (2008) Overview of recent developments in sustainable biomass certification. Biomass Bioenerg 32(8):749–780
149. Vandenberghe LPS, Valladares-Diestra KK, Bittencourt GA, Zevallos Torres LA, Vieira S, Karp SG, Sydney EB, de Carvalho JC, Thomaz Soccol V, Soccol CR (2022) Beyond sugar and ethanol: the future of sugarcane biorefineries in Brazil. Renew Sustain Energy Rev 167:112721. https://doi.org/10.1016/j.rser.2022.112721
150. Venkatachalam V et al (2018) LCA and eco-design: consequential and attributional approaches for bio-based plastics. Proc CIRP 69:579–584
151. Vimmerstedt LJ et al (2015) Maturation of biomass-to-biofuels conversion technology pathways for rapid expansion of biofuels production: a system dynamics perspective. Biofuels Bioprod Biorefin 9(2):158–176
152. Wasevar KL (2022) Process intensification and green engineering in process industry. In: Environmental sustainability and industries. Elsevier, pp 433–459

153. Wilson L et al (2023) Future perspectives in enzyme immobilization. In: Biocatalyst immobilization. Academic Press, pp 403–426
154. Wright MM, Brown RC, Boateng AA (2008) Distributed processing of biomass to bio-oil for subsequent production of Fischer-Tropsch liquids. Biofuels Bioprod Biorefin 2(3):229–238
155. Yadav A et al (2023) Development of lignocellulosic biorefineries for the sustainable production of biofuels: towards circular bioeconomy. Bioresour Technol 129145
156. Yamakawa CK, Qin F, Mussatto SI (2018) Advances and opportunities in biomass conversion technologies and biorefineries for the development of a bio-based economy. Biomass Bioenerg 119:54–60
157. Ye Z et al (2022) A review of the advances in catalyst modification using nonthermal plasma: process, mechanism and applications. Adv Colloid Interface Sci 102755
158. Yildirim O et al (2016) Opportunities and challenges for drug development: public–private partnerships, adaptive designs and big data. Front Pharmacol 7:461
159. You C, Han S, Kim J (2021) Integrative design of the optimal biorefinery and bioethanol supply chain under the water-energy-food-land (WEFL) nexus framework. Energy 228:120574
160. Zahraee SM, Shiwakoti N, Stasinopoulos P (2020) Biomass supply chain environmental and socio-economic analysis: 40-Years comprehensive review of methods, decision issues, sustainability challenges, and the way forward. Biomass Bioenerg 142:105777
161. Zetterholm J et al (2020) Economic evaluation of large-scale biorefinery deployment: a framework integrating dynamic biomass market and techno-economic models. Sustainability 12(17):7126
162. Zhu X et al (2023) Biorefinery. In: Future proteins. Academic Press, pp 447–468

Economic Feasibility of Biodiesel Production from Jatropha Oil

S. Sakthivel

Abstract This study aims to assess the economic viability of small-scale biodiesel production, known as fatty acid methyl ester (FAME), using the non-edible oil from *Jatropha curcas*. It is crucial to understand the economics of biodiesel production for its potential industrial application. The findings of this paper indicate that the cost of oil feedstock is the most significant expense, contributing 70% of the total production costs, making it a critical economic factor in biodiesel production. Other costs, including methanol, catalysts, utilities, labor, and capital cost (fixed), account for 8, 1, 9, 2, and 10% of the total costs, respectively. While capital expenditure (CAPEX) has a minimal impact on costs, operational expenditure (OPEX) significantly influences the overall production expenses. Currently, the cost of producing biodiesel is lower than that of conventional diesel fuel. However, the biodiesel offers the advantages of reduced pollution compared to petroleum diesel, enhances energy security, and ensures environmental protection. Additionally, glycerol as a by product enhances economic feasibility for producers by serving as a raw material for the production of value-added chemicals. The recovery and purification of crude glycerol (~80% w/w) for commercial sale can lower the total production costs by about 5%, offering an economic boost to biodiesel production.

Keywords Biodiesel · Economic analysis · Methanol recovery · Glycerol · *Jatropha curcas L. oil* · Production costs

1 Introduction

The Indian National Biofuel Policy of 2009 advises against the use of edible oils for biodiesel production due to social and economic concerns, recommending the utilization of non-edible oils, which are known as the second-generation feedstocks. These oils include non-food crops, algae, biomass, and forestry residues. In a move to further promote biofuel production, the Government of India introduced an updated

S. Sakthivel (✉)
Technology Group, TATA Consulting Engineers Limited, Mumbai, India
e-mail: ssakthivel@tce.co.in

S. Kumar et al. (eds.), *Clean Energy Transition-via-Biomass Resource Utilization*, Green Energy and Technology, https://doi.org/10.1007/978-981-97-6321-4_10

National Biofuel Policy in 2018, focusing on ethanol production. This new policy broadens the variety of raw materials eligible for ethanol production, including sugarcane juice, sugar beet, sweet sorghum, corn, cassava, and damaged food grains such as wheat and broken rice. It also supports the use of surplus food grains, non-edible seeds, used cooking oil, and crops with a short growth cycle for ethanol production. The policy aims to reduce import dependency, foster a cleaner environment, enhance public health, improve waste management, develop infrastructure in rural areas, and increase farmers' incomes. Various oilseed crops such as *Jatropha curcas, Madhuca indica (mahua), Pongamia pinnata (karanja), and Azadirachta indica (neem)* are available in India and suitable for biofuel production, which can thrive in arid and marginal lands with minimal water, maintenance, and soil fertility requirements. Additionally, microalgae oils are explored by academic and industrial researchers in India, especially in demonstration-scale microalgae processes.

Vegetable oils or animal fats, known chemically as triglycerides, undergo a reaction with short-chain alcohols in the presence of a catalyst, producing fatty acid methyl ester (commonly referred to as biodiesel) and glycerol as a by-product. The feedstock quality, particularly the presence of free fatty acids (FFAs), is critical in selecting the appropriate technology. High FFA levels in the feedstock can lead to soap formation when reacting with an alkaline catalyst, adversely affecting biodiesel yield and glycerol separation. It is noted that if the FFA content exceeds 3%, the transesterification process may not be successful [1]. For feedstocks with high FFA content, pre-treatment with an acid catalyst through an esterification reaction is required. Conversely, if the oil has low FFA content (less than 3%), it can directly undergo a transesterification reaction with methanol, producing biodiesel and glycerol. Non-edible oils typically have high FFA content, ranging from 8% to 35%. For example, *Jatropha curcas oil* contains over 12% of FFA [2]. High FFA oils require more alkali catalyst to neutralize the FFAs, leading to soap formation and making direct transesterification unsuitable. Acid pre-treatment followed by base-catalyzed transesterification is recommended for oils with high FFA content [3, 4]. This step is crucial for processing non-edible oils with high FFAs. The composition and types of fatty acids present are detailed in Table 1 [5].

The esterification reaction involves FFA reacting with methanol under the influence of an acid catalyst, such as diluted hydrochloric acid (HCl) or sulfuric acid, (H_2SO_4), resulting in the production of biodiesel and water.

Table 1 Fatty acid composition of *Jatropha curcas oil*

S. no	Fatty acid	Formula	Composition (%)
1	Palmitic acid, C16:0	$C_{16}H_{32}O_2$	17.62
2	Stearic acid, C18:0	$C_{18}H_{36}O_2$	6.78
3	Stearic acid, C18:0	$C_{18}H_{34}O_2$	36.76
4	Linoleic acid, C18:2	$C_{18}H_{32}O_2$	38.17
5	Linolenic acid, C18:3	$C_{18}H_{30}O_2$	0.67

Esterification Reaction (preferably in the presence of acid catalyst) (Eq.1) and transesterification reaction (Eq. 2)

$$\text{FFA} + \text{Alcohol} \rightarrow \text{FAME (Biodiesel)} + \text{Water} \quad (1)$$

$$\text{Oil/Fat (Triglycerides)} + \text{Alcohol} \rightarrow \text{FAME (Biodiesel)} + \text{Glycerol} \quad (2)$$

Biodiesel manufacturers prefer methanol for its cost-effectiveness compared to other alcohols like ethanol, propanol, butanol, and amyl alcohol. Researchers suggest an optimal biodiesel conversion using a molar ratio of 1:6 for oil to methanol and an alkaline catalyst concentration between 1 and 1.5% by weight (e.g., NaOH or CH_3ONa) relative to the oil feedstock [6, 7]. Various technologies are employed in biodiesel production, including homogeneous acid or alkali catalysis, alkali catalysis with acid pre-treatment, heterogeneous solid catalysis, supercritical methods, and enzymatic (biochemical) processes. Among these, the homogeneous alkaline catalysis method, with or without a pre-esterification step, in batch or continuous operation, is a preferred choice for many small-scale producers. This preference is due to its lower energy requirements compared to the heterogeneous catalyst process [2, 8] and the supercritical method [9, 10], which require high temperatures ranging from 150 to 320 °C.

Figure 1 llustrates the process flow diagram for producing biodiesel via a homogeneous alkaline method with an initial acid pre-treatment. The biodiesel production process includes several key steps: (a) pre-treatment of oil feedstock to address high FFAs, (b) conversion of triglycerides into biodiesel, (c) separation of the product, and (d) recovery of excess methanol and glycerol. Initially, oil feedstock with high FFAs and methanol, in a specified molar ratio, are introduced into reactor-1 along with an acid catalyst solution for the esterification step, transforming FFAs into esters, fatty acids, and water. This mixture is then separated, and the organic layer is combined with methanol in a 1:6 molar ratio and a sodium-based catalyst (e.g., $NaOH/NaOCH_3$) in reactor-2 for transesterification. This reactor, a continuous stirred-tank reactor (CSTR), operates between 50 and 65 °C at atmospheric pressure, with a residence time ranging from 15 to 60 minutes. Maintaining the reaction temperature close to methanol's boiling point enhances the reaction rate without causing excessive methanol vaporization. Post-reaction, the mixture is decanted to separate glycerol and to purify the crude biodiesel, which is then sent to an alcohol recovery unit to remove any excess alcohol. Following this, it undergoes water washing to eliminate catalyst traces, glycerol, and other impurities. The final step involves drying or distillation to remove moisture and residual methanol, yielding high-purity biodiesel. Glycerol separated in the process is treated in a neutralization or acidulation unit to neutralize leftover alkaline catalyst, followed by a separation process to remove water and methanol, which can be recycled. This results in glycerol of approximately 80% purity, representing a secondary revenue stream from the biodiesel production process.

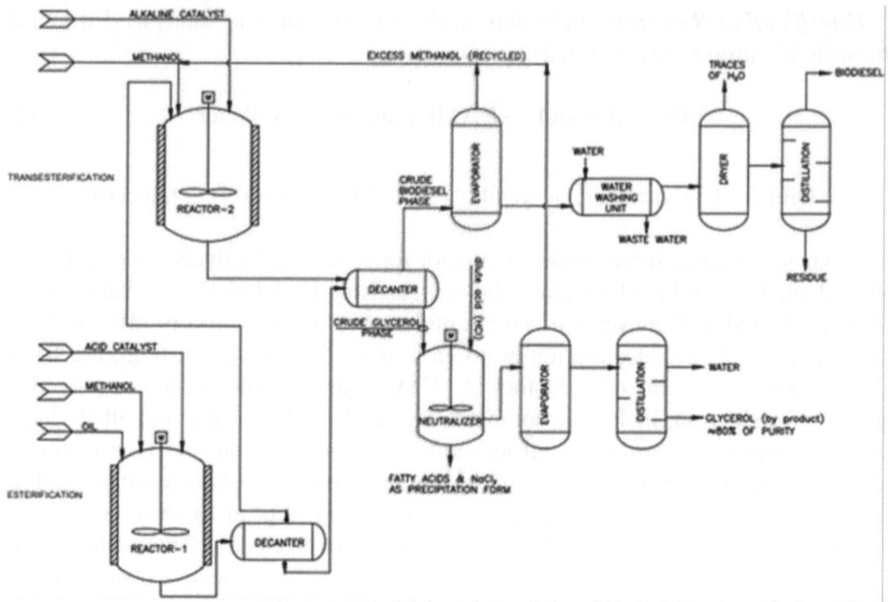

Fig. 1 Schematic diagram of homogeneous catalysts process with acid pre-treatment

2 Economic Approach

The economic and cost-effective production of biodiesel depends on various operational factors, including the type and concentration of catalysts, the molar ratio (MR) of oil to alcohol, and the efficiency of processes like soap separation, glycerol removal, water washing, and recovery of excess methanol. This article investigates into these crucial parameters.

2.1 Alcohol

According to the stoichiometric reaction, one-mole triglyceride reacts with three moles of alcohol to produce three moles of FAME and one mole of glycerol. Theoretically, methanol consumption in the transesterification reaction is about 12.6% (w/w) of oil feedstock. When alcohol is used as theoretical requirement (molar ratio of 1:3), the reaction reaches to equilibrium stage, which reduces the quality of biodiesel formation. The theoretical requirement of alcohol depends on the amounts fatty acids in the oil. Generally, the excess methanol is used for the transesterification reaction and its support to shift the equilibrium toward product formation, favoring the forward reaction in the reversible reaction. Typically, excess methanol is used between 60 and 100% of the stoichiometric amount of methanol. Thus, the excess methanol requirement depends on the composition of fatty acids. Sakthivel et al. [5] reported that the yield increases with an increased molar ratio until 1:18, after which further increases in methanol content show an insignificant improvement in biodiesel formation. However, the excess methanol does not become part of the final product and can be recovered and reused in the same process.

After the reaction, the excess methanol is present in both the crude biodiesel stream (i.e., organic phase) and the crude glycerol stream (i.e., aqueous phase). Typically, the crude biodiesel phase is contains 30–40% (w/w) of total excess methanol, while the glycerol phase contains around 60–70% (w/w) of total excess methanol. Methanol can be separated using various techniques, such as vacuum stripping, distillation column, or flash evaporation. The biodiesel ASTM D6751 and EN 14214 biodiesel standards allow 0.2 weight % of methanol in the final product. The methanol recovery process not only enhances product quality but also provides significant economic and environmental benefits [11]. Methanol, classified as a hazardous material due to its high flammability and toxicity, should not be disposed into wastewater streams. Moreover, with methanol prices ranging from approximately US$ 350–400/MT, it is economically sensible for producers to recover and reuse this valuable chemical rather than lose it with waste. The process of using a homogeneous alkaline catalyst with acid pre-treatment tends to consume less methanol compared to heterogeneous and supercritical methods, as reported by [12, 13] making it a more efficient and cost-effective approach for biodiesel production.

2.2 Catalyst

Alkali catalysts such as sodium hydroxide, potassium hydroxide, sodium methoxide, sodium ethoxide or potassium methoxide are commonly used along with methanol to enhance the rate of reactions. Typically, sodium-based catalyst (e.g., NaOH or $NaOCH_3$) is preferred for a small-scale production due to as inexpensive and easy to handle in transportation and storage. A few more additional reasons for using NaOH for commercial purposes include (a) the required quantity of KOH in the transesterification reaction is higher compared to NaOH to obtain the same yield and (b) the market price of KOH is about 2.5 times more expensive than NaOH [14]. The catalyst is never consumed by the reaction and remains present in the crude biodiesel phase and glycerol phases after the reaction. The majority of catalyst content is found in the glycerol phase as form of sodium glycerate, sodium methylate and sodium soaps. These sodium soaps are soluble in glycerol phase and are removed during neutralization step by addition of hydrochloric acid (HCl). During neutralization, the sodium salt reacts with HCl to form sodium chloride, which precipitates out as precipitation form.

2.3 Glycerol as By-product

Glycerol as high valuable by-product is generated during the tranesterification reaction. It can be separated after the reaction by using a centrifuge or gravity decanter, which is a density difference technique. Glycerin is produced as an approximately 10% (w/w) (i.e., 104 kg) in the transesterification of a metric ton of oil feedstock. In this glycerol phase, excess methanol act as a solubilizer and it can hold-up the separation efficiency. There are two steps involved for purification of glycerol like (a) an acid neutralization and (b) methanol removal after centrifuge or gravity decanter. Typically, the glycerol phase stream has around 50% of glycerol [15]. During the transesterification process, glycerol is produced as a valuable by-product, constituting about 10% (w/w) or 104 kg for every metric ton (MT) of oil feedstock processed. This glycerol can be initially separated from the reaction mixture using density-based techniques such as centrifugation or a gravity decanter. Excess methanol present in the glycerol phase can hinder the efficiency of this separation due to its solubilizing properties. The purification of glycerol involves two main steps: (a) acid neutralization to remove FFAs and salts, and (b) the removal of methanol. Typically, the glycerol phase contains about 50% glycerol, with the remainder consisting of excess methanol, traces of catalyst, and sodium soap [15]. The purity of glycerol can be enhanced from 50% to 75–90% through a soap splitting method applied during the neutralization stage. To achieve pharmaceutical-grade glycerol with purity levels of 99.5–99.7%, further processing is required, including vacuum distillation, ion exchange, and adsorption techniques. In the initial neutralization step, FFAs and salts are eliminated from glycerol through the addition of an acid, such as HCl, facilitating a soap splitting reaction.

2.4 Product Purification

There are two primary methods available for the product purification of biodiesel such as wet and dry washing. The water washing method is the most effective for removing traces of glycerol, excess methanol, and other contaminants such as catalyst residues, dirt. This step helps to meet the final product quality with ASTM/EN standard and doesn't allow more than 0.02 wt. % (200 ppm) glycerol content in the final product of biodiesel. The dry washing methods such as ion exchange resin, membrane and magnesium silicate powders (e.g., magnesol) are used to purify biodesel, with water for cleaning biodiesel. Dennis et al. [6] reported that magnesol process is more effective at removing methanol compared to the ion exchange resin process. Later, Michelle et al. [16] studied the effect of adsorbents (like cellulose and starch) on the purification of biodiesel produced from sunflower oil using dry cold washing method.

3 Results and Discussion

The costs associated with producing or manufacturing can be broadly categorized into operational expenditure (OPEX) and capital expenditure (CAPEX). Production costs typically contain direct operating costs, fixed charges, and plant-overhead costs. Direct operating or production costs include expenses for raw materials, chemicals and catalysts, utilities or service facilities, power, and supervisory labor. Fixed costs comprise depreciation, property taxes, insurance, and equipment, among others. Plant-overhead costs cover general plant maintenance, safety services, plant superintendence, control laboratories, salvage services, property protection, and facilities for warehousing and storage [17]. For the purposes of this study, several assumptions have been made: (a) equipment depreciation is calculated as 10% of the capital cost over a 10-year lifespan, (b) interest or escalation rate costs are added at 2% to the capital cost, (c) land costs are excluded, (d) economic considerations such as corporate tax rates, internal rates of return, salvage values, and construction interest rates are not included, (e) government subsidies are not factored into the calculations, and (f) the estimated production cost accounts for income from the sale of glycerol (with 80% (w/w) purity) as a secondary revenue source, assuming a production capacity of 1000 kg/day for a batch process.

In the transesterification process to produce biodiesel from *Jatropha curcas oil*, the stoichiometric requirements are 952 liters of oil and 120 liters of methanol, constituting 12.6% of the oil volume, to yield 1000 liters of biodiesel and 72 liters of glycerol as a by-product. However, in practical applications, to achieve 1000 liters of biodiesel, 952 liters of feedstock and 240 liters of methanol are needed, considering a 100% excess of methanol to optimize the process efficiency and depending on the catalyst used. Homogeneous catalysis typically results in conversion yields ranging from 90 to 99% for *Jatropha curcas oil*. These yields vary based on factors such as the type of catalyst, molar ratio, temperature, and residence time. Alkaline catalysts are used at a rate of 1–1.5% (w/w) of the oil feedstock, while acid catalysts range from 0.5 to 1% (w/w) for effective conversion and separation. Operating a biodiesel production facility requires utilities such as electrical energy, steam, process and cooling water, and nitrogen. According to Kiss et al. [18] the utility requirements for a commercial-scale homogeneous process plant to produce 1 MT of biodiesel include 145 kg of steam, 25 m^3 of cooling water, 38 m^3 of process water, 12 kWh of electrical energy, 10 kg of HCl (37%), 5 kg of sodium methylate as the catalyst, and 1.5 kg of NaOH, among others.

Feedstock: This analysis incorporates the costs of methanol (US$ 400/MT), oil feedstock (~US$ 600/MT), and glycerol (US$ 400/MT) to calculate the production expenses for biodiesel derived from *Jatropha curcas L. oil*. The comprehensive production cost estimation includes expenses related to raw materials (oil feedstock), chemicals and catalysts, utilities, labor, and investment costs (depreciation). The calculated cost for producing biodiesel is US$ 0.754/liter, excluding taxes, and including additional revenue from by-product of glycerol and methanol recovery (i.e., – US$ 0.065/liter). Otherwise, the cost of biodiesel is US$ 0.82/liter. The cost

Table 2 Economic analysis for biodiesel production from *Jatropha curcas L. Oil*

Description	US$/Liter
Capex	
Machine/Equipment	0.08
Repair and Maintenance	
Admin and overhead	
Opex	
Crude Jatropha Oil	0.65
Methanol	
Catalyst (NaOH)	
HCl/Chemicals	
Utility	
Cooling Water	0.09
Process Water	
Steam Consumption	
Electrical Energy	
Labours	
Capex + Opex + Utility Cost	0.820
Recovery of by-product / Excess CH_3OH	−0.065
Overall cost of biodiesel production	0.754

Note—Process efficiency, excess methanol, and glycerol recovery are considered as 96, 95, & 80%, respectively. 1 US Dollar is currently equivalent to 83 Indian Rupees

of the oil feedstock alone accounts for approximately 70% of the OPEX, aligning with findings from various researchers who have noted that raw material costs typically constitute the largest portion (about 70–80%) of total production expenses [19, 20]. Acevedo et al. [21] studied on the use of palm oil in a pilot-scale facility capable of processing 20,000 liters per day, or 850 liters per batch. This study highlighted that the raw materials needed for biodiesel production represented the largest share of expenses, accounting for 72.6% of the overall production costs. Reducing the production costs of biodiesel poses a significant challenge for producers, yet the environmental benefits, including the potential for carbon credits, offer substantial value. The operational expenditure can vary depending on factors such as the process adopted at the plant, its scale, and geographical location.

Figure 2 displays the cost distribution for biodiesel production. The analysis shows that 70% of the cost comes from raw materials, 10% from chemicals, 3% from utilities, 6% from labor, and 6% from fixed costs. A similar observation was reported by May and Tinia [22]. It indicates that the selection of low-cost oil feedstock must be essential to ensure the economic viability of biodiesel production. Typically, the depreciation cost is considered over a period of 10 years with 10% of fixed costs and its servicing thus accounted for 5–10% of total production costs [20, 23]. In

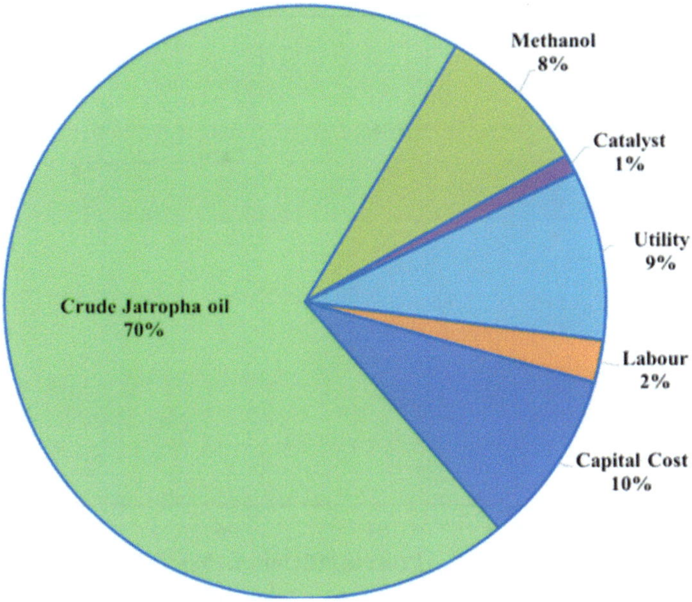

Fig. 2 Cost distribution of biodiesel production

this analysis, an estimated fixed cost of depreciation is accounted about 6% of total production costs. The fixed cost of biodiesel production is a relatively small proportion of the total production cost. The results indicate that the cost of servicing capex investments is not a major contributor to the overall production cost of biodiesel. It's inferred that the biodiesel industry is highly sensitive to feedstock price compared to opex and capex. Consequently, the biodiesel industry faces significant risk with increase in feedstock prices.

The overall costs of biodiesel production can be reduced by better utilization or higher conversion of raw materials, which can be achieved through increased process efficiency. Generally, the plant should operate at higher efficiency for better profitability. The process efficiency varies with the types of technology such as pre-treated alkali-catalyzed, heterogeneous acid catalyst, and supercritical process and generally, this process efficiency is 95%, 94%, and 98%, respectively. The present work analyzed the effect of production costs on process efficiency as shown in Fig. 3. The analysis resulted that if process efficient increases by 1% in biodiesel yield, (e.g., from 95 to 96%), the biodiesel price could be reduced by approximately 1% per liter of biodiesel.

Utilities and energy consumption: Operating a biodiesel plant requires various utilities such as electrical energy, steam, process and cooling water, and nitrogen. The costs associated with utilities and energy consumption typically account for 5–10% of the total biodiesel production costs. Marchetti [12] observed that, for a plant with an annual capacity of 36,036 MT,

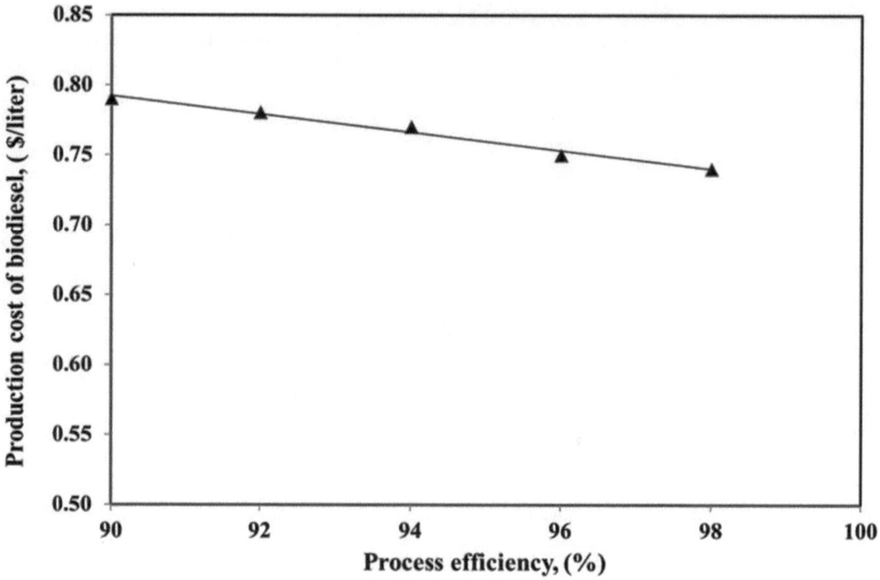

Fig. 3 Effect of process efficiency on biodiesel price

the utility costs for processes using heterogeneous catalysis were 6.1 times higher than those for homogeneous processes, largely due to the higher energy requirements of heterogeneous catalysis. This is because heterogeneous catalyzed processes operate at higher temperatures (approximately 150–350 °C) owing to the lower activity of the solid catalyst, thus consuming more energy. In contrast, West Alex [13] conducted an assessment of four biodiesel production processes with a capacity of 8,000 metric tons per annual (MTPA) using the HYSYS simulator, and their findings suggested that heterogeneous processes could have lower utility costs compared to homogeneous processes. Further analysis by Kiss et al. [18] on a larger industrial-scale plant capable of producing 100,000 MTPA highlighted the advantages of the heterogeneous process over the homogeneous process, including higher biodiesel yield, greater glycerin purity, and lower costs for catalyst and maintenance. This indicates that, while heterogeneous processes may initially seem more energy-intensive, they can offer economic benefits at larger scales or under certain conditions. Reducing production costs in biodiesel manufacturing is achievable through the recovery of glycerol produced during the transesterification reaction and by enhancing the process efficiency to approach 99%. Figure 4 illustrates how the efficiency of the glycerol recovery operation influences the overall production costs of biodiesel. As the performance of the glycerol recovery unit improves, the production cost of biodiesel correspondingly decreases. The findings indicate that optimizing glycerol recovery can lead to a reduction in production costs by approximately 5%. This observation aligns with previous research, which reported a potential reduction in production costs of about 6% during the transesterification

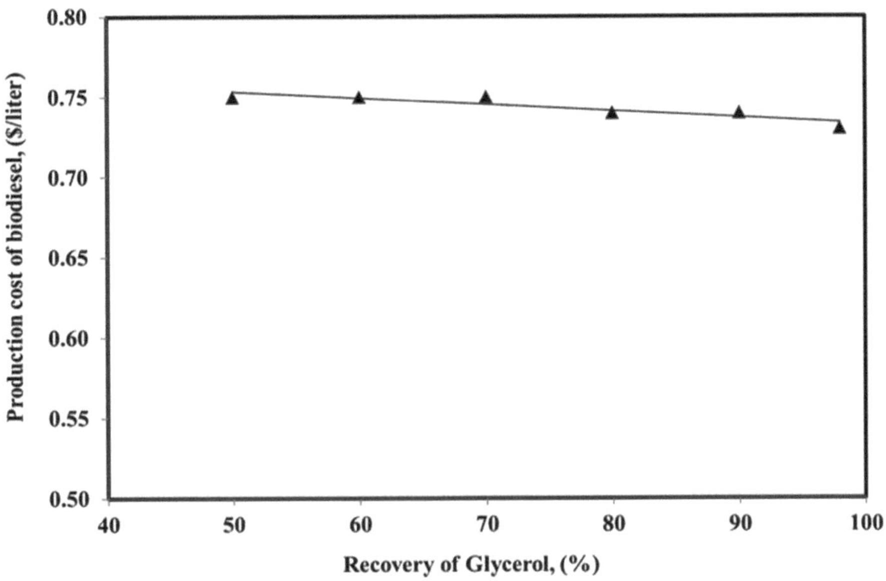

Fig. 4 Effect of recovery of glycerol on biodiesel cost

of degummed soybean oil [20] and a similar reduction during the transesterification of Ethiopian mustard oil [24].

4 Conclusions

The economic analysis of producing biodiesel from *Jatropha curcas L. oil* using a homogeneous alkaline catalysis process showed that the cost of feedstock is the primary challenge to the economic sustainability of biodiesel production, comprising 70% of the overall production costs. Costs associated with energy, chemicals and utilities, labor, and fixed charges are comparatively less significant. This highlights that the price of oil feedstock is a critical barrier and challenge to initiating large-scale commercial biodiesel production, surpassing even the impact of technology selection on economic viability. However, the process also yields glycerol as a valuable secondary source of revenue. The recovery of glycerol, with an 80% w/w purity from the aqueous phase, can reduce production costs by approximately 5%, offering a significant boost and incentive for biodiesel manufacturers.

References

1. Canakci M, Gerpen JV (1999) Biodiesel production via acid catalysis. Trans ASAE 42:1203–1210
2. Sakthivel S, Halder S, Gupta PD (2012) Influence of co-solvent on production of biodiesel in batch and continuous process. Int J Green Energy 10:876–884
3. Berchmans HJ, Hirata S (2008) Biodiesel production from crude Jatropha curcas L. seed oil with a high content of free fatty acids. Bioresour Technol 99(6):1716–1721
4. Ramadhas AS, Muraleedharan JSC (2005) Biodiesel production from high FFA rubber seed oil. Fuel 84:335–340
5. Sakthivel S, Halder S, Gupta PD (2012) Optimization of process variables for production of biodiesel in packed bed reactor using response surface methodology. Int J Ambient Energy 34(2):83–91
6. Leung DYC, Wu X, Leung MKH (2010) A review on biodiesel production using catalyzed transesterification. Appl Energy 87:1083–1095
7. Lu H, Liu Y, Zhou H, Yang Y, Chen M, Liang B (2009) Production of biodiesel from Jatropha curcas L. oil. Comput Chem Eng 33(5):1091–1096
8. Halder S, Sakthivel S, Jayaraj KM, Gupta PD (2014) Studies on transesterification of Karanja (Pongamia pinnata) oil in packed bed reactor. Chem Eng Commun 201(1): 88–101
9. Samniang A, Tipachan C, Kajorncheappun-ngam S (2014) Comparison of biodiesel production from crude Jatropha oil and Krating oil by supercritical methanol transesterification. Renew Energy 68:351–355
10. Salar-García MJ, Ortiz-Martínez VM, Olivares-Carrillo P, Quesada-Medina J, de los Ríos AP, Hernández-Fernández FJ (2016) Analysis of optimal conditions for biodiesel production from Jatropha oil in supercritical methanol: quantification of thermal decomposition degree and analysis of FAMEs. J Supercrit Fluids 112:1–6
11. Dhar BR, Kirtania K (2009) Excess methanol recovery in biodiesel production process using a distillation column: a simulation study. Chem Eng Res Bull 13:55–60
12. Marchetti JM, Miguel VU, Errazu AF (2008) Techno-economic study of different alternatives for biodiesel production. Fuel Process Technol 89:740–748
13. West AH, Posarac D, Ellis N (2008) Assessment of four biodiesel production processes using HYSYS. plant. Bioresour Technol 99:6587–6601
14. Sakthivel S, Suresh S, Selvaraju N (2018) Biodiesel—technical viability for India. In: Kumar S, Sani R (eds) Biorefining of biomass to biofuels. Biofuel and biorefinery technologies, Vol 4. Springer, Cham, pp 343–359
15. Van Gerpen J, Shanks B, Pruszko R, Clements D, Knothe G (2004) Biodiesel production technology. Department of Energy, Washington DC
16. Michelle GG, Douglas QS, Luis CM, Daniel P (2015) Purification of biodiesel by dry washing, employing starch and cellulose as natural adsorbents. Fuel 155:1–6
17. Peters MS, Timmerhaus KD (1991) Plant design and economics for chemical engineers, 4th Edition. McGraw-Hill, Inc. ISBN 0-07-0496137
18. Kiss FE, Jovanovic M, Boskovic GC (2010) Economic and ecological aspects of biodiesel production over homogeneous and heterogeneous catalysts. Fuel Process Technol 91:1316–1320
19. Demirbas A, Kinsara RA (2017) Cost analysis of biodiesel from kernel oil of tea seed. Energy Sources Part B 12(5):480–486
20. Haas MJ, McAloon AJ, Yee WC, Foglia TA (2006) A process model to estimate biodiesel production costs. Bioresour Technol 97:671–678
21. Acevedo JC, Hernández JA, Valdés CF, Khanal SK (2015) Analysis of operating costs for producing biodiesel from palm oil at pilot-scale in Colombia. Bioresour Technol 188:117–123. https://doi.org/10.1016/j.biortech.2015.01.071
22. May YK, Tinia IMG (2011) A review of biodiesel production from Jatropha curcas L. oil. Renew Sustain Energy Rev 15:2240–2251

23. Lim S, Teong LK (2010) Recent trends, opportunities and challenges of biodiesel in Malaysia: an overview. Renew Sustain Energy Rev 14(3):938–954
24. Dorado MP, Cruz F, Palomar JM, Lopez FJ (2006) An approach to the economics of two vegetable oil-based biofuels in Spain. Renew Energy 31(8):1231–1237

Insights into Next-generation Biofuel Strategies and Policy Considerations in India

Jyoti Verma, Yatish T. Shah, Suresh Sundaramurthy, and S. Arisutha

Abstract While India is the third largest consumer of energy in the world, its per capita energy usage is about one-fourth of China and one-tenth of the U.S. Both traditional and renewable energy including solar, wind, ocean, and tidal energy, biomass, hydrogen, etc. would need to complement each to meet India's growing demand for power in the future. However, this review particularly focuses on harnessing the renewable energy from biomass that is widely available in a country like India. India needs to develop and implement green and clean technologies suitable for indigenous conditions (a) to meet the growing demands, (b) to reduce the dependency on foreign imports of oil and coal, and (c) to reduce CO_2 emissions to help contain global warming. Green and clean technologies using renewable resources need to be developed, customized, and implemented on a massive scale. The review rather than just targeting the scientific community as the audience, also identifies industrial stakeholders as potential readers, both nationally and internationally hence, the Indian political perspective is also presented to encourage interaction between the industry and academia. This review presents a critical analysis of the National Policy on Biofuels (NPB) 2018, along with a comparison with biofuel policies of other nations, the Biodiesel Blending Program (BBP) and Ethanol Blending Petrol (EBP) programs, and various challenges and opportunities suggesting future reforms. Biofuels such as ethanol and biodiesel produced from different feedstocks as promoted in the National Biofuel Policy 2018 of India are critically discussed in this review. Attention is also paid to fuels derived from lignocellulosic biomass along with the microbial biofuel

J. Verma (✉)
Biotechnology for Energy, ONGC Energy Centre, 8th floor, core 3, Scope Minar, Lakshmi Nagar, Delhi 110092, India
e-mail: verma_jyoti@ongc.co.in

Y. T. Shah
Retired University Provost, Norfolk, VA, USA

S. Sundaramurthy (✉)
Department of Chemical Engineering, Maulana Azad National Institute of Technology, Bhopal, Madhya Pradesh 462 003, India
e-mail: sureshs@manit.ac.in

S. Arisutha
Eco Science and Technology, Bhopal, Madhya Pradesh 462 003, India

S. Kumar et al. (eds.), *Clean Energy Transition-via-Biomass Resource Utilization*, Green Energy and Technology, https://doi.org/10.1007/978-981-97-6321-4_11

production route, in addition to single-cell oil. The two major gaseous biofuels, biogas and biohydrogen, are also briefly discussed in the Indian context.

Highlights

1. Global picture on biofuels
2. Sources of biofuels and factors influencing biofuel production
3. Gaseous biofuel products: biogas and biohydrogen
4. Future directions and role of global energy players
5. Commercial and economic viabilities of processes
6. Policy considerations and suggestions

Keywords Pradhan Mantri JI-VAN Yojna · Lignocellulosic biomass · Green hydrocarbons · 2G ethanol · Pre-treatment · Simultaneous saccharification and fermentation

1 Introduction

Biofuels are derived from renewable biomass resources such as forest residues, municipal solid waste (MSW), agricultural residues, and surplus food grains, and have enormous potential to help in achieving renewable energy targets. Biofuel production from lignocellulosic biomass has the potential to improve India's national energy security by a significant reduction in dependence on foreign imports of oil, lowering greenhouse gas emissions, curtailing global warming, and also improving rural economies by creating employment opportunities [1]. Many researchers globally have worked on the biological production of liquid fuels from non-food biomass and coal [2]. This technology has existed for the past three decades in developed nations like the US and others. India being a developing country is slowly stepping forward in this direction by implementing National Energy policies. If the aspirations of Indians regarding development and standard of living are to be realized, energy production and imports will need to increase significantly. For example, the electricity generated in 2017 reached 1200 Billion units, and further, it is likely to reach 1600 Billion units by 2022 and spiral up to 2100 Billion units over the next decade. The current global experience suggests that the lack of cost-effective technology for the efficient conversion of biomass to liquid fuels is the key bottleneck in the production of lignocellulose-derived biofuels [3, 4]. Advances in agriculture and biotechnology in countries like the US, Brazil, China, and European countries have made it possible to provide subsidies/tax benefits for corn to ethanol production from 2G processes using waste biomass.

India is aiming for significant investment in technologies for biofuel production, principally ethanol, and research is underway to overcome the biological, chemical, and engineering barriers for diesel, gasoline, and jet fuel hydrocarbon production.

Meeting jet fuel HC (hydrocarbon) specifications for properties like combustion characteristics, energy content, stability, density, corrosivity, fluidity, electrical conductivity, and lubricity is crucial because aircraft frequently cross international borders as compared to regional vehicles. Hydrocarbon fuel production from biomass has the following advantages: Firstly, the existing infrastructure (engines, pipelines) need not be modified as "green hydrocarbon fuels" are similar to the existing petroleum products, except for their source [5]. Secondly, the energy equivalence of fuels from biomass-based hydrocarbons is similar to that of petroleum-derived fuels. Thirdly, heterogeneous catalysts are more economically viable in comparison to biological catalysts as they are inherently recyclable hence, they could be used over longer time periods. Fourthly, the high-temperature requirement for hydrocarbon production from biomass accelerates the conversion process in production plants. Lastly, energy-consuming distillation processes could be eliminated for the production of immiscible hydrocarbons from lignocellulosic biomass [5]. Bio-reactors can be placed in close proximity to the biomass source, thereby minimizing the transportation cost.

Presently, India being the world's third largest energy importer, has aspired to a 10% lowering of crude oil imports by 2022. A depreciating INR (Indian National Rupee) value has resulted in an increasing fiscal deficit because of large crude oil imports, suggesting that local production of biofuels requires considerable attention. Biofuel policies have been framed for the production and utilization of biofuels (ethanol and biodiesel) [6]. Initially, India started with a 5% ethanol blending pilot program in 2001, followed by the launch of a biofuel mission in 2003 to achieve 20% biodiesel blending in a diesel by 2011–12. With the advent of the National Policy on Biofuels (NPB) in 2009 and its revision in 2018, an indicative blending target of 20% ethanol in petrol and 5% biodiesel in diesel by 2030 was proposed. The objective of such a policy was to facilitate the optimal development of domestic biomass feedstock for the production of biofuel. The policy strategized the vision and mission toward attaining sustainability by using drylands for biofuel development with financial, institutional, and technological interventions, also increasing employment opportunities. The USA is the largest biofuel producer in the world (38,088 thousand tons of oil equivalent (ktoe) in 2018, accounting for 39.9% of global production, followed by Brazil (21,375 ktoe accounting for 22.4% of global production), while India holds a global share of 1.1% (1023 ktoe) [7]. Demand-side policies focus on encouraging investments in biofuel crop cultivation to ensure mandatory proportionate biofuels blending, whereas the supply side policies envisage rebates on excise duties for biofuels and compatible fuel vehicles [8].

The biofuel policy of 2009 envisaged the choice of any wasteland oilseed-bearing plant as the feedstock, unlike the initial biodiesel mission program for which *Jatropha curcus* was chosen as the primary feedstock for biodiesel production [8]. A study by Arjune (2017), on the Indian biofuel policies outlined the use of wasteland for cultivating oilseed-bearing plants but not the use of agricultural waste for biofuel production [9].

Presently, in petrol, the blending percentage of ethanol is 2.0%, and in diesel, biodiesel blending is less than 0.1%. The NPB 2018 aims to increase the ethanol blending in petrol to 20% under the EBP program and the biodiesel blending in diesel to 5% under the BBP by 2030.

NPB 2018 policy classifies biofuels as:

1. First Generation (1G) "Basic Biofuels"—Ethanol from molasses and biodiesel from non-edible oilseeds.
2. Second Generation (2G) "Advanced Biofuels"—Ethanol from Municipal Solid Waste (MSW).
3. Third Generation (3G)—Bio-CNG.

1.1 Global Picture of Biofuels

1.1.1 Environment and Technology

The environmental damage stemming from the burning of stubble (crop residue) is an area of concern due to its contribution to poor air quality worldwide. Usually, the burning of biomass in the northern hemisphere intensifies in late March to May, releasing many chemically and radiatively important trace gases and aerosols into the atmosphere. Satellite images revealed that large areas of Central Africa and South America, over the tropical Atlantic, and the Indian Ocean showed elevated levels of O_3, CO, and aerosols due to long-range transport of pollutants emitted from biomass burning. Jain et al. (2014) estimated that the largest amount of crop residues were burned in the Indian states of Uttar Pradesh (22.3 Mt) and Punjab (21.3 Mt), followed by Haryana (9.2 Mt) and Maharashtra (6.8 Mt) [10]. Uttar Pradesh was the highest contributor to the burning of sugarcane trash, followed by Karnataka. Oil seed residues were burned in Rajasthan and Gujarat while the burning of fiber crop residues was dominant in Gujarat (28.6 Mt), followed by West Bengal (24.4 Mt), Maharashtra, and Punjab. Of the different crop residues burned, the major contribution (93%) came from rice (43%), followed by wheat (21%), and then, sugarcane (19%).

1.1.2 Gaseous and Aerosol Emissions

It was further estimated that on-farm burning of 98.4 Mt of crop residues led to the emission of 8.57 Mt of CO, 141.15 Mt of CO_2, 0.037 Mt of SOx, 0.23 Mt of NOx, 0.12 Mt of NH3, 1.46 Mt NMVOC, and 0.65 Mt of NMHC, 1.21 Mt of particulate matter for the years 2008–09. CO_2 accounted for 91.6% of the total emissions. Burning of agricultural residues resulted in the conversion of 70, 7, and 0.66% of the carbon present in rice straw to CO_2, CO, and CH_4 emissions, respectively, while 20 and 2.1% of the nitrogen in straw were emitted as NOx and N_2O, respectively, and 17% of the sulfur in straw was emitted as SOx upon burning. This practice not only adds to pollution but also results in the loss of nutrients present in crop residues. Jain et al. (2014) suggested that the maximum loss of nutrients was due to sugarcane trash burning followed by rice and wheat straw [10].

1.1.3 Challenges and Opportunities

India provides a great investment opportunity for international players in the biofuel market. The profit-related challenges in the biofuel production process include a low tolerance to inhibitory compounds from pre-treatment, lack of useful co-products, low yield, lower productivity and concentration of lipids, and lack of harvesting and dewatering technologies [11, 12]. As methanol is derived from fossil resources, using ethanol as a blending agent rather than methanol, can reduce fossil fuel use. As hexose sugars are easily fermented to ethanol and pentose sugars can be used for diesel production, co-production of ethanol and biodiesel is feasible. Glycerol is obtained as a by-product from biodiesel production, and could be used as a feedstock for biogas and biodiesel production [13, 14]. In a lipid accumulation reactor, glycerol feedstock increases biodiesel production and substrate concentration in the hydrolysate. Thereby, decreasing energy usage in lipid accumulation and lowering water content in extraction. Figure 1 shows a schematic of 2G ethanol production from lignocellulosic biomass.

Futuristic scenarios for next-generation biorefineries should necessarily examine and introduce subsequent reforms based on the following points:

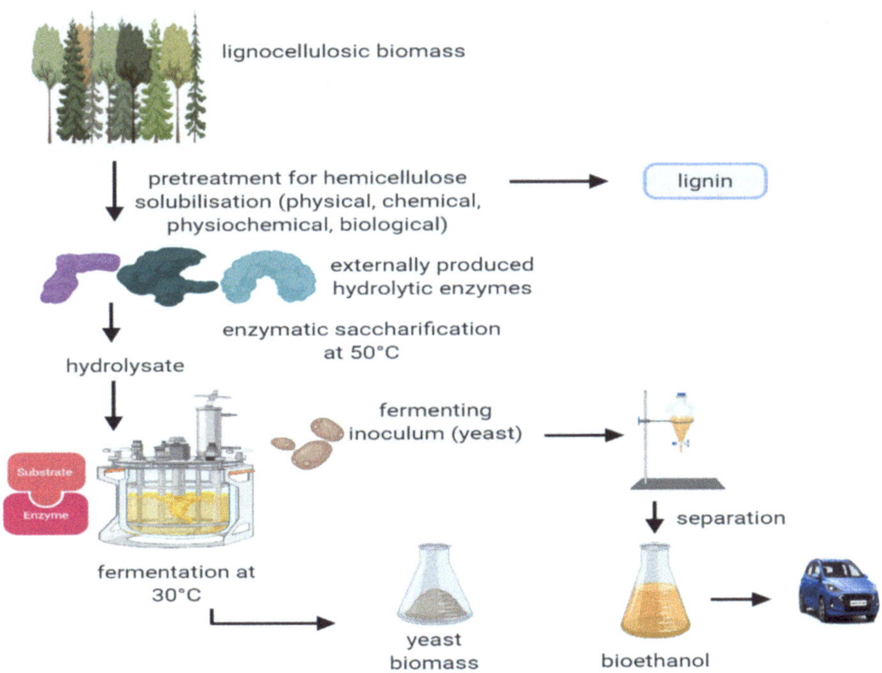

Fig. 1 Schematic of 2G ethanol production from lignocellulosic biomass

i. Use of agricultural waste for biofuel production.
ii. Cost estimation should be such that the production cost for biodiesel should be less than the purchase price.
iii. The availability of raw material feedstock as the use of edible feedstock or arable land for biofuel production is discouraged.
iv. The biofuel policy measures should not be biased toward the supply side and the promotion of flex fuels for biofuel demand generation is encouraged. Flexible fuel vehicles (FFVs) have an internal combustion engine and are compatible with gasoline and any blend of gasoline and ethanol up to 83%. Depending on geography and season, gasoline-ethanol blending can be up to 51–83% ethanol (E85 or flex fuel). As per IHS Markit records of 2017, the United States had more than 21 million FFVs.
v. Measures need to be taken to overcome the restrictions prevalent on the inter-state movement of ethanol and molasses.
vi. Creating awareness about tree-borne oilseeds (TBO) cultivation for biodiesel production.
vii. The impact of potential feedstock such as Used Cooking Oil (UCO) on biofuel production.

1.1.4 Current and Future Prospective of Biofuel Production in India

The national policy on biofuels was first introduced in 2009 to strengthen India's energy infrastructure and sustainability, which currently aims at achieving 20% blending of biofuels with fossil-based fuels by 2030. Although presently, India is highly dependent on exhaustible fuel resources due to its huge reliance on imported crude oil to quench its domestic consumption needs, the government has set ambitious targets to exploit bioenergy resources from a fairly under-explored sector. The goals include electrification of all census villages by 2019, 24×7 electricity and 175 GW of renewable energy capacity by 2022, energy emissions intensity reductions of 33–35% by 2030, and more than 40% of electricity production from non-fossil fuels by 2030. Moreover, the Ministry of New and Renewable Energy (MNRE) had strategized to contain the country's dependence on foreign gas and oil imports by elevating domestic production, endorsing biofuels and renewables, introducing efficient energy norms, refinery process improvement, and demand substitution.

The policy facilitates national energy security, green fuel adoption, encouragement of biofuel technological advancements, generation of employment, and combating climate change.

The government of India (GoI) is actively promoting 2G ethanol production via effective implementation of the Pradhan Mantri JI-VAN Yojna (Pradhan Mantri Jaiv Indhan-Vatavaran Anukool fasal awashesh Nivaran Yojana) scheme to facilitate integrated projects for bioethanol production using lignocellulosic biomass and other renewable feedstocks. The scheme was articulated to limit the current levels of fossil fuels usage by 10% by 2022 while targeting 20% blending in petrol and 5% in diesel by 2030. The initiative supports 12 commercial-scale 2G ethanol projects and 10 demonstration-scale projects based on lignocellulosic biomass. There is a provision for supporting 6 commercial and 5 demonstration projects in Phase I (2018–19 to 2022–23) and Phase II (2020–21 to 2023–24). The research and innovation aspect is based on the viability enhancement of 2G ethanol projects including enzyme cost reduction, lignin valorization, etc. [3]. Currently, oil marketing companies (OMCs) are blending petrol with ethanol up to 10% as per the current BIS (Bureau of Indian Standards) specifications. Existing first-generation routes for ethanol production, may not be sufficient for achieving the said targets, hence to maximize the production, other petrochemical routes need to be explored.

ONGC ENERGY CENTRE (OEC), set up by ONGC, an exploration and production company in India, has initiated Research, Development, and Demonstration (RD&D) in several areas of renewable energy beyond hydrocarbons to address green and clean energy alternatives, having biotechnology as one of the key RD&D areas for indigenous development of affordable technologies in relation to contemporary developments taking place in India and abroad.

In 2019, CSIR-IIP (Council for Scientific and Industrial Research-Indian Institute of Petroleum) produced bio-jet fuel for commercial trials of two aircraft including a SpiceJet flight from Dehradun to Delhi and an Indian Airforce (IAF) fighter jet plane. The SpiceJet's Bombardier Q-400 aircraft was powered by a blend of 75% air turbine fuel and 25% bio-jet fuel. IAF's formidable workhorse, the Russian-made AN-32 aircraft was formally fleet certified to fly on blended aviation fuel containing up to 10% of indigenous bio-jet fuel.

Bio-jet fuel production scale-up at suitable existing refineries like MRPL (Mangalore Refinery and Petrochemicals Limited), which is an ONGC subsidiary is underway. According to IIP, bio-jet fuel can be produced from used cooking oil, animal fat, waste dairy fat, and sewage sludge. The freezing point of such oil needs to be below −47 °C so it doesn't freeze at altitudes at which aircraft fly, should not be flammable on the ground during landing, must have the same density as ATF, should not choke the filters and have the desired calorific value.

After the adoption of the National Policy on Biofuels (NPB) in 2018, India has been constantly aiming at building a sustainable biofuel ecosystem. However, the Biodiesel Blending Program (BBP) has not been a great success so far due to its cost being higher than that of conventional diesel. The implemented Goods and Services Tax (GST) since July 2017 and gaps in supply chain management of feedstock, as most of the Indian states do not allow free inter-state movement of molasses, are a few of the factors responsible. Moreover, the policy is not economically viable as it does not promote biofuel import and export to other nations. In the following section, the possible sources of biofuel and factors influencing indigenous biofuel production have been discussed.

1.2 Sources of Biofuels and Factors Influencing Biofuel Production

1.2.1 Usage of Lignocellulosic Biomass

Lignin is a highly complex crosslinked aromatic polymer that is covalently linked to hemicellulose and stabilizes the mature cell walls. Efficient lignin degradation/biodegradation would result in higher fuel yield and enhanced economic outlook for lignocellulose utilization [15]. Enormous amounts of lignin are produced as by-products in the paper and pulp industry, therefore lignin products like lignosulfonates/sulfonated lignin used as plasticizers in making concrete and plasterboard are cheaply available. Soltanian et al. (2020) have explicitly reviewed the effects of pre-treatment methods on the exergetic aspects of lignocellulosic biofuels [16].

In a typical plant cell, the dry weight of the cell wall comprises approximately 35–50% cellulose, 20–35% hemicellulose, and 10–25% lignin [17]. Other studies report that woody biomass accounts for 40–50% of cellulose, with 20–30% of each lignin and hemicellulose. The cell walls of grasses (cereals residues and sugarcane bagasse) have very different cell wall composition to that of the woody biomass from trees indicating the need for specific processing technologies to achieve optimal outcomes. Lignin, with a higher calorific value (CV) than cellulose and hemicellulose, is capable of contributing 40–50% of the overall heating value or total available energy of the biomass. Thereby, chemical/biochemical conversion of lignin resources to produce fuel/energy is a favorable strategy. When the biomass undergoes combustion, it leaves behind ash due to various plant-specific chemicals in the biomass matrix including extractives like resins and phenolics and minerals like calcium, magnesium, and potassium [18]. Biodiesel from lignocellulosic biomass can be produced either through biochemical or thermochemical process routes, the latter having a lower potential for cost reduction due to the developed and optimized FT (Fischer–Tropsch) process as compared to the former route [19]. Further research emphasizing process design, optimal pre-treatment, and useful by-product formation along with improving the yield via strain identification and employing genetic engineering is likely to optimize the overall mass and energy balance for lignocellulosic ethanol or biodiesel production. According to a press release on October 2018 by the Press Information Bureau, GoI-MoPNG (Ministry of Petroleum and Natural Gas), survey studies suggest that lignocellulosic biomass surplus availability in India is in the range of 12–16 crore tons per annum which, if exploited commercially, could yield 2500–3000 crore liters of ethanol per annum. GoI is already coming up with schemes with a mandate for 2G bioethanol production from leftover agricultural residues, organic wastes, and by-products. Figure 2 depicts the energy path for the bioconversion of solid waste biomass to fuel, heat, and power.

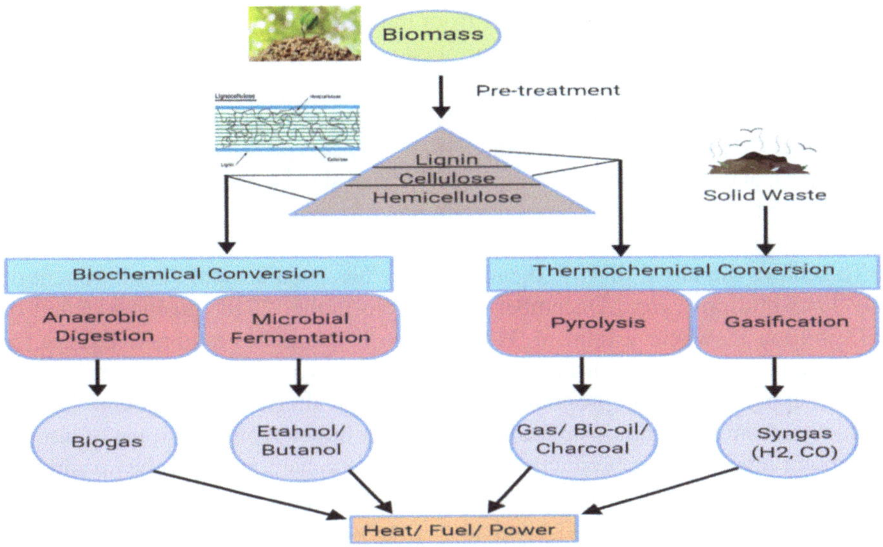

Fig. 2 Energy path for bioconversion of solid waste biomass to fuel, heat, and power

1.2.2 Effects on Biological Processes Due to Diversity of Lignocellulosic Biomass Along with Geological/climatic Conditions

Based on the soil-climate conditions, feedstock for ethanol varies largely from continent to continent and country to country. Ethanol production from sugarcane is possible via four different modes—directly from sugarcane juice, B-molasses, and C-molasses and from the bagasse. India produces ethanol largely from C-molasses. Per ton of crushed sugarcane produces sugar (107 kg) and molasses (46 kg), which in turn can yield 11 L of ethanol [20]. Ethanol could also be produced from the sugarcane bagasse, which is the residue after sugar extraction. Research has identified improved pre-treatments to allow biochemical conversion by enzyme hydrolysis and fermentation. Ethanol production from whole stalks without separation of the sugar is yet another option.

In a country like India, where geological or climatic conditions vary largely, there exists a diversity of lignocellulosic biomass feedstock for ethanol production. Every year, there is uncertainty in the amount of agricultural produce generated based on unpredictable climatic conditions that are changing abruptly due to global warming. For instance, there had been a bumper sugarcane crop production of 300 million tons (MT) during 2017–18 resulting in 32.1 MT of sugar and 13.8 MT of molasses as compared to 20.26 MT of sugar and 9 MT of molasses in 2016–17. Due to this, sugar mills witnessed huge losses due to sugar prices falling below the production cost. As a solution to this, the Indian government permitted ethanol production directly from sugarcane juice or B-molasses by amending the sugarcane control order of 1966 in July 2018. This initiative enabled the sugar mills to produce ethanol during periods of surplus sugarcane production apart from indirectly supporting the sugar prices. Before this, ethanol production was permitted only from C-molasses formed during the sugar processing of raw cane juice.

The ethanol purchase price for the OMCs from sugar mills for the EBP program was fixed at INR 43.70, INR 47.49, and INR 59.13 per L produced from C-molasses, B-molasses, and sugarcane juice, respectively. This step by the government ensured that sugar mills could divert the sugarcane juice for ethanol production during the years of surplus sugarcane production [20].

To achieve the target of 10% blending, there has been a supply commitment of only 1763 million L from December 2017 to November 2018 causing a deficit of 1373 million L, where a total of 3136 million L ethanol is required. Except for Uttar Pradesh and Maharashtra, all the Indian states have witnessed this shortfall in ethanol supply for the EBP program. To overcome this shortfall of indigenous biofuel production from lignocellulosic feedstock, another biotechnological route of biofuel production involving the production of lipids from sugars may also be considered and is discussed in the following section. Furthermore, the bagasse could also be used to produce biofuel by biochemical or thermochemical routes. Conversion of the bagasse to sugars could allow this to be used as feedstock for the production of either ethanol or lipids.

1.2.3 Microbial Biofuel Production

Production of microbial oil using oleaginous microbial fermentation is an area of great interest in the biofuel industry. Oleaginous organisms including bacteria, yeast, filamentous fungi, and algae have the ability to accumulate more than 20% of their dry weight as lipids. Presently, the utilization of alkanes as carbon sources by microorganisms is being aggressively researched [21]. For microbial oil production, various yeast strains like *Rhodotorula glutinis, Rhodosporidium toruloides, Cryptococcus albidus, Yarrowia lipolytica,* and *Lipomyces starkeyi* are the most promising candidates. Long ago, xylose fermentation was a challenge, but recently suitable yeasts have been discovered [22, 23] and also genetically engineered for xylose fermentation to ethanol [24]. To elevate the domestic energy supply security, biofuel production from lignocellulosic materials appears to be an explorable opportunity. Biodiesel being a bio-degradable and non-toxic renewable energy source has a high production cost as the only limitation [25]. Biochemical conversion of lignocellulosic biomass to sugars facilitates the options of either ethanol production or biodiesel production using oleaginous yeast.

Oleaginous microorganisms accumulate lipids due to cell starvation from nitrogen or other key nutrients such as iron, zinc, magnesium, and phosphorous [26]. Lipid is accumulated as triacylglycerol contributing to more than 20% of cell biomass. Different chemical routes employing biochemical and thermochemical processes are employed to convert lignocellulosic materials to fuels, worldwide. The biomass is hydrolyzed to produce sugar that could be used as feedstock for fuels and chemical production in biochemical conversion. Whereas, the biomass is gasified followed by catalytic conversion of syngas to different fuels and chemicals in the thermochemical process. Feedstocks like lignocellulosic biomass fractions such as straw and forest residues have advantages like easy availability [27], high growth rate, high oil content, the resemblance of TAG to plant oil [28], lower environmental impact [29], also do not require additional land for cultivation thereby preventing food-to-fuel debate [30]. Cellulose and hemicellulose are two different types of polysaccharides contained in lignocellulosic biomass, which upon hydrolysis yields glucose and a mixture of pentose and hexose sugars, respectively. Conjugation of systems biology techniques and metabolic engineering for the development of more efficient and tolerant microbial communities are the recent industrial approaches [31]. There is significant potential to engineer crops such as sugarcane to have a composition that is better suited as a substrate for biochemical conversion. This could involve developing crop varieties with lower or altered lignin composition or with a higher proportion of C-6 sugars in their cell walls. Another alternative route of biofuel production, single-cell oil, can also be explored and is briefly discussed in the next section.

1.2.4 Single Cell Oil

From an economic viewpoint, a microbial application using environmental lignocellulosic waste for conversion to high-value products like single-cell oil (SCO), is an explorable opportunity. SCO has industrial application in biodiesel production [32] with polyalcohols having applicability in pharma, food, and cosmetic industries. The cost of industrial production of microbial SCO depends on the composition of the culture media and the efficiency of the lipid extraction process. *Rhodotorula mucilaginosa* yeast isolated from Benjamin leaves is reported to produce 9.7 g/L SCO in a nitrogen-limiting medium with a dry biomass of 16.1 g/L [33]. Isolates can be evaluated by Sudan black B staining and lipids can be extracted by the Bligh and Dyer method. Additionally, 49.3 g/L of xylitol production was reported in a medium with 140 g/L of xylose, as a result of NADH accumulation that inhibits xylitol dehydrogenase enzyme under an anaerobic environment [33]. It was reported that higher xylitol production in *R. mucilaginosa* was obtained when the xylose concentration was increased from 80 to 140 g/L, leading to higher fermentation time while further increase of xylose concentration decreased the xylitol production. In the first stage of lipid production in oleaginous microorganisms, cell growth occurs and in the second stage, the extra-cellular carbon is accumulated in intra-cellular lipid form as triacylglycerol [34]. The obtained oil can then be trans-esterified using 80% sulfuric acid as a catalyst, based on the molecular weight of the oil. The presence of some inhibitors like furfural and formic acid can inhibit lignocellulosic hydrolysate, limiting yeast growth and decreasing lipid production as compared with the lipid yield in a xylose medium [35].

SCO serves as both a functional oil supplier and a feedstock for biodiesel production but high fermentation costs are a challenge in industrial application [14]. So cheap, hydrophilic, and hydrophobic lignocellulosic biomass sources are utilized as substrates for SCO production. López et al. (2016) reported the extraction of microalgal saponifiable lipid (SL) fractions consisting of polar lipids (glycolipids and phospholipids), from wet *Nannochloropsis gaditana* microalgal biomass using seven extraction systems, and the polar lipid contents of some fractions were reduced by low-temperature acetone crystallization [36]. The polarity of the first solvent in the extraction system affected the polar lipid content in the extracted lipids. Methanolysis, allowed transformation into fatty acid methyl esters (FAMEs, biodiesel), followed by lipases and *Rhizopus oryzae* catalysis in the tert-butanol medium.

Dalmas et al. (2019) studied SCO production from yeast *Rhodosporidium thoruloides* grown in a sugarcane medium [37]. The biomass was concentrated by sedimentation and the extraction of acid-hydrolyzed cells was done with recurrent extraction (RE) and wet extraction (WE) processes. The solvent extraction with hexane and chloroform in RE and WE, respectively, resulted in high lipid yields (m/m). Both extracts depicted similar density, iodine value, saponification number, FAME compositions, moisture, and acid value, other than the difference in peroxide content, which was lower in chloroform WE extract. The WE process consumed 13% of the energy used in the RE process and energy conservation analysis revealed that around 65% of the energy from molasses could be transformed into microbial lipids. The highly efficient methodology was intended for industrial-scale use because of the minimum energy requirement and desirable characteristics of the bio-oil achieved by WE [37]. So far, it has been shown that lipids play a very important role in the production of bio-oils via a biotechnological route, therefore, the optimal requirements for lipid production have been described in the next section.

1.2.5 Optimal Requirements for Lipid Production

The effect of different carbon sources such as glucose, sucrose, and fructose on lipid production in *Rhodotorula glutinis*, *Rhodosporidium toruloides,* and *Lipomyces starkeyi* was studied by Vijayakumar et al. [38]. *Rhodosporidium* sp., *Rhodotorula* sp., oleaginous yeast strains accumulate 70% of their biomass as intra-cellular lipids under nutrient-limiting conditions [39, 40]. It was reported that glucose produced higher biomass and lipid content, and *Rhodotorula glutinis* exhibited a higher lipid production of 2.43 g/L and biomass of 10.21 g/L on glucose. A few micro-algae have been reported for their ability to accumulate lipids utilizing CO_2 as a carbon source [41], but yeasts have comparatively faster growth rates, can accumulate lipids in separate lipid bodies, and utilize low-cost fermentation media.

Increasing glucose concentrations result in a decrease in dry biomass due to intolerance of the yeast cells to high glucose concentrations causing an intrinsic increase in the osmotic potential as reported by Syed et al. [42]. The highest lipid production was reported at a glucose concentration of 80 g/L by Leesing and Baojungharn [43] and this study revealed that an additional increase in glucose concentration inhibits lipid production. *R. mucilaginosa* showed the highest lipid production at pH 5, 25 °C in 72 h. The oleaginous yeast *Yarrowia lipolytica* has the tendency to degrade its storage lipid after 72 h, irrespective of the presence of an unconsummated carbon

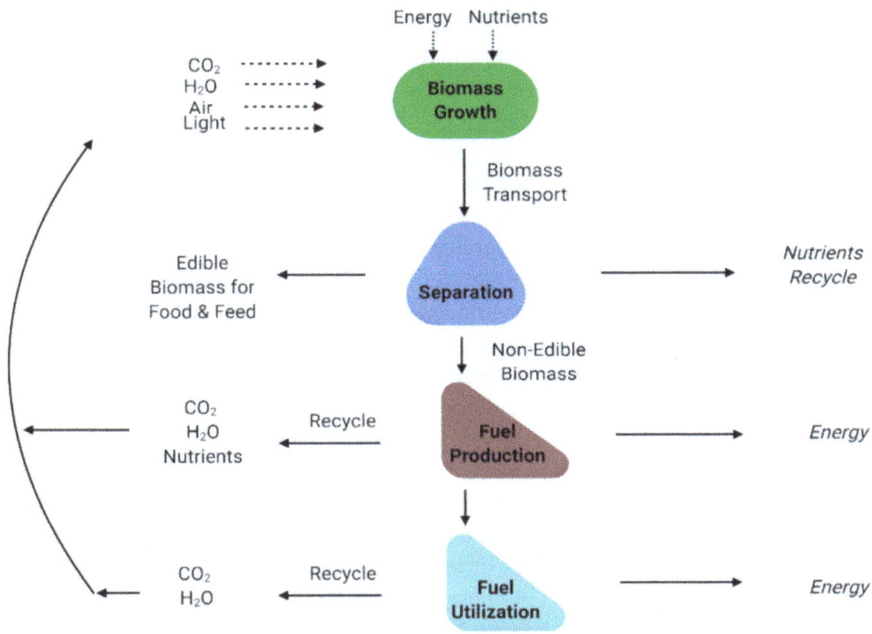

Fig. 3 Primary process requirements of the biofuel economy

source in the culture medium, causing lipid turnover [44, 45]. Figure 3 shows the primary process requirements of the biofuel economy that are usually implemented worldwide.

Leesing and Baojungharn (2011) reported *Torulaspora globosa* YU5/2 in an N_2-limiting medium supplemented with 80 g/L glucose, showed a volumetric lipid production rate of 0.520 g/L/d, specific yield of lipid as 0.472 lipid/g cells, lipid concentration as 4.16 g/L, and specific rate of lipid production as 0.156 g/L/d [46]. When sweet potato tubers hydrolysate was used as a carbon source, a maximum cell yield coefficient of 0.728, maximum specific yield of lipid of 0.237, and volumetric lipid production rate of 0.619 were reported via the Monod model by the group. From an economic point of view, parameters like cell yield based on a substrate, specific growth rate, volumetric substrate consumption rate, specific substrate consumption rate, product yield based on substrate, specific product yield and volumetric product formation rate in fermentation processes are of considerable interest for scaling up the process. The group also reported that in the case of *Trichosporon fermentans*, glucose in the medium is initially utilized for cell growth, increasing biomass and lipid yield but it slightly decreases when the N_2 source is exhausted, leading to high glucose consumption and a decrease in pH of the medium. For microbial lipid production, a fed-batch fermentation process is widely used [47].

The use of woody trees like willow, poplar, and eucalyptus as feedstock in energy production requires structural polysaccharides (lignin and cellulose) to be hydrolyzed to produce soluble sugars that can be used as nutritious substrates for oleaginous microorganisms. Wang et al. (2014) reported that the fatty acid composition of lipids was altered in *Lipomyces starkeyi* when a high carbon-to-nitrogen ratio was used in glucose media, facilitating C16:0 fatty acids and saturated C16:0 and C18:0 fatty acids accumulation [48]. A two-step dilute sulphuric acid hydrolysis was used. In the first step of hydrolysis, xylose (15.5 g/L) was the major sugar product and in the second step, glucose (21.9 g/L) was the major product. However, C18 fatty acids and C16:1, C18:1, and C18:2 unsaturated fatty acids were restricted. Using crude glycerol as the sole carbon source in fermenting *L. starkeyi*, the dry cell weight, lipid content, and lipid productivity were reported as 9.1 g/L, 46.2%, and 4.2 g/L, respectively, for glycerol, and using willow wood sawdust hydrolysate as carbon source, the dry cell weight, lipid content and lipid productivity were reported as 8.2 g/L, 42.7% and 3.5 g/L, respectively. Xylose is converted into xylulose via a two-step reduction and oxidation and is further converted into glucose via the pentose phosphate pathway [49]. The substrate acid hydrolysis results in the formation of degradation products like acetic acid, water-soluble lignin, furfural, and 5-hydroxy methyl furfural, which could inhibit the cell growth of *L. starkeyi.* The concentrated raw hydrolysate solution can be detoxified of cell growth inhibitors via active charcoal adsorption, also resulting in the loss of fermentable sugar concentration. The main by-product of methyl ester-based biodiesel production is crude glycerol, having efficient reuse being questionable. Figure 4 shows the types of biofuels that could be derived from lignocellulosic biomass by adopting various pathways. The availability of desired biomass and its supply chain management is a crucial parameter in the techno-economic feasibility of biofuel production, which has been briefly studied in the following section.

1.2.6 Biomass Availability, Suitability, and Supply Chain Management

The value chain of 1G ethanol production involves the use of feedstocks like crushed sugarcane, bagasse, molasses, sugarcane juice, sugar beet, sweet sorghum, corn, cassava, rotten potatoes, damaged food grains, and others. The value chain of 2G ethanol production involves the use of feedstocks like surplus biomass, sugarcane bagasse, agricultural waste, grasses, algae, forestry waste, twigs, sawdust, and others.

Various feedstocks, such as non-edible oilseeds used in cooking oil (UCO), animal tallow, algal feedstock, acid oil, etc., are invariably used for biodiesel production for blending with diesel for automobiles. In 2008, India became the world's leading producer of Jatropha with cultivation of 4,07,000 ha (47% of global cultivation area). Presently, even less than 0.5% biodiesel blending is being practiced, while the Biodiesel Blending Program (BBP) under NPB 2018 aims to increase the biodiesel blending percentage to 5% in diesel by 2030. With the advent of GST on July 1, 2017, the excise duty on biodiesel significantly increased from 6 to 18%, which severely impacted the nascent biofuel industry growth in India. Consequently, biodiesel is costlier by INR 10–12 per L than conventional diesel. The tax benefit for biodiesel producers was considerably reduced and no incentives were there for the end consumers to use biodiesel. For instance, the 18% tax resulted in the loss of

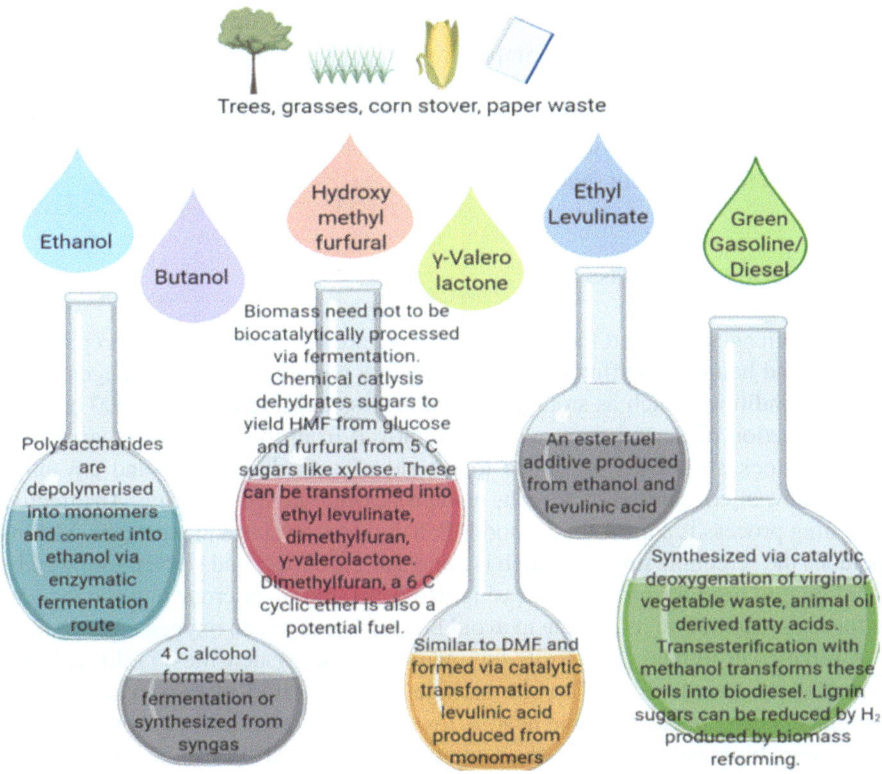

Fig. 4 Types of biofuels derived from lignocellulosic biomass

major biodiesel consumers such as Indian Railways, OMCs, and other companies that are running at underutilized capacity. An import duty of 25% on refined bleached deodorized palm stearin, which is a raw material for biodiesel production, with an additional 18% GST on biodiesel made it economically unviable as compared to diesel. Subsequently, OMCs were impacted by incurring under-recoveries followed by shut down of many biodiesel-producing companies. The bulk biodiesel feedstock palm stearin is usually imported; hence, there exists an insufficiency of domestic raw material. It is suggested that the large quantities of used cooking oil generated annually should be extensively diverted to be used as raw material for biodiesel production. Subsidies should be provided to lower the biodiesel retail price as compared to that of conventional diesel. Other feedstocks such as Jatropha and other tree-borne oilseeds (TBOs) like Pongamia, Calophyllum, and Madhuka should be exploited for biodiesel production by All India Coordinated Research Projects (AICRP). The TBO's productivity needs to be enhanced via better cultivation techniques on wasteland, and the development of improved plant varieties, thereby generating indigenous cheaper raw material. The in-house R&D units of all the biodiesel producer companies should be recognized by the Department of Scientific and Industrial Research

(DSIR) as suggested by Das [20]. Financial support in the form of incentives should be given to biodiesel machinery imports, flexi vehicles, and suppliers of UCO for biodiesel production. International import and export of biodiesel may be considered to promote competition. Additionally, to promote indigenous biofuel production, storage and transportation of biomass feedstock also requires attention that is briefly outlined in the subsequent sections.

1.2.7 Storage of Biomass

Major challenges in mixed feedstock processing are biomass feedstock preparation, storage, and handling. Different kinds of biomass require a specific arrangement of storage conditions. Such as storage of corn in silos holding up to 350,000 bushels, post inspection of dented corn, which is removed from the cob during harvesting. The feedstock shipments are made to get rid of cob, dust, chaff, and any other foreign unwanted materials before the next processing stage of steeping. An effective screening process is a must for a smooth transition to the subsequent stages. Milling includes the screening process which is done using a blower and screen, followed by grinding into flour for further processing. However, the sugar crops must be reduced to form a syrup before its storage to avoid biomass decay in an oxygen environment.

Unlike coal, biomass due to its low mass and energy densities, is difficult and expensive to transport. The on-site storage of unprepared biomass can result in biological decay releasing some heat. A low density of biomass necessitates a relatively large biomass storage space as compared to coal. MSW can be collected and delivered to a common location, depending on the municipality's size. Biomass pre-treatment reduces the cost of transportation, storage, and handling. Table 1 summarizes the effect of various biomass pre-treatment processes on the storage of biomass feedstock.

1.2.8 Economic Forecast of Biomass Production, Land Consideration, Transport, and Conversion Costs

Biomass rich in carbon content is considered to be useful for biofuel production. The NBP policy 2018 permits using sugarcane juice, sugar-containing materials, for example, sugar beet, sweet sorghum, starch-containing materials such as corn, cassava, and damaged food grains (wheat and broken rice), and rotten potatoes for ethanol production. Subject to the approval of the National Biofuel Coordination Committee (NBCC), surplus food grains may also be used for ethanol production. The policy proposes a higher purchase price for 2G biofuels vis-a-vis 1G biofuels along with a VGF scheme for the 2G ethanol biorefineries set up of INR 50,000 million in 6 years to promote extensive research and development across the value chain. Although NBP 2018 permits the feedstock import and 100% Foreign Direct Investment (FDI) in the biofuel area, however, the import and export of biofuels are prohibited. To support these projects financially, tax credits, prior depreciation on

Table 1 *Source* Modified from Shah and Gardner [50]

S. no	Biomass pre-treatment	Effect on storage
1.	Sizing (grinding, chipping, chunking, milling)	Should be done before transportation but storage of sized materials increases dry matter losses and microbiological activities leading to GHG (CH_4, N_2O) emissions
2.	Drying	Reduces dry matter losses, decomposition, self-ignition, and fungi developments during storage
3.	Bailing	Better for storage and transportation; higher density and lower moisture content
4.	Briquetting	Higher energy density, the possibility for more efficient transport and storage Possibility for utilization of coal infrastructure for storage, milling, and feeding; rate of combustion comparable with coal Easy moisture uptake leads to biological degradation and losses of structure. Require special storage conditions
5.	Pelletizing	Higher energy density leads to better transportation, storage, and grinding and reduced health risks. Demanding with regard to storage conditions
6.	Torrefaction	Improved hydrophobic nature—easy and safe storage, biological degradation almost impossible
7.	TOP process	Better volumetric energy density leading to better storage and cheaper transportation

plant expenditure, and large funding for 2G biorefineries set up are provided by the government. As per the policy, the biofuel storage, distribution, and marketing would be taken care of by the oil marketing companies (OMCs).

To encourage second-generation (2G) ethanol production, the government has incentivized the relevant schemes; however, the existing (1G) ethanol production that comes from the molasses route, contributes significantly under the EBP program. The NBP 2018 acknowledges that due to the competitive use of molasses in potable liquor and chemical industries, its availability has limitations.

To avoid the food-to-fuel conflict, the policy encourages the OMCs to sign Ethanol Purchase Agreements with 2G ethanol suppliers for 15 years of period to give assured market to private stakeholders support 2G ethanol initiatives and to prove the technology on a commercial scale.

The policy encourages the nascent Advanced Biofuel industry stakeholders by providing fiscal incentives in the form of tax credits, advance depreciation on plant expenditure, and differential pricing vis-a-vis 1G Ethanol, VGF, and others to set up 2G Ethanol Biorefineries. The EBP program could have been even more accelerated in case, these incentives had been extended to 1G ethanol production from molasses as India was a record holder for molasses production in 2018 and also being the world's second-largest sugarcane producer. Though the policy allows the use of surplus food grains as feedstock, such as corn, sugar beet, sweet sorghum, corn, cassava, and rotten potatoes, for ethanol production as per the decision of the National Biofuel Coordination Committee (NBCC) to promote the first generation fully developed technologies, this may likely to impact the food prices as it happened in 2006 in USA, where 20% of the total US corn production was used as feedstock for ethanol production, thereby causing a hike in global food prices [51]. Hence, there needs to be a check on this and also GHG emission-balance to limit the biofuel production from food crops grown on agricultural land, as restricted by the Energy Independence and Security Act (EISA) in the USA. It is worthwhile to mention that the USA being the world's largest ethanol producer utilizes 80% of the global corn produced as the major primary feedstock [52], and therefore, its corn-ethanol program affects the developing nations where corn is the staple food source [53].

1.2.9 Transportation

An export tax of INR 2 per L has been charged by the Uttar Pradesh state on all outbound non-potable ethanol, which is largely responsible for obstructing its free inter-state movement via artificially depressed prices [54] and negatively affecting the EBP program. The ethanol availability for the EBP program for 2017–18 has been 1400 million L [55] while the ethanol deficit is likely to be 2604 million L by 2022 [56].

Moreover, a few states have mandated "No Objection Certificates (NOCs)" from the respective state excise department for inter-state transport of ethanol. The sugar mills need to furnish details of the input tax credit (ITC) for the ethanol supplied to the OMCs for the EBP program, since the goods and services tax (GST) has been imposed from July 1, 2017. As a consequence, the OMCs deduct the ITC amount from the ethanol price payable to the respective supplier. The stand-alone ethanol suppliers and sugar mill suppliers buying molasses from other mills for ethanol production to utilize their excess capacity are severely impacted by this ITC deduction.

The applicable GST rate on molasses is 28% presently, but there is no GST on molasses used to manufacture ethanol for captive consumption. In India, around 20% of ethanol production is affected by the ITC directive of the OMCs.

The excessive sugarcane production during 2018 caused a sudden collapse of molasses cost to almost zero after March 2018 in Uttar Pradesh (UP) to such an extent that a few mills were left with no option other than storing molasses in open brick-lined pits in their premises to avoid paying an excise duty to the state government. This led to the quality degradation of unsold molasses and posing an environmental risk. While a few other mills bore the transportation cost of molasses to distilleries for processing into ethanol, the so-called "free molasses" (after a 12% reserved quota for liquor manufacturers) was prevented from being sold outside UP without permission from the excise commissioner; in the majority of the cases, there is a delay in excise permissions as reported by Das [20].

The Rangarajan committee reported that a few states also impose regulatory obstructions by imposing a quota system on the various end use of molasses and suggested the government to remove such restrictions on the sale of molasses by facilitating its free transport across the Indian states [57]. The committee also recommended that the molasses sale/purchase price should be determined by the market for creation of a unified national market leading to better price recovery of molasses aiding the EBP program.

Apart from biomass feedstock logistics and supply chain management, the stability of the produced biofuel and mass energy balance is another aspect to be considered while analyzing the techno-economic feasibility of the applied biotechnological process which is briefly discussed in the subsequent sections.

1.2.10 Biofuel Stability

McCormick and Westbrook [58] reported that if the B100 stability is higher than a 3 h induction time, blends from such fuel tend to be stable on the induction time. During storage, most biodiesel samples, regardless of initial induction time begin to oxidize immediately. If induction time <= 3 h limit, B100 surpasses the specification limits for either stability or acid value within 4 months (4 weeks on the D4625 test which is a standard test method for middle distillate fuel storage stability at 43 °C). Even B100 samples not showing a significant change in acidity or in deposit formation with induction times greater than 7 h will cross the oxidation degradation specification limit at only 4 months. To maintain the oxidative stability for B5 and B20 blends for up to 12 and 4 months, respectively, in storage, the 3 h B100 induction time limit is adequate. The stability of B5 and B20 blends is affected by B100 stability and could be controlled by synthetic antioxidants, irrespective of diesel fuel aromatics, sulfur level, or stability.

1.2.11 Mass-energy Balance

Process economics plays a vital role in biomass conversion process development, whether it is chemical/enzymatic, or both. One important factor is that supply chain logistics investment up to 15 miles must be kept to an absolute minimum, which consumes approximately 35% of the expected energy budget. Currently, organizations are directing their work patterns by acquiring the approach of green supply chain management (GSCM) to combat harmful environmental concerns [59]. Needless to say, a giant and far distant energy production plant source from the starting material would not serve the purpose. In such processes, right from the feedstock processing up to the biorefining, energy budget credits for by-products formation, blending, and the energy gained from the increase in octane points are also considered to generate a positive net energy value.

For lignocellulosic biofuels, factors like the feedstock diversity and availability, use of inedible crops, limited fertilizer usage, use of barren land, and conversion process technology that is evolving every day make such an analysis more difficult and less significant. Hence, considering only process energy balance in the context of biomass to biofuels conversion would be misleading [60]. Da Silva et al. (1978)

analyzed the energy budget for the production of ethanol from different agricultural biomass [61] and concluded that the industrial phase usually consumes about 60–75% of the entire energy which includes various energy forms used in agricultural and industrial processing, other than solar energy for photosynthesis. Different steps in the industrial stage, such as extraction, hydrolysis, alcohol fermentation, and distillation, use about 6.5 kg steam/L alcohol. In contrast to cellulose-to-ethanol technology, sugarcane-to-ethanol technology is way more direct in terms of energy balance evaluation. For example, in Zimbabwe, sugarcane-to-ethanol production has depicted the energy ratio as 1.52 if all the major products are considered and 1.15 if only ethanol is considered as the reaction output. Similarly, the reported net energy ratio for Brazil is 2.41 [61], and Louisiana, United States is 1.85 [62]. The low energy ratio accounts for (i) the energy-intensive agricultural phase, due to higher fertilizer demand and (ii) larger fossil fuel consumption in biomass processing. Seemingly, factors like climate change and soil quality, process conversion technology, agricultural technology, logistical issues, etc. affect the energy balance statistics.

Karlsson et al. (2016) reported that for energy and mass balance, using extraction methods without drying the yeast, increased the lipid yield and decreased the lipid accumulation residence time [63]. Vegetable oils such as rapeseed oil, palm oil, and soybean oil are potential sources for FAME production. The group reported process modeling using Aspen Plus™ and designing the heat exchanger network using Aspen Energy Analyzer™ Ethanol production from CS using NREL model as described in the process defined by Humbrid, formed basis for pre-treatment modeling, hydrolysis, biogas, wastewater treatment, and CHP processes. Pre-treatment of dilute acid-impregnated wheat straw is done using steam explosion (190 °C for 10 min). Lipid production using simultaneous saccharification and fermentation (SSF) reduces the inhibition of product during hydrolysis and increases the yield of sugar [12]. Relatively low acetic acid concentration can be utilized as a carbon source for *L. starkeyi*, whereas concentration beyond 3.9 g/L could be inhibitory [64]. *L. starkeyi* assimilates lignocellulosic hydrolysate compounds like glucose, xylose, and acetic acid and accumulates fatty acids with high degree of saturation favoring biodiesel production [65]. In many studies, inhibitor concentrations representative of biomass hydrolysate can be tolerated by oleaginous yeasts [66]. Usually, the residence time for yeast propagation is 2 days and for lipid accumulation is 5 days. The group reported the yeast propagation volume as 23.8 m^3/h and lipid accumulation volume as 88.4 m^3/h. As the yeast drying process is more energy consuming, hence wet yeast process is preferred for lipid extraction [67]. However, yeast drying decreases lipid losses during extraction, which helps to maintain the overall energy balance. After lipid extraction, the remaining yeast cell mass can be fed to the anaerobic digester to calculate the theoretical methane yield and efficiency [68]. High-pressure water scrubbing is reported to upgrade the biogas methane content to 96% (mole fraction) using Cozma's model [69]. Bio refinery's internal heat demand was met by lignin and other residues' combustion. Lipid accumulation and yeast propagation in agitation and aeration reactors are known to be the most energy-consuming process. Additionally, residence time and concentration of sugar in the hydrolysate affect the energy usage. Lignocellulosic hydrolyzing enzymes are constantly under improvement for better results and for the same conversion efficiency, the current enzyme products with significantly low energy usage are being introduced. However, using residual yeast cell mass for producing biogas, could serve as a nutrient source for yeast cultivation, thereby, decreasing nitrogen usage and improving the energy balance [26].

Studies suggest that the production cost of biofuel is inversely proportional to the capacity of the production facility, very similar to most of the petrochemical industries. However, the inflection point determines the minimum total cost corresponding to production cost for every scale increase in the plant capacity [70]. The empirical relationship between the above-discussed factors has been examined using production function variants and the computed F values at a 5% significance level [71]. Advanced fermentation technology employing recombinant microbes has the potential for reducing the overall cost by 25% or more [72]. Lignin being a phenolic polymer, cannot be fermented to sugar, rather the pathway could be diverted to the production of chemicals such as methyl aryl ethers, which are chemically compatible with gasoline as a high-octane enhancer. A competitive price for biofuels could be ensured by the adaptation of the above strategies. The various bioconversion processes for biofuel production reported by the researchers have been studied in the next section. Such processes can be optimized further for obtaining maximum biofuel production. Figures 6 and 7 illustrate the desirable attributes of "green gasoline" and "green diesel", respectively.

1.2.12 Bioconversion Process

Many pre-treatments have been developed to improve the enzymatic digestibility of lignocellulosic biomass. Mechanical processing to reduce particle size aids digestion. Hot water or steam treatment is used. Chemical treatments include various types of acid and alkali. More recent approaches have used ion liquids and eutectic solvents.

Ethylenediamine (EDA) pre-treatment had evolved as a novel method for bioethanol production. Qin et al. (2018) reported an optimized Simultaneous Saccharification and Co-Fermentation (SSCF) process making use of EDA-pretreated corn stover (CS) to study the constraint factors in the production of bioethanol [89]. The group achieved the highest ethanol concentration at 6% glucan loading: 12-h pre-hydrolysis, 34 °C, pH 5.4, and 5 g dry cell/L inoculum size. Upon increasing glucan loading from 6 to 9%, the subsequent increase in the concentration of ethanol from 33.8 to 48.0 g/L was reported due to Tween 20 and BSA which enhanced the enzymatic efficiency. Also, a decrease in ethanol yield by 7% was reported due to a reduction of glucan enzymatic conversion and xylose metabolism in the strain. Enzyme dosage was decreased by 40% (from 20 to 12 mg protein/g glucan), in case of solid-recycled SSCF process for achieving ethanol concentration (~40 g/L) as compared to the conventional SSCF. At high solid loading, glucose enzymatic yield and yeast viability are the main factors that affect ethanol production. It is studied that pre-treatment significantly affects the biorefinery costs and downstream processing [90]. Figure 5 shows the schematic of the biomass-based biorefinery recovery process that depicts the various steps involved in the methodology. Ethylenediamine (EDA) pre-treatment provides high enzymatic digestibility at high solid-to-liquid ratios for the "dry-to-dry" process and ambient pressures [91, 92] while preserving maximum biomass sugars. EDA boils at 117 °C to react with biomass and separate from it, simultaneously. Post EDA pre-treatment, maximum cellulose, and hemicellulose remain in water insoluble solid fraction. Lignin gets relocated to the biomass surface with ether and ester bonds in the lignin–hemicellulose complex getting cleaved, and partial crystal cellulose getting transformed into amorphous cellulose [91]. To prevent enzymatic hydrolysis inhibition by soluble oligosaccharides and phenolics derived from pretreated CS, washing is done prior to enzymatic hydrolysis to detoxify the process and reduce enzyme dosage [92, 93]. Glucose enzymatic yield of pretreated CS is estimated to be over 90% [92]. SSCF emerges to be a promising bioethanol refinery process. In this process, pretreated biomass is enzymatically hydrolyzed simultaneously with the co-fermentation of hexose and pentose sugars (glucose and xylose) carried out by recombinant *Saccharomyces cerevisiae*, thereby reducing cost of investment, energy saving, and increasing ethanol productivity by reduction in inhibition of end-products in comparison with Separate Hydrolysis and Co-Fermentation (SHCF) [94–96]. High ethanol concentration can be achieved by high solid loading which also reduces the distillation cost. However, ethanol yield tends to decrease due to high solid loading because of deficiency of cell viability, lack of enzyme activity, and other unknown factors that could be studied by mass balance data of SSCF to understand the corresponding tactics [96–98]. Qin et al. (2018) reported that solid loading affects the glucose enzymatic yield and xylose utilized by the strain [89]. It was studied that recycling of solid residues in SSCF, recycles the unhydrolyzed carbohydrates, yeast cells, and adsorbed enzymes. For enzyme recycling, alkaline-pretreated biomass having lower lignin content is considered to be more suitable than acid-pretreated biomass [99], thereby reducing enzyme dosage by 30% [100–102]. Table 2 provides an overview of the methods used in the bioconversion process for improving the fungal lignocellulolytic activity.

Recombinant xylose-fermenting strain *S. cerevisiae* SyBE005 was reported to produce higher ethanol concentration (33.7 g/L) at 96 h at 34 °C in the case of SSCF of EDA-pretreated CS [103]. 12-h pre-hydrolysis resulted in maximum (5 g dry cell/L), ethanol concentration at 96 h and 24-h pre-hydrolysis resulted in even higher ethanol concentration with initial 12-h SSCF. An increase in inoculum size, pH (5.4), and solid loading (9%) further tend to enhance the ethanol concentration. However, ethanol yield and glucan conversion were reported to decrease at 9% glucan loading as compared to 6 and 7.5% glucan loading. Enzymatic hydrolysis of glucan is inhibited by an increase in solid loading but the same is not the case for xylan and xylose metabolism is also inhibited. It was implied that additives do not have a significant effect on yeast metabolism and Tween 20 and BSA promote cellulose hydrolysis by enhancing the stability of cellulase and specific adsorption of cellulase to cellulose, in lignin's presence [103, 104].

Catalytic trans-esterification under mild reaction conditions can convert algal triglycerides into biodiesel via a chemical conversion route. The trans-esterification effects on selected fuel properties had been examined revealing that cetane number

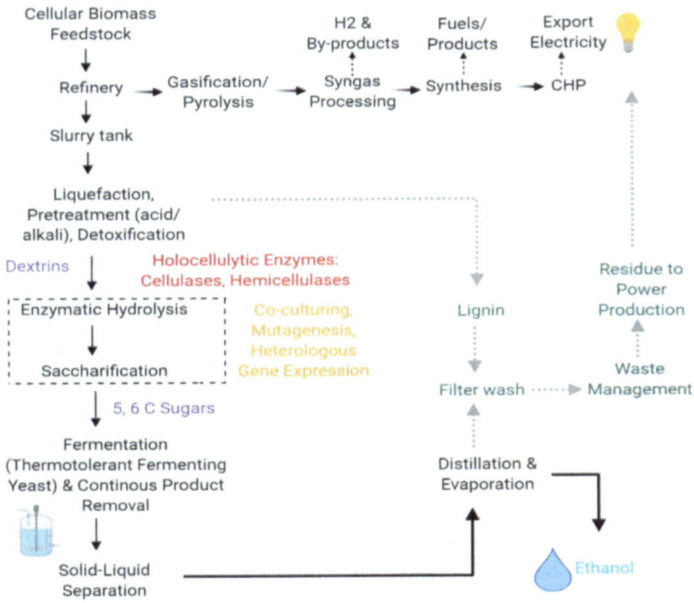

Fig. 5 Schematic of biomass-based biorefinery recovery process

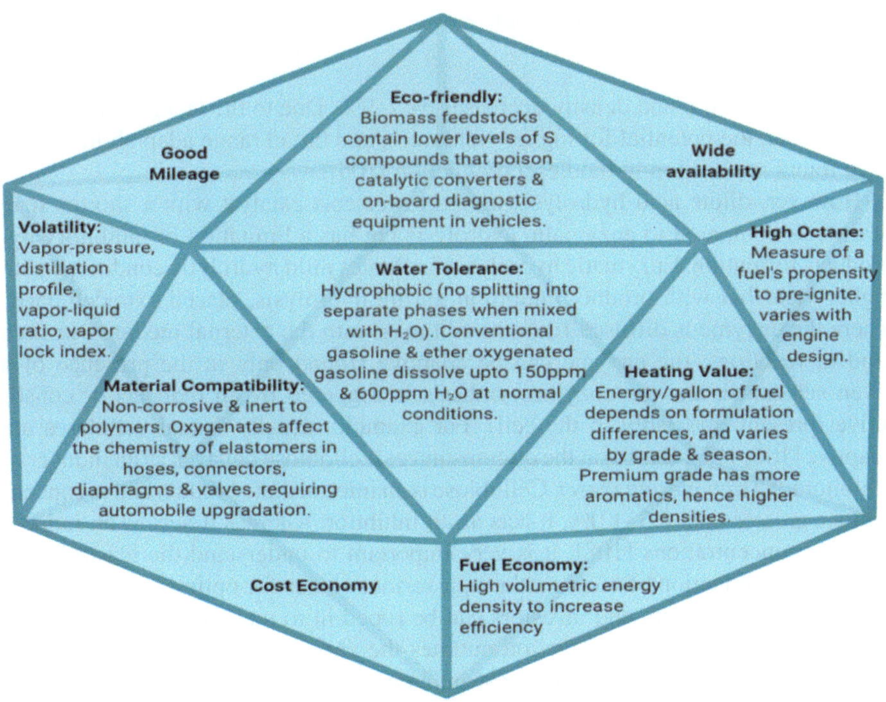

Fig. 6 Desirable attributes of "green gasoline"

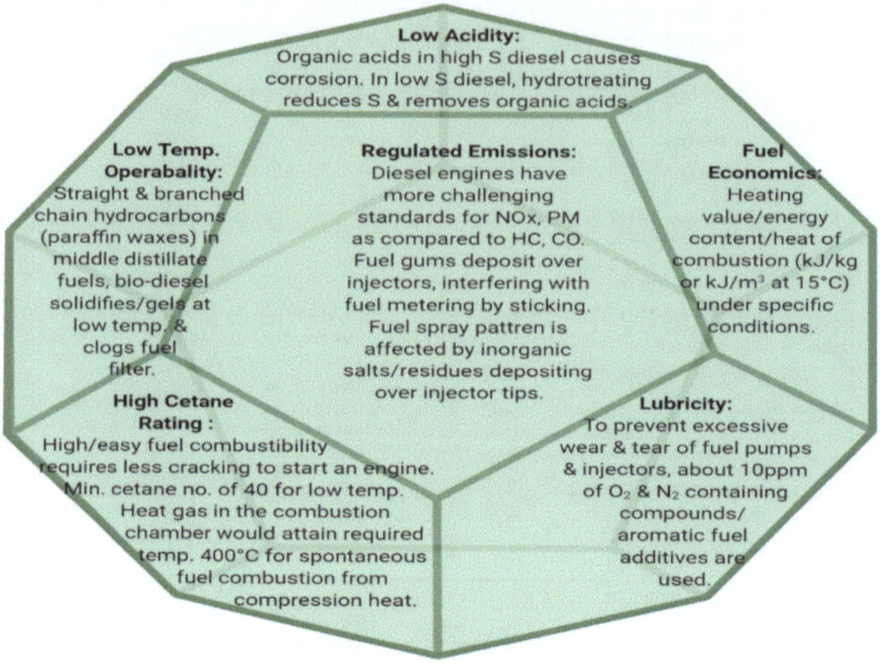

Fig. 7 Desirable attributes of "green diesel"

goes up by 60–70% and density decreases by 7–9%. Due to the fast growth rates of micro-algae, the potential for higher yield rates, and broad range adaptability, algae have found application in biodiesel production [105].

However, dilute acid hydrolysis uses a lower cost catalyst with a shorter span of time as compared to enzymatic hydrolysis but has a limitation of inhibitory by-products formation. Enzymatic hydrolysis facilitates mild hydrolytic conditions with higher yield but with product inhibition during hydrolysis, placed extracellularly, where the enzyme is diffused from the cell mass into the external medium. In such kind of reactions, the enzyme can be adaptive (found only in the presence of a given substance, and the enzyme synthesis is triggered by an inducer) or constitutive (present invariably in the cell). For example, usually fungal cellulases are adaptive [106, 107]. Based on the circumstances, cellobiose behaves as an inducer or inhibitor for *Trichoderma reesei*. Cellobiose is an intermediate product and at concentrations greater than 0.5–1.0%, it acts as an inhibitor, whereas it acts as an inducer at lower concentrations [108]. It is very important to understand the feasibility of large-scale production of biofuels through various routes for optimized production, so that the stakeholders and investors can be roped in to make India self-reliant in energy production. The next section outlines the various Indian organizations that are actively pursuing efforts to increase indigenous biofuel production.

Table 2 Some methods that have been used to improve the fungal lignocellulolytic activity or stability are mentioned in the following table

Methods	Fungal strain	Enzyme	Altered feature	Technique	Reference
Mutagenesis	*T. reesei* RUT-C30	Cellulases	Activity	UV treatment > 2 rounds of NTG treatment	Peterson and Nevalainen [73], Seidl et al. [74]
	T. reesei P9	Thermophilic endo-1,4-β-xylanase (XynII)	pH stability (alkalinity), thermostability	SDM (using PCR and synthetic oligonucleotide primers) (N97R + F93W + H144K)	Fenel et al. [75]
	T. atroviride TUB F-1724	β-glucosidase	Activity	2 rounds of NTG treatment > UV treatment	Kovács et al. [76]
	T. reesei (T2C:T28C mutant)	Endo-1,4-β-xylanases II (XynII)	Thermostability	PCR and synthetic oligonucleotide primers (Engineering a disulfide bridge at N-terminal region)	Fenel et al. [77]
	P. verruculosum 28 K mutants	Cellulases and xylanases	Activity	Four cycles of UV mutagenesis > two-stage fermentation process	Solov'eva et al. [78]
	T. reesei (Variants L218H, Q139R/N342T)	Endo-β-1,4-glucanase II	Catalytic efficiency, pH optimum	Saturation mutagenesis > random mutagenesis > DNA shuffling	Qin et al. [79]
Heterologous gene expression	*H. jecorina* (P201C)	Cel12 A	Thermostability	Mutation > heterologous expression in *A. niger*	Sandgren et al. [80]
	T. reesei RUT-C30	Thermostable β-glucosidase (*cel3a*) from thermophilic fungus *T. emersonii*	Activity	Heterologous gene expression using *T. reesei* *cbh1* promoter	Murray et al. [81]

(continued)

Table 2 (continued)

Methods	Fungal strain	Enzyme	Altered feature	Technique	Reference
	Acidothermus cellulolyticus and *T. reesei*	Endoglucanase and cellobiohydrolase	Bi-functional endo- and exo-acting cellulase	Chimeric protein, expressed in *T. reesei*	Bower et al. [82]
	T. reesei	Cellobiohydrolase (I & II)	Activity	Overexpression using *T. reesei cbh1* promoter	Miettinen-Oinonen et al. [83]
Mixed-culturing	*A. ellipticus* and *A. fumigatus*	β-glucosidase	Activity	Solid-state fermentation using pretreated sugarcane bagasse	Gupte and Madamwar [84]
	T. reesei D1-6 and *A. wentii* Pt 2804	Cellulases, xylanases	Activity	Mixed culture medium (M3) supplemented with trace metal and vitamin solutions	Panda et al. [85]
	T. reesei RUT-C30 and *A. phoenicis*	β-glucosidase	Activity	Shake flask culture	Duff [86]
	T. reesei RUT-C30 and *A. niger* LMA	β-glucosidase	Activity	Fed-batch fermentor on a cellulose-yeast extract medium	Ahamed and Vermette [87]
	T. reesei LM-UC4 and *A. phoenicis* QM329	Cellulases, hemicellulases	Activity	Solid-state fermentation using ammonia-treated bagasse	Dueñas et al. [88]

1.2.13 Feasibility for Large-scale Production Through Fermentative Routes

Praj industries have established a technology center, Praj Matrix, for the formulation of bio-products using useful bacteria, yeasts, fungi, enzymes, anti-microbials, and nutrition biomolecules that caters to bio-products requirements to over 12 countries. The expertise is based on the design and operation of continuous, fed-batch, and batch-type fermentation processes. Effytone is a developed bio-nutrient with complete yeast nutrition to achieve better fermentation and a high rate of yeast metabolism. Effytone provides vital micro-nutrients facilitating healthy yeast growth, therefore ensuring rapid conversion of sugars to alcohol, with higher yields and minimal bacterial contamination and by-product formation. Praj offers variants based on feedstock and application.

The DBT-ICT Lignocellulosic Ethanol Technology is based on innovative pretreatment strategies for low and high lignin biomass, radical intensification for the enzyme process, and intensification of fermentation steps. The technology produces separate enzyme amenable cellulose and hemicellulose fractions along with lignin, with more than 90% yield of sugars from biomass using high ethanol tolerant strains for C5 and C6 fermentation in high cell density column fermenters. The technology is sustainable as it has low CAPEX and a low plant footprint. The large-scale fermentation technology has targeted the following approaches: modification of growth phases of engineered microorganisms, media engineering, fermenter design, extractive fermentation, metabolomics, and metabolic flux modeling. DBT-ICT has developed a rapidly growing and high oil-yielding micro-algae and microorganisms for the production of bio-oil/bio-crude as feedstock for energy.

The efforts are also being made to develop sustainable strategies for the production of algae as a source of biofuel feedstock/biodiesel/value-added products. Growth and media engineering, consortia designing, CO_2 mitigation, and algal strain improvement by genetic modification/metabolic engineering/hybridization methods are a few approaches that are being considered. Photo bioreactor/Raceway pond designing for efficient scale-up of algae biotechnology is also underway.

The use of synthetic biology techniques for the synthesis of drop-in biofuels (butanol, biodiesel, biohydrocarbons), large-scale bioproduction of amino acids, and synthesis of furanics from biomass is also being studied. Construction of synthetic metabolic pathways for the production of high-value compounds, redirecting fluxes toward biofuel production by pathway analysis, vector construction for shuttle/transient/integrative cloning and expression of genes, recombinational methods for overexpression/silencing of genes, and alleviating product toxicity in biofuel production by directed evolution for tolerant strains are the approaches that are being followed.

The various biofuel products that are being produced by the Indian energy players/institutions and the comparison of biofuel versus petro-fuel along with future perspectives regarding the feasibility of bio-refineries have been briefly studied in the following sections.

1.3 Biofuel Products

1.3.1 Gaseous Biofuels

Biogas

The IndianOil (IOCL) has developed a two-stage anaerobic process for converting various types of organic wastes to biogas with the ability to produce 15–20% more methane than other contemporary technologies and higher gas yield. BPCL is considering setting up Compressed Biogas (CBG) plants using agricultural waste residues in Bina (Madhya Pradesh) and Bhandara (Maharashtra). Besides this, BPCL is facilitating investors to set up CBG plants based on Biomass/Municipal solid waste, etc. under the SATAT (Sustainable Alternative Toward Affordable Transportation) scheme of GoI and Biodiesel Plants based on Used Cooking Oil (UCO).

Praj RenGas technology has developed renewable gas plants with over 40 industrial installations having capacities of up to 1500 M^3/hr. Renewable Natural Gas is the biomethane produced by anaerobic digestion of agri-residues/organic wastes. The technology offers the provision of advanced gas cleaning (H_2S and CO_2 removal). The technology uses multi-feedstocks such as agri-residues (rice/wheat/corn straws, cane bagasse/beet pulp), sugar processing co-products (press-mud/spent-wash (vinasse)), composite or mixed industrial waste streams. The technology claims to enhance the value of press-mud (solid residue co-product from cane juice clarification) by 3–4 times. The multiproduct technological applications are:

1. Compressed biogas (CBG) for transport and grid applications.
2. Combined Heat and Biopower (CHP) applications.
3. Value adding organic soil conditioner co-products.

PRAJ has developed unique technologies for high gas yields and purity in renewable gas plants such as

1. Feedstock pre-treatment for preservation and high yields:
    - Hydrolysis and Preservation of TVS (total volatile solids) in convertible form.
    - Proprietary Microbial Consortium development.
2. High efficiency and fast biomethane production:
    - Proprietary rumen microbe consortium development.
    - High biogas and methane yields with CO_2 reduction.
3. Biogas cleaning to CBG:
    - Efficient low-cost chemical H_2S removal.
    - Efficient water based CO_2 removal.
    - Cleaning and drying through the Molsieve technique.

The benefits of CBG plant are:

1. Continuous operation.
2. Low cost of CBG production.
3. High co-product credit from organic soil conditioner.
4. Customized multi-application reactor designing.
5. CBG/Biopower/Industrial gas production.

The DBT-ICT technology aims to produce biogas from agri-waste and MSW for more than 5% LPG/CNG replacement. The CTL or fuel oil can be anaerobically digested to produce biogas, which can be purified to produce methane or bio-CNG. The DBT-ICT Rapid-Anaerobic Digestion Technology is reported to enhance the production of biogas and bio-CNG from lignocellulosic biomass or MSW, which is first converted into bio-crude oil via catalytic liquefaction within 2 h. This bio-crude oil undergoes acidogenesis/acetogenesis where organic acids are formed, which undergo methanogenesis in anaerobic digester for greater than 90% biogas production with no solid waste as a by-product. The process duration is 24 h. The technology has been scaled up to 1-ton MSW/day Demonstration Plant and 200-ton Bio-waste/day Commercial Scale Plant.

Bio-hydrogen

Verma et al. (2021) have reviewed the biohydrogen production from microbial fuel cells using electricigens [109]. DBT-TERI Centre of Excellence (CoE) in Advanced Biofuels and Bio-commodities, supported by the Department of Biotechnology (DBT), Govt of India is an initiative by The Energy and Resources Institute (TERI) in collaboration with IIT Guwahati, Indian Agricultural Research Institute, and Transtech Greenergy [110]. Efforts are being made to generate advanced biofuels (algal biodiesel, pyrolytic oil, Biohydrogen; and Bio-commodities; 2, 3 Butane Diol, acrolyn, high-value pigments, bioplastic, aqua feed, and animal feed) through the marine algal route and from selected third generation feedstock (algae and aquatic macrophytes). Apart from assessing their sustainability, focus is on the development of technologies for the production of value-added bio-commodities including bio-plastics, fish feed, nutraceuticals, and novel specialty and industrial platform chemicals, through a biorefinery approach.

The center is working toward the demonstration of large-scale algal bioreactor with the capacity of 100,000 L with the selected indigenous saline algal strain being adapted to grow in the coastal site. Large-scale cultivation of *Azolla microphylla* in domestic gray water is under the demonstration phase, aiding in the wastewater treatment and clean hydrogen production from generated value-added biomass feedstock (aquatic plant biomass and deoiled algae). Crude glycerol being the co-product of biodiesel production, is explored for the production of 2, 3 Butane Diol (by novel strain *Enterobacter tabaci*) and Acrolyn, the value-added industry platform chemicals. Hydrogen production by *Azolla* and water hyacinth biomass through the use of select C5 and C6 sugar co-fermenting microbe is also being studied.

Patel et al. (2015) (DBT-IOC Centre for Advanced Bio-Energy Research) have reported the biohydrogen production from a novel alkalophilic isolate *Clostridium* sp. IODB-O_3, isolated from sludge and identified by 16S rDNA gene analysis [111]. Maximum hydrogen yield of 2.54 ± 0.2 mol-H_2/mol-reducing sugar from low-cost wheat straw pre-hydrolysate (WSPH) and 2.61 ± 0.1 mol-H_2/mol-reducing sugar from pretreated wheat straw enzymatic-hydrolysate (WSEH) were obtained at 37 °C and pH 8.5. The cumulative H2 production (mL/L), 3680 ± 105 and 3270 ± 100, H_2 production rate (mL/L/h), 153 ± 5 and 136 ± 5, and specific H_2 production (mL/g/h), 511 ± 5 and 681 ± 10 with WSPH and WSEH were obtained, respectively. Biomass pre-treatment via steam explosion produces large quantities of WSPH that remains unutilized for bioethanol production due to the non-availability of efficient C5-fermenting microorganisms. This study aimed at reducing economic constraints on the overall cellulosic ethanol process and also establishing a sustainable biohydrogen production process.

Kumari and Das (2017) have reported a "biohythane" process for the maximization of gaseous energy recovery from organic wastes using a 10 m^3 biohydrogen pilot plant and, both mesophilic and thermophilic microorganisms [112]. The technology license agreement of the process has been transferred to M/s. Dhampur Sugar Mills Ltd. The research group has successfully conducted the pilot plant studies on the biomethanation process. High-yielding hydrogen producing bacterial strain *Enterobacter cloacae* IIT-BT 08 was involved during the development of a commercially competitive and environmentally benign bioprocess via fermentation. Das research group also worked on the development of a novel two-stage fermentation process (dark fermentation followed by photofermentation) for improving hydrogen productivity. The pilot plant studies for the commercial exploitation of the process using cane molasses, rice mill wastes, distillery effluent, sewage sludge, etc. to synchronize the wastewater bioremediation along with clean energy generation were also conducted.

Biohydrogen can be produced either directly or indirectly by photolysis of water by blue-green algae and micro-algae, oxidation of organic acids by photofermentation, and dark fermentation (using mesophilic or thermophilic bacteria). The two groups of microflora: acidogens and methanogens are mainly involved in the biogas generation process. The dark fermentation process is a promising method for biohydrogen production among other processes. The spent media of the dark fermentation process contains large volumes of volatile fatty acids (acetate, butyrate, propionate, etc.) which can be used as substrates for methanogens. Therefore, integration of the biohydrogen with biomethane processes such as "biohythane" developed by Das research group, could help in the improvement of gaseous energy recovery.

Kumari and Das (2017) have reported the improvement of biohydrogen production using acidogenic culture [112]. Yadav et al. (2021) have reported enhanced co-production of biohydrogen and biobutanol under optimized conditions with maximum biohydrogen production and yield of 9.25 L/L and 231 mL/g, respectively [113]. Maximum biobutanol concentration and yield of 16.75 g/L and 0.41 g/g, respectively with maximum substrate energy recovery of 78% was reported. Yadav et al. 2021 have studied the Acetone-butanol-ethanol (ABE) fermentation process for the development of a sustainable route for biohydrogen and biobutanol production using *C. saccharoperbutylacetonicum* [113]. Process temperature (32.65 °C), headspace (58.21% (v/v)) and butyric acid supplementation (9.16 g/L) led to maximum substrate energy recovery of 78% with biohydrogen and biobutanol production of 5.9 L/L and 16.75 g/L, respectively. Das research group has also studied the feasibility of power generation and hydrogen production in a microbial electrolysis cell using organic wastes like sewage sludge. The maximum hydrogen yield of 4.5 mmol H_2 g-1COD reduced was obtained at 1.0 V in Microbial Fuel Cell [112].

1.3.2 Biofuel versus Petro-fuel

The composition and property differences of biofuel vs. petro-fuel are tabulated in Tables 3, 4, and 5 below. This is dependent on how biofuel is produced. Direct liquefaction of biomass will produce liquid very different from petro-fuel. On the other hand, biofuel produced via FT process will be very similar to petro-fuel generated by the FT process. In general, biofuel tends to be less aromatic than petro-fuel, etc.

Other challenges are linked to poor or incompatible fuel quality including fuel feeding, co-milling, deposit formation, increased corrosion and erosion, and the need for new fly ash (slag) utilization strategies. Mixed feedstock with low biomass concentration has less of these issues, but as the low-quality biomass concentration elevates, significant issues are encountered affecting the plant economics. For instance, herbaceous biomass and coal are not as good a mixed feedstock as wood chips and coal. Besides, there are many process scale differences between biofuel generation and petro-fuel generation from coal or oil. The low density of biomass is a negative factor and leads to smaller scale operation than coal or oil refining processes. A workable strategy for biofuel production is a distributed strategy. It is to be noted that one can never produce biofuel at the scale of coal and oil refining

Table 3 Coal and biomass compositions of NUON/IGCC Plant

Feedstock type	Lower heating value (MJ/kg)	Feedstock input (1000 metric tons/Yr)	Feedstock (% by weight)	Feedstock (% by energy input)
Waste wood	15.4	130	22.5	14
Dried sewage sludge	8.2	40	7	2.3
Other biomass	10.2	10	0.5	0.7
Total biomass	13.6	185	30	17
Coal	29	400	70	83
Total feed	24.4	577	100	100

Source Ratafia-Brown et al. [114]

Table 4 Polk IGCC plant coal/coke and biomass combined feedstock

Feed composition (Wt%)	Coke + Coal	Biomass	Combined feed	Recycle solids to gasifier
C	82.88	49.18	82.02	66.26
H	4.50	5.78	4.71	0.29
N	1.85	0.24	1.81	0.95
S	2.99	0.06	3.13	2.31
O	3.53	39.42	3.92	0.00
Ash	4.25	5.32	4.4	30.19
Total	100.00	100.00	100.00	100.00
HHV, BTU/lb dry	14,491	8,419	14,470	9,698
% of original feed recycled	–	–	–	48.6
Lb carbon/Million BTU	57.2	58.41	57.21	68.32
Effective Lb carbon/ Million BTU[a]	57.2	1		56.71

Source: Ratafia-Brown et al. [114] and McDaniel [115]
[a]Accounts for biomass carbon recycle and carbon released during biomass preparation

Table 5 Polk power station biomass co-gasification test results

Property	Base fuel	Biomass	Total/Average
Fuel feed rate (Lb/Hr, As-Received)	164,840	1945	166,786
Moisture content (Wt%)	7.82%	46.8%	8.27%
Higher heating value (BTU/lb, As-Received)	13,322	4424	13,218
Higher heating value (MMBTU/Hr)	2196	8.6	2,205
Net power production (kW)	219,640	860	220,500

Source: Ratafia-Brown et al. [114] and McDaniel [115]

processes. Such parameters of course affect the economy of the process. Biomass is difficult to transport over long distances unlike coal, oil, or gas.

1.4 Future Directions and Role of Global Energy Players

1.4.1 Feasibility of Biorefinery

Even, the lack of required distillation capacity in large sugarcane producing states for processing of surplus molasses coupled with higher crop production like in 2018 and restriction controls on sale and transportation of molasses has affected the EBP program in a big way. Of late, Central Pollution Control Board (CPCB) has imposed stringent norms on the discharge of effluents, leading to a considerable restriction on the issuance of new licenses for the setting up of distilleries and sugar mills particularly in UP which is the largest sugarcane producing state in India.

Praj industries has set up the Bio-MobilityTM platform technology for renewable transportation fuel production across all modes of mobility (surface, air, and water). The company has set up 1 MLPA integrated biorefinery in Pune, India. The technology aims for "Rengas" (Renewable Natural Gas) production from agri-feedstock, such as biomass, press mud, etc., and Biodiesel named as "Ecodiezel" produced from Used Cooking Oil, Crude Palm Oil, Palm Fatty Acid Distillate, palm stearin, tallow, corn oil, and acid oil using enzymatic technology. Marine Biofuels are produced from lignin-based feedstock and Sustainable Aviation Fuel (SAF) from sugars, starch, and biomass from renewable sources.

The IndianOil Corporation Limited (IOCL) has developed a new and economical enzymatic pre-treatment process at the pilot level using agriculture waste and is now scaling it up in a demo plant at Panipat. IOCL has proposed a 2G ethanol project worth Rs. 909 Crore at Panipat Refinery. The project envisages the production of Ethanol from multi-feedstock lignocelluloses using indigenous technology. The company is also working on the project "Off-gas to Ethanol (3G)" technology worth Rs. 598 Crore at Panipat Refinery. This project envisages the conversion of Off-gas to Ethanol using LanzaTech Technology. IOCL R&D is also working on a bio-route to convert the CO_2 from refinery flue gases to ethanol by installing India's first-of-its-kind CO_2 to Ethanol production plant at Panipat. These projects help in meeting the Government of India's policy on ethanol blending to motor spirit.

The Bharat Petroleum Corporation Limited (BPCL) after commissioning its first 2G Ethanol Plant at Bargarh District in Odisha, plans to set up a second (2G) Ethanol Bio-Refinery in Odisha in early 2022. The first plant produced around 3 Crore Liters of Ethanol per annum using agricultural waste surplus (rice straw) available in the vicinity (https://www.bharatpetroleum.com/blog/Blog-Detail.aspx?id=21) [116].

Three demonstration/pilot-scale high-value-multiproduct biorefineries based on DBT-ICT technology were proposed to be operational in 2018. The techno-economics and life-cycle assessment of any technology is an important aspect to be considered in the commercialization of the technology, which is discussed in the next section.

1.4.2 Techno-economics and Life-cycle Assessment of Technology

It is suggested that the financial incentives laid out in the NPB 2018 for the promotion of 2G ethanol production should be extended to 1G ethanol also in a country like India where surplus sugar production takes place. This aspect has been vividly discussed in the subsequent sections of this paper. As suggested by an excellent review by Das (2020), the concept of differential pricing of 2G ethanol as compared to 1G ethanol should be abolished [20]. The provision for 12% reserved quota of molasses policy of states like UP whereby non-captive sugar mills are mandated to sell 12% of their molasses production to potable liquor companies needs re-consideration. The approval process for issuance of new licenses should be made more robust for set up of distilleries and sugar mills especially in states with large feedstock production, ensuring that norms on the discharge of distillery effluents are followed to speed up the process. As per the requirement, the import of low-grade ethanol may be considered and purchase of flexi vehicles should be encouraged by providing tax incentives for the same. Financial incentives based on the life-cycle assessment of sugar mills and distilleries (including non-captive sugar mill owners) involved in non-potable 1G ethanol production and molasses should be considered by the government.

For a commercially viable project, after accounting for feedstock, decreasing the investment cost via reduction in the auxiliary equipment is the second step in the process. The main challenge in the industry is location flexibility for availing more viable technical and economical projects.

Valdivia et al. (2016) classified the projects as Greenfield/Stand-alone, Colocation/Bolt-On, and Hybrid/Integrated based on the cost of production of a second-generation plant [117]. A Greenfield/Stand-alone plant has its own complete value chain management (core process, biomass handling, pre-treatment, enzymatic hydrolysis, fermentation, and distillation). Such a plant is the most expensive among all. The Colocation/Bolt-On project has pre-existing infrastructures and operations that can be shared with proximally related industries. If anything is required, that would be the construction of the process area, feedstock procurement value chain, and logistics that have to be completely in place. Construction of auxiliary operations units is very low. In a greenfield/stand-alone project, a 30% higher initial investment is there for a non-core area such as cogeneration. Without synergies with external utilities, extra costs are incurred with storage equipment and other logistical issues. In comparison with a stand-alone plant, a co-located plant requires less than 50% capital investment. The Hybrid/Integrated plant has the lowest cost among all the classifications. The entire value chain is integrated within a 1G facility, enabling synergies in feedstock supply and product logistics.

Economics of the technological process can be improved by enzyme cost reduction, improving catalyst activity, valorization of the lignin present in the raw material, enhancing the pre-treatment efficiency, or genetically improving the yeast microbial organism. There is a consensus that the final enzyme cost contribution should be stabilized at around \$0.4/gal [118].

In the entire process, the pre-treatment step accounts for an investment of 30–50% of the total equipment cost. Due to the intrinsic recalcitrant property of the lignocellulosic biomass, biomass to sugar yields are usually low resulting in high pre-treatment costs [119].

Common pre-treatment methods include (i) physical pre-treatment, involving mechanical processing and extrusion to reduce particle size but increase surface area, (ii) chemical pre-treatment in acidic, neutral, or basic conditions, (iii) physiochemical pre-treatment involving steam explosion (in the presence or absence of SO_2) and CO_2 explosion (using super critical CO_2 that produces carbonic acid), and (iv) biological pre-treatment using microorganisms like brown, white, and soft-rot fungi to pretreat biomass [120].

It is noteworthy to mention that not all pre-treatment methods are equally commercially viable and depend on the following factors:

I. How efficient the process is in opening of the cell wall to expose lignin to the surface in decentralized biomass processing? This can make the biomass durable for long-term storage without the addition of any external binding agents. Pre-treatments like alkaline hydrogen peroxide and ozonolysis that have the tendency to degrade lignin are less viable.
II. If the densified, pretreated biomass has dual applications such as fertilizer, soil amendments, animal feed, and biomass composites in addition to biorefinery feedstock, enhances its economic viability.
III. Quantity of water requirement to get rid of toxins from the pretreated biomass, especially in water-stressed regions.
IV. Less energy intensive (moderate temperature) with cheaper chemicals and easy scale-up to meet the biorefinery needs of dealing with larger volumes of feedstock, over 2000 tons per day.
V. Processes that use supercritical fluids (water and CO_2) operate at a very high pressure and require additional cost.

Let us briefly understand the various platforms for the production of biofuels that have been reported by researchers worldwide and have been practiced globally in the next section.

1.4.3 Thermochemical, Chemical, and Syngas Platforms for the Production of Biofuels

Lee and Shah (2012) have discussed the crop oils, biodiesel, and algae fuels in great detail, also examining the fast pyrolysis and gasification of biomass [121].

The production of 2G biofuels from lignocellulosic feedstocks can be achieved via two ways. First is biochemical—enzymes and other microorganisms convert cellulose and hemicellulose feedstocks components to sugars, prior to ethanol fermentation. Second is thermochemical—pyrolysis/gasification technologies produce a synthesis gas ($CO + H_2$) from which biofuels (synthetic diesel or aviation fuel), can be reformed. Additional pathways to produce 2G biofuels are still being researched in laboratories and pilot plants, and the R&D investment is significantly low.

Ganguli et al. (2018) reviewed that the biochemical and thermochemical routes have similar potential yields in terms of energy due to higher energy density by volume of synthetic diesel than ethanol [117]. However, the yields vary in terms of liters per ton of feedstock [118]. The final product in the two production routes is different. While the biochemical route produces ethanol only, the thermochemical route produces a range of longer-chain hydrocarbons including biofuels suitable for the aviation and marine sector. The preferred production technology out of the two depends on various factors.

The DBT-TERI Centre of Excellence in Advanced Biofuels and Bio-commodities is exploring the production of three forms of advanced biofuels—algal biodiesel, biohydrogen, and pyrolytic bio-oil—from next-generation feedstock through the following methods:

i. Biodiesel production from the wet algae harvested for lipid recovery.
ii. De-oiled algae biomass for pyrolytic bio-oil production, followed by upgradation to drop-in fuel forms through specific catalysts.

1.4.4 Biomass to Liquid Fuel via Fischer–Tropsch and Related Syntheses

Shah (2012) stated that biomass can be converted to biofuels in various ways [122]. For bioenergy generation from lignocellulosic biomass feedstock, thermochemical methods are usually used for the production of biosyngas and bioliquids. Gasification, reforming, pyrolysis, extraction, and liquefaction technologies convert biomass into biogas and bioliquids. Biomass gasification is largely practiced globally, followed by reforming and gas cleaning to produce clean syngas. The syngas can be converted to bioliquids via well-known FT and related syntheses. This versatile method can produce biofuels for heat, electricity, or transportation as well as for chemical feedstock.

The FT synthesis process can be made generally selective with a proper combination of operating conditions and catalysts. Global efforts are underway to prevent the reaction $2CO \rightarrow CO_2 + C$ and for improving the catalyst activity, selectivity, and stability. Apart from extensively studying multiple forms of catalysts such as iron, nickel, cobalt, and ruthenium, Exxon and other oil companies have studied both cobalt and ruthenium catalysts with different types of promoters. In recent years, MOF (metal–organic framework) has been tested to improve the selectivity of FT reactions. This work is being carried out at NETL (National Energy Technology Laboratory), Pittsburgh. The production cost of the purified synthesis gas can be as high as 70% of the total cost of the FT process.

Shah (2012) reported that another process isosynthesis used for the branched hydrocarbons production is crucial because (a) at optimized conditions, it generates saturated branched-chain aliphatic hydrocarbons with 4–8 C atoms, (b) in contrast to other synthesis reactions, it makes use of challenging reducible oxides as catalysts, and (c) it produces isobutane for high-octane gasoline [122]. The catalysts used for synthesis are thorium oxide, aluminum oxide, tungstic oxide, uranium oxide, and zinc oxide. The efficacious catalysts for isosynthesis are the tetravalent oxides, thoria, ceria, and zirconia. Of all, thoria or promoted thoria has proved to be the most promising one. Unlike catalysts of the iron group, thoria is not poisoned by sulfur compounds. Activity decline due to carbon deposition can be resolved by passing the air over the catalyst to generate CO_2 at the synthesis temperature.

Thoria catalyzes the formation of relatively high C number hydrocarbons from synthesis gas. Additionally, gasol hydrocarbons comprising of C_3–C_4 also contain huge volumes of isobutane. In contrast, the trivalent oxide alumina largely produced methane and carbon with low amounts of "gasol" and isobutane traces. The most active divalent oxide, zinc oxide does not produce liquid hydrocarbons but methane and alcohols. The optimum temperature for thoria catalyst has been reported between 375 and 475 °C. Alcohol production prevailed below 375 °C and methane and LPG were the main products in the pressure range of 300–600 atm. Below 300 atm, gas conversion was low; above 600 atm, methane and dimethyl ether were obtained. The alumina has proved to be the most potent promoter.

1.4.5 Processes for Syngas Production and Syngas to Liquid Conversion

The transformation process of fossils (crude oil, heavy oil, coal, shale oil, tar sand, bitumen, etc.) or biomass to liquid fuels (methanol, biogasoline, biodiesel, bio-jet fuel, etc.) primarily comprises pre-treatment, gasification, gas cleaning and conditioning, FT process, and final product separation and upgrading. Pre-treatment steps may include sizing (grinding, chipping, chunking, milling), drying, bailing, briquetting, washing/leaching, pelletizing, and torrefaction. The various merits and demerits of such processes are briefly reviewed by Shah [122]. For combustion and gasification, biomass particle size, water content, decay resistance, and inorganic impurities are important aspects to be considered. Therefore, leaching and torrefaction are important pre-treatment steps. Leaching process tends to get rid of salts and other inorganic impurities from the biomass and is carried out with hot water or steam.

Advantages of torrefied biomass for gasification and other applications [123]:

i. Torrefied product has lower moisture content and higher heating value.
ii. Storage and transport become easy.
iii. Hydrophobic; does not absorb moisture in storage and transportation.
iv. Resistant to fungal attack.
v. Easy combustion with less smoke and faster ignition.
vi. Significant biomass chemical energy conservation.
vii. Has heating value (11,000 Btu/lb) comparable to coal (12,000 Btu/lb).
viii. Electricity generation with a similar efficiency as from coal (35% fuel to electricity).
ix. Has grindability similar to that of coal.
x. Requires grinding energy 7.5–15 times less than that for untreated biomass for the same particle size.
xi. Mill capacity 2–6.5 times higher as compared to untreated biomass.
xii. Better gasifier fluidization properties.
xiii. Suitable for applications in heating, fuel, steel, and new materials manufacturing industries.

Numerous reactor technologies for the gasifier have also been tested. Earlier studies used fluidized bed (both homogeneous and turbulent), fixed bed, moving bed, and circulating fluidized bed (CFB) technologies. The CFB allowed a better control of the feedstock residence time in the gasifier. An entrained bed gasifier (or even a fast fluidized or transport bed) is also used and is recommended for new gasification processes. Extensive R&D in the design of gasifier internals is being done to enhance the energy efficiency, reducing generation of harmful gaseous impurities, and better handling the slag formation in the reactor. Several devices have been developed (e.g., Stamet) to effectively feed dry biomass in the gasifier. Coal gasification is operated at a higher temperature than biomass gasification. This is because biomass is more reactive and easier to thermally decompose. For biomass, a pressurized oxygen-blown entrained-flow gasifier is favorable. This technology has been identified as an optimum technology for biosyngas production as it has the merits of (a) high efficiency of biosyngas production [124], (b) fuel flexibility for all types of biomass, e.g., wood, straw, grassy materials, and coal with the possibility to use coal as backup fuel, and (c) suitability for scales of several hundreds to a few thousand megawatts [125]. Biomass requires considerable pre-treatment to facilitate stable feeding to the gasifier without excessive inert gas requirement. Torrefaction has efficiency up to 97% and torrefied biomass can be handled and fed to the gasifier with existing coal infrastructure. Entrained flow gasification for coal is a well established and commercial technology that can be readily adapted for biomass, especially for India.

The raw syngas from the gasifier requires considerable cleaning, conditioning, and treating before further catalytic synthesis. The major problem during gas cleaning is the presence of 7 g/m^3 of tars in the gas. Tars are condensable organic compounds with boiling points between 80 and 350 °C. When the system temperature goes below 350 °C, tars condense resulting in fouling and system failure. Inorganic biomass impurities are NH_3, HCl, and H_2S and minor quantities of COS, CS_2, HCN, and volatile metals. Also, 2 g/m^3 of solids are present in raw biosyngas. A typical gas conditioning lineup comprises gas cooling, water–gas shift (or other reforming), CO_2 removal, and impurities removal (e.g., H_2S, COS, HCN, volatile metals). Cooling can be done by a cooler or a water quench. While a cooler can utilize latent heat in the gas, for biomass there is an increased risk of fouling due to relatively high alkaline and chloride concentrations in comparison to coal. Fouling issues are prevented in a gasifier using water quench. Other than gas cooling, biosyngas conditioning and the treatment of fossil fuel-based syngas are similar. Biosyngas can be cleaned using the available technologies for commercial coal-to-liquid operations. No biomass specific impurities require a totally different gas-cleaning approach [125]. Significant efforts are being made to examine CBTL (mixture of coal and biomass to liquid) process.

An integrated process for biomass to biodiesel oil production via Fischer–Tropsch synthesis by replacing conventional gasification with a hydrogasification reactor to produce biosyngas from biomass is being developed globally [126]. The economics of an integrated BTL (biomass to liquid) process has been vividly discussed by Shah [122]. Bottlenecks in the BTL (Biomass to liquid fuels) technology are as follows:

Commercializing BTL plants to generate green fuel has several drawbacks pertaining to the use of biomass alone as a feedstock for either power generation or synthetic oil production. No consistent biomass supply is existing to operate a commercial-sized power or fuel plant. For fuel production, the economic viability scale requires 500,000 Bbl (barrels)/day of oil production. At the present time, the largest BTL (biomass-to-liquids) plant that may be built will most likely produce 50,000 Bbl/day [127]. Pure biomass gasification process is limited to a small scale, has high capital (fixed) cost, has lower thermal efficiency, and carries shutdown risk. All of these are alleviated by the use of coal. The mixture of coal-biomass provides a stable and reliable feed supply that generates less CO_2. Coal allows continuous plant operation when biomass feedstock is not easily available. Co-gasification reduces the cost associated with fossil fuel consumption although some types of biomasses can add significant cost to the overall fuel production. An effective CBTL (Coal-Biomass to Liquid Fuels) process will certainly be a way forward to reduce fossil imports.

1.4.6 Aqueous-phase Reforming (APR) and Bio-forming Process

Global technology processes to convert biomass to liquid fuels include (i) sugar fermentation to ethanol, (ii) pyrolysis or high-pressure liquefaction of biomass to bio-oils, (iii) gasification of biomass to syngas followed by Fischer–Tropsch (FT) synthesis to alkanes, (iv) AD of cellulosic waste to produce H_2 and CH_4, (v) Mobil process of conversion of carbohydrates to aromatic hydrocarbons and coke with Zeolite Socony Mobil (ZSM)-5 catalyst, and (vi) supercritical water extraction or biomass gasification to H_2 or liquid fuels. A selective process to generate H_2, syngas, alkanes, and monofunctional groups using low-pressure catalytic process in an aqueous environment is briefly reviewed by Shah [128]. Monofunctional groups can also be further upgraded to various liquid fuels (such as diesel, gasoline, and jet fuel) using a selective catalytic "bioforming process". The selective APR reactions occur in the temperature range of 215–265 °C. The upgrading of monofunctional groups requires a higher temperature. The nature and quality of products strongly depend on the feedstock and the nature of the catalyst, support, and promoter along with operating conditions (temperature, pressure, acidity of slurry and catalyst) and solid concentration in the feed slurry.

1.4.7 Aqueous-phase Reforming

Tao et al. (2012) and Alonso et al. (2010) showed that carbohydrates and polyols such as methanol, ethylene glycol, glycerol, and sorbitol can be efficiently converted to H_2 and CO_2 at 500 K by reforming under aqueous conditions [129, 130]. This is applicable to all carbohydrates in wastewater from biomass processing of cheese whey, beer brewery, sugar processing, and carbohydrate streams from agricultural products, such as corn, sugar beets, and hemicellulose from any biomass [131]. The produced H_2 can be used to hydrogenate components of lignocellulosic biomass to produce glycols and other polyols, thus enlarging the feedstock possibilities for APR. The H_2 can also be used to produce ammonia and fertilizer, an additive to gasification products to produce liquid fuels via FT synthesis, and a fuel source for polymer electrolyte membrane (PEM) fuel cells.

Primary feedstock examples are water-soluble oxygenated hydrocarbons such as sugars, sugar alcohols, saccharides, and other polyhydric alcohols. Secondary feedstocks such as sugar crops, grain crops, agricultural waste (cornstalks, straw, seed hulls, sugarcane leavings), bagasse, nutshells, manure (from cattle, poultry, and hogs), wood materials (wood or bark, sawdust, timber slash, mill scrap), municipal waste (waste paper, yard clippings), and energy crops (poplars, willows, alfalfa, switchgrass, prairie bluestem, corn, soybean) are first converted to primary feedstock.

While APR can only be used for selective feedstock without their pre-treatments, it is faster than AD process for generating H_2 from cellulosic waste. The operating conditions and the nature of the catalyst not only affect the selectivity between H_2 and alkanes but also the level of CO production. A low CO concentration in the product requires an ultra-shift operation where a water–gas shift reaction is favored. The lowest level of CO requires the lowest partial pressure of CO_2 and H_2 in the gas phase so that the forward water–gas reaction is thermodynamically favored. These conditions are achieved by operating the reactor at the saturation pressure for water (at the reaction temperature) and using a low feed concentration of oxygenates [132]. Very few studies have been done on the novel reactor design to carry out the APR process.

1. The alkanes reforming is only favorable at high temperatures. However, the reforming of oxygenated carbon (with a C/O ratio of 1:1) and the water–gas shift reaction is possible at low temperatures. This facilitates APR in the liquid phase.
2. The activation energy required to break up the C–C bond in oxygenated compounds is lower than that required in alkanes. Thus, H_2 and CO_2 from oxygenated compounds are produced in a single reactor. This can be accomplished in the liquid phase only for high-boiling compounds (glucose and sorbitol), whereas for low-boiling compounds (glycerol, ethylene glycol, and methanol), the reactions can occur in both gas and liquid phases.
3. The catalyst choice can affect the products. Pt, Pd, and Ni–Sn alloys show high selectivity for H_2, whereas Ni catalysts tend to make more alkanes. Ru and Rh catalysts also make alkanes with very little H_2. More acidic support favors alkanes production, whereas more basic/neutral support (alumina) favors H_2 production. The acidic aqueous solution similarly promotes alkanes production due to acid-catalyzed dehydrogenation reactions (followed by the hydrogenation of the metal). The basic aqueous solution favors H_2 production. The promoters such as Re add acidity to the catalyst, thereby reducing H_2 formation.
4. The feed type and its concentration affect the product distribution. Sorbitol gives higher selectivity for H_2 than glucose. Within polyols, H_2 selectivity decreases with an increase in the C number of the feed and an increase in feed concentration due to an increase in side reactions. APR of platform chemical glycerol has been very widely studied. APR can also be applied to the secondary feedstock as long as they are properly pretreated by hydrolysis (either acid or enzymatic) and/or hydrogenation depending on the feedstock nature.
5. Davda et al. (2005); Davda and Dumesic (2003) have outlined a number of different pathways that can occur in the APR reactor depending on the nature of the catalyst, its acidity, and acidity level of aqueous solution, temperature, and pressure to obtain the desired product distributions [132, 133]. Generally, a higher C number in the feed and more acidity on the catalyst or aqueous solution

favor C–O scission and more alkanes production. The reverse conditions promote C–C bond cleavages to form H_2 and CO_2. The latter compounds can undergo undesirable methanation and FT reactions to produce more alkanes. Some metals (Ru and Rh) favor C–O scission and form more alkanes. Pt and Pd favor C–C scission. More bifunctional catalysis can occur by the combination of metal, support, and solutions. In general, high H_2 selectivity requires high C–C scission, low rates of C–H scission, and low rates of methanation and F–T reactions. Low CO level can be obtained by operating the reactor with low partial pressures of H_2 and CO_2. The use of microchannel reactors has been found to have positive effects on the APR process [134, 135].

1.4.8 APR Versus Steam Reforming

The low-temperature APR to produce H_2 has significant advantages over conventional steam reforming that are mentioned below [136]:

1. The process occurs in one liquid phase eliminating the energy requirement to vaporize water and carbohydrates. Steam reforming requires high temperature and is accompanied by a phase change.
2. The raw materials for APR are non-inflammable and non-toxic allowing them to store and handle safely and conveniently.
3. The temperature and pressure used in APR favors the thermodynamics of water–gas shift reaction facilitating high conversion of CO in one reactor. This produces a nearly pure H_2 stream (with very low CO concentration).
4. The conventional PSA, cryogenic separation and membrane technologies are easily applicable to the product stream to separate CO_2 from H_2 since pressures used in APR vary from 15 to 50 atm. Steam reforming is carried out at low pressure, thus requiring product pressurization to carry out effective separation. Pure H_2 can be produced more easily by the APR process.
5. Low temperatures in APR minimize the decomposition reactions for carbohydrates and catalysts coking. Coking of the catalyst is an issue in conventional steam reforming.
6. APR can produce H_2 in a single reactor as opposed to a conventional steam reforming process that will generally require a multi-stage process.
7. Since APR produces H_2, syngas, lower alkanes, and monofunctional groups (which can be further processed to generate liquid fuels), the operating conditions and catalysts can be manipulated to obtain the desired selectivity among various products. More product possibilities are offered than the conventional steam reforming process. APR is a selective process that can only be used for a certain type of feedstock. However, steam reforming can be used for all carbonaceous feedstock.

1.4.9 Hydrothermal Processing of Biomass and Waste in Supercritical Water

This area has gained tremendous interest worldwide in the production of hydrogen and methane. Supercritical extraction of biomass is heavily pursued to generate hydrogen. Since biomass is more reactive than coal or oil, it is an ideal material for producing hydrogen. Currently, more than 96% of hydrogen is produced from the reforming of fossil fuels. Generation of hydrogen from a renewable fuel is an important greens technology.

As water behaves differently with an increase in temperature and pressure, the exploitation of its properties in subcritical regions for thermochemical transformations is getting prevalent in R&D. This facilitates organic reactions involving hydrocarbons to occur in the aqueous medium, as water becomes non-polar with an increase in temperature. Water can be used as a reaction medium for the transformation of raw fuels (coal, biomass, waste, and others) to produce gaseous, liquid, and solid synfuels. The properties of water at 370 °C are similar to those of acetone at 25 °C.

Geological transformation of biomass waste, human and animal remains buried underground results in the formation of fossil fuels with higher C content and lower hydrogen and oxygen content than the original biomass, taking millions of years. The youngest lignite coal contains H/C and O/C similar to that of biomass and peat.

Shah (2014) has profoundly illustrated that the aging and geological transformation of biomass can be accelerated by the hydrothermal conversion processes [136]. High thermal and pressure forces exerted during hydrothermal carbonization processes rapidly convert biomass into more coal (hydrochar), oil (bio-crude), or syngas similar to natural gas, due to strong thermochemical interactions between the biomass and water. Biomass produces cleaner hydrochar exhibiting properties like subbituminous coal, used as a raw material in the production of gaseous and liquid fuels.

Under subcritical conditions via hydrothermal liquefaction, biomass can be converted to oils, such as bio-crude, which can be processed further. Moreover, hydrothermal gasification produces methane, hydrogen, or syngas. India certainly needs to look into these processes, being integrated with advanced technologies.

While the affinity of coal for water is not as pronounced as that of biomass, the coal-water chemistry can be further explored: (1) the pre-treatment effect of coal by water in coal liquefaction, (2) coal liquefaction in high-temperature and high-pressure water, and (3) the use of coal-water slurry in combustion processes. Although low-rank lignite coal is amenable to liquefaction in high-temperature water, liquid product is poor quality and its upgradation is a daunting task. Coal-water slurry can be used as a combustion fuel in boilers, gas turbines, and diesel engines.

1.4.10 Methane Production by Anaerobic Co-digestion of Biomass and Waste

Both hydrothermal and anaerobic digestion (AD) processes can produce hydrogen and methane from biomass. Shah (2016) has deliberated methane production by anaerobic co-digestion of biomass and waste, as being the most energy-efficient biochemical process [137]. AD is a multistep biological process for the biodegradation of organic material by anaerobic microorganisms for conversion into biogas/methane and digestate in the absence of oxygen. The nature of the microbial commodity may vary widely. The process is applicable to animal waste, food and organic waste, various wastewater treatment plants, crop and crop residues (CRs), energy crops, industrial sludge, municipal solid waste (MSW), and municipal sewage sludge to produce H_2. Conversion of this waste into fuel/energy appears to be a more sustainable strategy in the long run. Hydrolysis/liquefaction, acidogenesis, acetogenesis, and methanogenesis are the four stages involved. These four phases occur in two parts.

The co-digestion process is reported to improve the efficiency of the AD to such an extent that in the recent decade, co-digestion has become a norm (and not an exception) in the AD process industries worldwide, thereby enhancing the process stability for cost-effectiveness. The advantage of using co-digestion is that the available free capacities of the existing digesters can be utilized merely by adding another substrate to a homogeneous mixture of two or more substances, at only a marginal additional cost.

Co-substrates such as maize or grass are added to livestock waste products (manure), which is an economically feasible method for enriching a digester with energy crops. However, such non-waste products that are explicitly grown for biogas production tend to increase the potential for atmospheric anthropogenic greenhouse gas (GHG) emissions as leakage of CH_4 during the process contributes to the GHG. GHG balance of essential parameters such as CH_4, N_2O, and CO_2 have been evaluated for co-digesters by various researchers.

It has been examined that a farm based on 100 cows can reduce approximately 118,000–138,500 kg CO_2 eq/year depending on the manure management system applied. This is equivalent to the CO_2 emission of an average passenger car traveling 1 million km [138–143]. This fugitive leakage can be as high as $5.1 \pm 1.8\%$ of total gas produced and should be reduced to the extent possible.

1.4.11 Advantages of Co-digestion

The equipment sharing is facilitated for waste that is versatile in nature, hence minimizing the cost burden due to the transportation of individual waste. An optimized choice of multiple substrates with specific proportions with the right nutrient balance can significantly boost methane production through synergistic reaction [141, 142].

Co-digestion of organic fraction of municipal solid waste (OFMSW) and vegetable oil (83/17 on dry weight) is reported to produce a higher specific methane yield of 686 N m3/tons volatile solid (VS). Moreover, OFMSW with animal fat (83/17 on dry weight) and cow manure (CM) with fruit and vegetable waste (FVW) (60/40 on weight) can generate very high methane yield of 490 and 450 N m3/tons VS, respectively [138–143]. Co-digestion can be synergistic, antagonistic, or neutral based on the co-substrates used.

1. Co-digestion produces superior quality fertilizer as the digestate and also improves the nutrient balance ratio of total organic carbon/nitrogen/phosphorous, which should be ideally 300:5:1. Moreover, maintaining a reasonable proportion of minerals (Na, K, Mg, Mn, and Fe) and balanced composition of trace metals, stabilizes the digestor performance.
2. Use of multiple substrates to avoid overloading and underloading of a single seasonal waste, maintaining a constant digestion rate throughout the year.
3. In agricultural digesters, co-digestion significantly improves the overall economics (payback time) of the plant. The provider pays a lower price at a farm-scale AD plant as compared to an incineration or composting facility (usually by a factor of 3 or 4). The farmer is credited with the income from gate/tipping fee.
4. Co-digestion lowers fossil fuel consumption and minimize carbon footprint by recycling challenging liquid "wastes."

You et al. [144] reported that anaerobic co-digestion of corn stover with swine manure (SM) can be largely improved with NaOH pre-treatment that reduced digestion time (18 to 12–13 days) and improved biogas yield. A C/N ratio of 25:1 in the substrate of corn stover and SM showed that lignin removal rates of 54.57–79.49% were achieved through the NaOH pre-treatment. The best yield was observed at 6% NaOH at 35 °C for 3 h. This was 34.59% higher than that from the untreated corn stover. The increase of methane yield was from 276 to 350 mL/g VS.

According to Braun and Wellinger [142], co-digestion is practiced in Germany since 2002, as well as in other parts of Europe, such as Austria, Switzerland, Italy, Denmark, Portugal, Sweden, France, Spain, and England. An American company CH_4 Biogas, uses Bigadan technology of a Danish corporation, for building co-digesters using manure and organic waste in the US. It recycles digestate to farmland and uses biogas production for power generation or upgraded it as a fuel. An economic analysis of an AD facility installed on a 700-cow dairy in northwest Washington state revealed that co-digestion with 16% organic wastes enhanced biogas production by two-folds and annual digester revenues by four-folds in comparison to a manure-only baseline [145–147].

It is recommended that India should certainly explore the application of such a process in small and medium-sized farm-scale digesters using mixed co-substrates with manure.

1.4.12 Co-gasification and Co-liquefaction of Coal-biomass Mixture

Co-gasification of multi-fuels, such as coal, oil, or biomass, has drawn tremendous interest globally during the last two decades as deliberated by Shah [148]. The gasification process involves partial oxidation of solid/liquid fuels under limited oxygen conditions. It is an exothermic reaction producing a mixture of synthetic fuel gas, which has significant heating value and can be used for heat/power generation in a downstream process, if required. Synthetic gas has numerous industrial applications in producing liquid fuels and chemicals by processes such as Fischer–Tropsch synthesis, oxysynthesis, and isosynthesis. While combustion/complete oxidation produces heat and flue gas, largely comprising of CO_2, water, and other gaseous impurities, the gasification process produces gaseous fuel, majorly containing methane, hydrogen, and carbon monoxide along with CO_2, water, and other impurities.

Large-scale coal gasification plants are associated with huge greenhouse gas (GHG) emissions and smaller scale plants are not economically viable. Large-scale biomass and waste gasification plants have demerits of unsustainable raw material supply, expensive storage and transportation, and energy-intensive feed preparation, whereas smaller scale plants are deprived of economic advantage. Therefore, to moderate the effects of both, co-gasification of a coal-biomass mixture is the way forward. In comparison to coal, biomass has a lower content of sulfur, N, and heavy metals, and if produced sustainably, it is also carbon neutral. Thus, the co-gasification of biomass and coal can considerably contribute to mitigating GHG and other emissions. Moreover, considering the fact that carbon capture and storage (CCS) can incur high energy penalties, which can range from 15 to 40% of the energy output, co-gasification of biomass-coal feedstock is relatively economical.

Biomass usually has a high content of hydrogen (H), enabling it to blend easily with low H-content coal. However, biomass as a gasification feedstock has the disadvantage of low energy density due to its high oxygen and moisture contents. All such limitations including seasonal biomass availability and higher tar release (due to excessive volatile release and low gasification temperature from biomass gasification) are compensated during coal-biomass co-conversion. Blending with higher-energy-content coal increases the temperature and enhances tar cracking, thereby boosting the efficacy of the overall system.

Syngas can be used for heat/power generation along with liquid fuels and chemicals needed for transportation and in industrial sectors. The existing infrastructure for natural gas can be used for synthetic gas as well. As a matter of fact, steam gasification of coal is not thermodynamically favorable, steam gasification of biomass is indeed a lucrative way for hydrogen production. Hence, co-gasification of coal and biomass in the presence of oxygen and steam may enhance the role of renewable energy input in the fossil-dominated industry. Such a process is likely to be more socially acceptable by pro-environment advocates and politicians. The large-scale Lurgi plant in the United Kingdom using coal and municipal solid waste is a successful example of its kind with minimal presorting or refuse-derived fuel (RDF) that has had significant pre-treatment such as mechanical screening, shredding, and

Table 6 Biomass feed preparation can be achieved by the following alternatives [50]

Drying/Grinding	Torrefaction/Grinding	Pyrolysis
Physical process	Physics/chemical process	Chemical process
Lowest energy	Medium energy	High energy
Dry feed/high energy for grinding	Dry feed/low energy for grinding	Slurry feed
Pelletization is possible	Pelletization is possible	Not applicable
Storage and transportation are problematic because of moisture absorption and fungal attack	Storage and transportation are easy	Slurry storage and transportation require an inert atmosphere
Grindability, fluidization properties, and combustibility for entrained-flow fluidization are superior	Grindability, fluidization properties, and combustibility for entrained-flow fluidization are superior	Pyrolysis can be considered as a first stage of two-stage gasification process

torrefaction. Table 6 shows the various ways by which biomass feedstock preparation can be done.

Coal-biomass gasification plant units can have a number of configurations for various degrees of flexibility, operability, and reliability depending on:

1. Overall scale of the plant and feedstock chemistry.
2. Desired biomass capacity and its geographical presence relative to the central gasification facility.
3. Biomass pre-treatment requirement.
4. Levels of integration of various equipment:

    a. At pre-treatment stage.
    b. Within the gasification stage.
    c. Within the product cleaning and separation stage.

Two global commercial plants employing co-gasification of coal and biomass mixtures are:

1. 250 MWe IGCC Plant, Nuon Power, Buggenum, B.V.-Willem-Alexander Centrale, Netherlands operational since 1993 and using Shell dry feed gasification technology. The gasifier is an oxygen-blown, continuous slagging, entrained-flow reactor.
2. 250 MWe IGCC Plant, Tampa Electric's Polk Power Station operational since 1996 and using slurry-fed GE (formerly Texaco) gasification technology with oxygen-blown, continuous entrained bed gasification reactor.

1.4.13 Co-liquefaction of Coal-biomass Mixture

Other than gasification, direct liquefaction (and extraction) of coal, biomass, and waste for producing synthetic liquids is an indeed valuable and commercially viable thermochemical transformation technology. In contrast to combustion, gasification, and pyrolysis, liquefaction occurs with multiple fuels, solid (coal, biomass, or waste), and liquid (oil or water), generally in the presence of a donor solvent. Liquid products are produced depending on the quality/nature of feedstock, operating conditions (temperature, pressure, reaction time), and the catalyst. Donor solvent quality influences the quantity and quality of products as well as the economic feasibility of the process. Synergy among mixture components affects the co-liquefaction or co-processing of coal, heavy oil, and other substances (polymers, rubber tires, oil shale, and other wastes) to produce liquid fuel.

Coal can be liquefied in the presence of oil or another donor solvent. An economically competitive method for liquid oil production has not been devised, yet. However, other laboratory-scale approaches have been investigated to produce oil from coal, biomass, and waste using heterogeneous fuels, with successful pilot-scale operations. A few of them are tested on a commercial scale also. Needless to mention, how attractive the lab-scale process appears for biofuel production, the success of a certain technology depends on the commercial and economical viabilities of the developed processes, which is briefly discussed below.

1.5 Commercial and Economical Viabilities of Processes

1.5.1 Cost Comparison of Biochemical and Thermochemical Routes

Wright and Brown (2007) studied the comparison between advanced biofuel production methods and grain-based ethanol production. It was reported that the total costs per gallon would be 44% higher for the biochemical cellulosic ethanol process as compared to the grain-based process and 48% higher for the thermochemical cellulosic biofuel process as compared to grain-based ethanol production [149]. The total cost considers the feedstocks, operation and management, credits, and capital charges.

Rismiller and Tyner (2009) estimated the net present value (pre- and post-tax) to compare the profitability of the three biofuel production types—grain-based ethanol, cellulosic biochemical ethanol, and cellulosic thermochemical biofuels [150]. Given input and output prices and other technical and financial assumptions, it was deliberated that all three production types are unprofitable in the absence of subsidies and mandates. In the presence of subsidies, cellulosic biofuels were predicted to have higher net present values than grain-based ethanol. It was also reported that when compared on an energy equivalent basis, the estimated cost of producing grain ethanol was $114/bbl crude oil equivalent, biochemical ethanol $141/bbl, and thermochemical gasoline $108/bbl.

Although there is no favored choice of technology, the ethanol produced via the biochemical route has the greater potential for cost reductions in contrast to liquid fuels produced by the thermochemical route. This is because the major technology for biomass-to-liquid plants (based on Fischer–Tropsch conversion) is mature and the process development usually involves assembling various proven components together [118].

The status of the various actively pursued Indian technologies for 2G biofuel production is mentioned below.

1.5.2 Indian Technologies for 2G Biofuel Production

Praj Industries being India's leading Ethanol plant manufacturers and providers of 2G bio-based solutions is actively involved in developing several state-of-the-art technologies for converting agricultural waste into biofuels and biochemicals. The 2G ethanol produced is supplied to major Indian oil companies like IOC, BPCL, HPCL, and MRPL. Besides ethanol, agri-waste feedstock can also be used as a raw material for compressed biogas (CBG) production. This technology has solved the problem of disposal of Agri residue, by controlling the "burning" issue to some extent. 2G feedstocks include agri-residues like rice and wheat straw, cane trash, corn cobs and stover, cotton stalk, bagasse, empty fruit bunches (EFB), etc.

The major challenges in the mass commercialization of 2G technology in India include hurdles at various levels of statutory clearances and permissions such as environmental and land acquisition, making the set up of bioethanol plant, a time-consuming process. Other measures like funding and mechanization of farming in India are required to facilitate quick and easy leftover straw collection post-grain harvesting.

To develop a robust ecosystem for the smooth functioning of supply chain feedstock management, farmer awareness is required about wrapping straw in bales or man-made bundles to be processed further for biofuel production. Usage of modern farm equipment should be promoted to extract the straws during harvest and shaping them into round or square bales to facilitate easy storage and transport. Farmer co-operative societies, rural entrepreneurs and women farmers need to be supported financially. The State Government is required to fix the minimum support price (MSP) of the straw for farmer assurance of a long-term buy-back agreement for collected straw.

Praj's 2G cellulosic ethanol pilot plant facility called "Enfinity" has been operational since 2009. More than 450 MT of biomass such as corn cob, cane bagasse, corn stover, empty fruit bunches (EFB), rice straw, etc. have been tested. The technology has been scaled up to 1 Million L per annum capacity through rigorous testing. PRAJ's end-to-end 2G "Smart Biorefinery" aims to process multiple feedstocks (lignocellulosic biomass/MSW) to produce multiple products like bioethanol, biobutanol, biochemicals, power, Bio-CNG, liquid CO_2, bio-fertilizer, etc. The technology claims to have optimized capex and opex with lower water and energy consumption and higher fuel-grade ethanol yields in compliance with Indian specifications. The technology meets statutory compliance with the Zero liquid discharge (ZLD) system and also produces premium grade extra neutral alcohol for beverage application as well as industrial grade ethanol (for perfumery, pharma, solvent, and rectified spirit).

Praj Industries also helps existing ethanol plants by way of modernization and retrofit services via Greenfield plants.

The DBT-ICT (Department of Biotechnology—Institute of Chemical Engineering) Centre has successfully developed a continuous processing 2G Ethanol technology from agricultural residue feedstock (bagasse, rice straw, wheat straw, bamboo, cotton stalk, corn stover, wood chips, etc.). The advantages over its competing technologies are the conversion of biomass feed to alcohol within 24 h, in contrast to 3–5 days from other developed technologies, unique plant design, advanced reactor design, and separation technologies with rapid reaction regime operations. The technology was proven at a 1-ton/day scale during Phase 1 at the India Glycols Ltd., Kashipur site. This was later scaled up to a demonstration-scale plant of 750,000 L annual alcohol capacity and Biotechnology Industry Research Assistance Council (BIRAC) supported pre-commercial scale at 10 tons biomass per day. This was also the world's first multi-feedstock continuous 2G ethanol plant. Larsen and Toubro Hydrocarbon Engineering being the engineering partner for the technology commercialization, the DBT-ICT Centre, along with its industrial partners, Bharat Petroleum Corporation and Hindustan Petroleum Corporation, are working to design and scale up the technology to 400 ton/day scale. The zero discharge (no vinasse) technology claims to have low OPEX due to the reuse of recycled chemicals and enzymes.

Lali (2018) has reported that the DBT-ICT technology converts complex carbon sources (solid, liquid, and gas) into simple sugars, lignin monomers, and CHO molecules for the production of fuels like ethanol, methane, methanol, "green diesel" and hydrogen, along with chemicals such as green PET, PLA, acrylate, butadiene, solvents, organic acids, and other value-added chemicals [151]. The technology basically uses three platforms: lignocellulosic sugars, catalytically thermo-liquified CTL oil, and lignin hydrolysate. The "action heroes" such as *E. coli, S. cerevisiae, Corynebacteria, Clostridia sp., Yarrowia sp., Pseudomonas sp., Cyanobacteria, Algae* and Methanotrophs have been reported for the conversion of terrestrial/marine/aquatic biomass into proteins, lipids, alcohol, organic acids, fatty acids, amino acids, terpenoids and hydrocarbons.

The other DBT-ICT technology aims for the production of Catalytic Thermo-Liquefaction (CTL) oil from wet/dry biomass at 150 °C, 15 Bar Pressure with oil yield greater than 90% and about 100% catalyst -efficiency due to its reusability. The calorific value of fuel/CTL oil was reported as 22–25 MJ/Kg with elemental composition as Carbon: 5 5%, Hydrogen: 6%, Nitrogen: 0.8% Sulfur: 0.22% Oxygen: 38%, and 0.5–1% of ash content. This CTL or fuel oil can be either directly used as furnace oil shipping fuel or boiler oil to produce power, or it can be catalytically upgraded via dehydration or hydrotreating followed by cracking or distillation to produce gasoline or diesel.

Three demonstration/pilot-scale high-value-multiproduct biorefineries based on DBT-ICT technology were proposed to be operational in 2018.

1.5.3 Role of ONGC Energy Centre

ONGC Energy Centre (OEC) via its in-house R&D projects at KDMIPE Dehradun, is working toward studying the feasibility of microbial oil production from oleaginous yeast and fungi. Bioconversion of unrecovered oil from matured oilfields for methane generation by using anaerobic methanogens is also being explored. The collaborative efforts of OEC, TERI, and ARI Pune are also underway for the development and demonstration of a bioconversion process for methane generation from subsurface lignite deposits. A microbial process for the enhancement of CBM (coal bed methane) in coal seams at Jharia wells has been developed in collaboration with TERI via *in-situ* biostimulation. Field tests of the developed technology were carried out successfully in a few depleted ONGC CBM wells, where a two to fourfold increase in gas production was observed after the demonstration of the developed processes. The tested wells are still in the production stage. The innovation has benefited the environment by extending the lifespan of coal seam gas wells. Post successful demonstration of these processes, a DST (Department of Science and Technology) sponsored project, in collaboration with TERI, has been undertaken for developing microbial processes for non-producing CBM wells with poor to marginal prospects and where no biogenic methane has been reported so far. OEC via its collaboration with ICT Mumbai is jointly working for the development of a process and prototype reactor using a multi-functional catalyst system for conversion of carbon dioxide to gaseous hydrocarbons such as methane.

Apart from strengthening the technological front of the Indian energy players, reforms in the existing national biofuel policy considering various perspectives outlined in this review need to be considered and are briefly discussed in the next section.

1.6 Policy Considerations

1.6.1 Transformation of the Indian Industry

While India is constantly striving hard to adopt the advanced proven global technologies that fit in Indian prospects for 2G biofuels production, the country needs to introduce future reforms that various other nations ahead in the biofuel area have acquired. This would indeed be a step forward for the effective implementation of advanced policies making it self-sufficient in the energy sector.

The Renewable Fuel Standard (RFS) in the USA is a federal program accountable to the Environmental Protection Agency that mandates the blending of a certain amount of renewable fuel in the locomotive fuel being enhanced every year. The RFS regulations were part of the Energy Independence and Security Act of 2007 (EISA) that stressed upon annual enhancement of biofuel volumes by 9%—2022, in addition to lowering of gasoline consumption by 20% until 2017 [152].

The policy directed 8–11% renewable fuel blending from 2010 to 2017 and classified them into four categories (Conventional Biofuel, Advanced Biofuel, Biomass-Based Diesel, and Cellulosic Biofuel), with a predefined feedstock and carbon intensity (CI) reduction (of at least 20%, 50%, 50%, and 60%, respectively) compared to the transportation fuel including indirect land use GHG emission [153]. The revised RFS2 promotes cellulosic biofuels vis-a-vis corn for producing ethanol given the latter's contribution to price rise in the international corn market and limited reduction in GHG emission [154].

1.6.2 Impact of Global Policies on Technology Development

The market segmentation strategy in conventional/advanced/cellulosic biofuels used in the United States market is based on granting price premiums for the production of cellulosic ethanol. Low interest rates and a venture capital culture were another strategy for advancing the deployment of 2G biofuels in the US. The drivers of 2G ethanol were developed under the Energy Policy Act of 2005 and were published as the Renewable Fuels Standard (RFS), followed by the updation to the Energy Independence and Security Act of 2007. The RFS's objective is to increase the biofuel blend up to 36 billion gallons (Bgal) by 2022 from 9 Bgal in 2008. In addition to blending legislations, the inclusion of several value-generating aspects at the federal level, (a case in point being the cellulosic waiver credit (CWC) which is a tax exemption that inversely correlates with gasoline prices) have helped in the adoption of 2G technology.

Implementation of the blending mandates and adoption of a roadmap (in 2011) for a "Competitive Low Carbon Economy" in the European Union led to 40%, 60%, and 80% reductions in GHG's emissions by 2030, 2040, and 2050, respectively, via low-carbon technologies and robust energy efficiency scheme.

China has shown considerable rapid growth in the advanced cellulosic ethanol industry. The National Development Bank in Brazil has strongly supported the sector to provide power to the global industry.

2 Conclusion

Undoubtedly, India's current energy demand–supply gap is accelerating exponentially and the myriad environmental, economic, and social bottlenecks linked with the global climatic crisis are urgent. With the increasing uncertainty in the world's oil supply and evidential rise in crude oil prices, the search for renewable substitutes is increasing day by day. The 2011 International Energy Outlook [155] presaged by its Reference Case Scenario that world consumption of marketed energy in totality will be incremented by 42% by 2035 from 2010 with an increase in liquid fuels consumption of 30% by 2035 from 2010 in the transportation sector. Biodiesel could be the potential alternative, provided the high cost of conversion, hence there is a need to explore cheaply available substrates that can bring down the production cost and increase the efficiency. The use of agro-industrial residues like sewage sludge, whey permeate, sugarcane molasses, waste glycerol, and monosodium glutamate wastewater as low-cost culture media, could be a solution as studied in different scientific reports. Microbial oil fatty acids are similar in composition and energy value to plant oils and animal fat, other than offering unique advantages over them. Inexpensive lignocellulosic biomass has universal abundance and good calorific value and does not compete with the production of agricultural food crops, hence can be effectively used for the effective preparation of oils. The same biomass could be used for producing bioethanol and other biofuels like drop-in fuels, pyrolysis oil, biogas, etc. Decentralized Compressed Biogas (CBG) production from biomass or MSW has significant potential for augmenting gas supply specifically in rural areas where cooking gas demand is growing substantially. To bridge the current/future energy demand–supply gap and confront the expense of negative geopolitical challenges posed by crude oil, it can serve as fungible fuels source. However, the biomass cost needs to be worked out as it varies according to region and type and is estimated as per barrel of oil energy equivalent. An innovative biomass supply chain model with direct payment to the biomass producers could be a solution to the challenge. Producing value-added co-products like furfural, acetic acid, etc. could lower ethanol production and utility costs. Globally, lignocellulose usage is economically viable as compared to other food crops such as sugarcane, corn starch, soybeans, etc. due to quick production at a significantly lower cost than agricultural crops. Also, they form the non-edible portion of the plant that is presently underutilized and could be derived from both agricultural and forest resources. To sum up, lignocellulosic biomass possesses the potential to meet the world's basic needs for sustainable production of transportation fuels, with limited impact on the nation's food supply. The role of nanotechnology should also be explored in the context of biofuel production. By the middle of the twenty-first century, surplus domestic non-food lignocellulosic biomass could be available for large-scale bioenergy and biorefining industries, if new energy policies are implemented seriously. The major question to be addressed here for biofuel production is to efficiently convert a large fraction of biomass energy into liquid fuels using low-cost processing technologies, which do not yet exist.

In the above study, we have seen that there is a huge potential in a wide diversity of biomass feedstocks including MSW, waste gases, etc., and simultaneously conversion technologies are at different maturity levels and expected to improve in the future. India definitely needs a national mandate for domestic biofuels production and folks are overwhelmingly supporting the development of advanced technologies for alternative fuels production, promoting renewable energy and reducing greenhouse gas emissions. The environmental perturbation resulting from the nation's dependency on fossils along with diminishing resources has spurred our society to broaden its outlook to search for renewable energy sources by utilizing the massively available biomass feedstock. To improve project viability and facilitate wider participation, government policies may consider including "Brownfield projects" and "Bolt on" plants in the designed schemes. Hence, the national policy may take its course in the direction that is biomass feedstock agnostic along with promoting biofuels on equal footing.

Production of ethanol from industrial wastes which is also a renewable feedstock, may be considered for viability gap funding (VGF). Based on the calorific value, bio-crude production, and drop-in-fuels production from MSW/biomass/plastic waste may be considered on the basis of ethanol equivalence. The implementation of the scheme may allow broader options for biofuels development and competition with each other, so that the best may prevail.

Acknowledgements The figures were created and exported using a paid subscription from BioRender.com. We are thankful to Dr. Nishant Joshi, from the School of Natural Sciences, Shiv Nadar University, Greater Noida, India for his support in the study.

Declaration of Competing Interest The authors declare no conflict of interest. The views expressed in this review are entirely of the authors and do not necessarily represent the collective views of the affiliated organizations.

References

1. International Energy Agency (IEA) (2011) Bioenergy. What is biomass? www.aboutbioenergy.info/de_nition.html. Accessed 29 April 2021
2. Hawkes GL, O'Brien JE, McKellar MG (2011) Liquid bio-fuel production from non-food biomass via high temperature steam electrolysis. In: Proceedings of the ASME 2011 international mechanical engineering congress and exposition IMECE2011. Denver, Colorado, USA
3. Bioenergy Feedstock Information Network (BFIN) (2011) Glossary of useful bioenergy feedstock terms. https://bioenergy.ornl.gov/main.aspx. Accessed 03 May 2021
4. Kent MS, Andrews KM (2007) Biological research survey for the efficient conversion of biomass to biofuels. Technical Report. SAND2006-7221. https://doi.org/10.2172/900852. Corpus ID: 55407028
5. Gerpen J (2005) Biodiesel processing and production. Fuel Process Technol 86:1097–1107
6. Kumar S, Shrestha P, Salam PA (2013) A review of biofuel policies in the major biofuel producing countries of ASEAN: production, targets, policy drivers and impacts. Renew Sustain Energy Rev 26:822–836
7. Dudley B (2019) BP statistical review of world energy 2019. BP, London
8. Pohit S, Biswas PK, Ashra S (2011) Incentive structure of India's biofuel programs: status, shortcomings and implications. Biofuels 2(3):355–369
9. Arjune V (2017) Perceptions and realities of promoting biofuels in India: a look at the national policy on biofuels. J Biofuels 8(1):8–14
10. Jain N, Bhatia A, Pathak H (2014) Emission of air pollutants from crop residue burning in India. Aerosol Air Qual Res 14:422–430. https://www.ictmumbai.edu.in/uploaded_files/16%20DBT-ICT%20CEB%20%202016-17.pdf
11. Sitepu IR, Garay LA, Sestric R, Levin D, Block DE, German JB, Boundy-Mills KL (2014) Oleaginous yeasts for biodiesel: current and future trends in biology and production. Biotechnol Adv 32(7):1336–1360
12. Jin M, Slininger PJ, Dien BS, Waghmode S, Moser BR, Orjuela A, da Costa SL, Balan V (2015) Microbial lipid-based lignocellulosic biorefinery: feasibility and challenges. Trends Biotechnol 33(1):43–54
13. Li Q, Du W, Liu D (2008) Perspectives of microbial oils for biodiesel production. Appl Microbiol Biotechnol 80(5):749–756
14. Huang C, Chen XF, Xiong L, Chen XD, Ma L, Chen Y (2013) Single cell oil production from low-cost substrates: the possibility and potential of its industrialization. Biotechnol Adv 31(2):129–139

15. Lee S, Shah YT (2012) Biofuels and bioenergy: processes and technologies. In: Green chemistry and chemical engineering. CRC Press, 341 pp. ISBN 1420089560, 9781420089561
16. Soltanian S, Aghbashlo M, Almasi F, Hosseinzadeh-Bandbafha H, Nizami AS, Ok YS, Lame SS, Tabatabaei M (2020) A critical review of the effects of pretreatment methods on the exergetic aspects of lignocellulosic biofuels. Energy Convers Manage 212:112792. https://doi.org/10.1016/j.enconman.2020.112792
17. Saha BC (2004) Lignocellulose biodegradation and applications in biotechnology. In: Saha BC, Hayashi K (eds) Lignocellulose biodegradation 2–34
18. Cohen D (2004) Form and distribution of trace elements in biomass for power generation, Technical Report. QCat Technology Transfer Center, Australia
19. Sims RE, Mabee W, Saddler JN, Taylor M (2010) An overview of second-generation biofuel technologies. Bioresour Technol 101(6):1570–1580
20. Das S (2020) The national policy of biofuels of India—a perspective. Energy Policy 143:111595
21. Matatkova O, Gharwalova L, Zimola M, Rezanka T, Masak J, Kolouchova I (2017) Using odd-alkanes as a carbon source to increase the content of nutritionally important fatty acids in *Candida krusei, Trichosporon cutaneum*, and *Yarrowia lipolytica*. Int J Anal Chem 9, Article ID 8195329. https://doi.org/10.1155/2017/8195329
22. Bettiga M, Bengtsson O, Hahn-Hagerdal B, Gorwa-Grauslund MF (2009) Arabinose and xylose fermentation by recombinant *Saccharomyces cerevisiae* expressing a fungal pentose utilization pathway. Microb Cell Fact 8:1–12
23. Lin Y, He P, Wang Q, Lu D, Li Z, Wu C, Jiang N (2010) The alcohol dehydrogenase system in the xylose-fermenting yeast *Candida maltose*. PLoS ONE 5:1–9
24. Sharma S, Arora A (2020) Tracking strategic developments for conferring xylose utilization/fermentation by *Saccharomyces cerevisiae*. Ann Microbiol 70:50. https://doi.org/10.1186/s13213-020-01590-9
25. Yang X, Jin G, Gong Z, Shen H, Bai F, Zhao KZ (2015) Recycling microbial lipid production wastes to cultivate oleaginous yeasts. Bioresour Technol 175:91–96
26. Fakas S, Papanikolaou S, Galiotou-Panayotou M, Komaitis M, Aggelis G (2008) Biochemistry and biotechnology of single cell oil. University of Patras, pp 38–60
27. Balat M (2011) Production of bioethanol from lignocellulosic materials via the biochemical pathway: a review. Energy Convers Manag 52(2):858–875
28. Meng X, Yang J, Xu X, Zhang L, Nie Q, Xian M (2009) Biodiesel production from oleaginous microorganisms. Renew Energy 34:1–5
29. Wiloso EI, Heijungs R, de Snoo GR (2012) LCA of second-generation bioethanol: a review and some issues to be resolved for good LCA practice. Renew Sustain Energy Rev 16(7):5295–5308
30. Escobar JC, Lora ES, Venturini OJ, Yáñez EE, Castillo EF, Almazan O (2009) Biofuels: environment, technology and food security. Renew Sustain Energy Rev 13(6):1275–1287
31. U.S. DOE Office of Science and Office of Energy Efficiency and Renewable Energy. Breaking the biological barriers to cellulosic ethanol: a joint research agenda. A research roadmap resulting from the biomass to biofuels workshop, 2006; December 7–9, 2005, Rockville, MD, Technical Report. DOE/SC-0095
32. Pirozzi D, Ausiello A, Yousuf A, Zuccaro G, Toscano G (2014) Exploitation of oleaginous yeasts for the production of microbial oils from agricultural biomass. Chem Eng 37:469–474
33. Granstrom T (2002) Biotechnological production of Xylitol with Candida yeasts. Available in the electronic publications archive of Aalto University. ISBN 951-22-5992-3
34. Sandhya M, Aravind J, Kanmani P (2013) Production of polyhydroxyalkanoates from *Ralstonia eutropha* using paddy straw as cheap substrate. Int J Environ Sci Technol 10(1):47–54
35. Zhao X, Peng F, Du W, Liu C, Liu D (2012) Effects of some inhibitors on the growth and lipid accumulation of oleaginous yeast *Rhodosporidium toruloides* and preparation of biodiesel by enzymatic transesterification of the lipid. Bioprocess Biosyst Eng 35(6):993–1004

36. López EN, Medina AR, Moreno PAG, Cerdán LE, Grima EM (2016) Extraction of microalgal lipids and the influence of polar lipids on biodiesel production by lipase-catalyzed transesterification. Biores Technol 216:904–913. https://doi.org/10.1016/j.biortech.2016.06.035
37. Dalmas Neto CJ, Sydney EB, Candeo ES, de Souza EBS, Camargo D, Sydney ACN, de Carvalho JC, Letti LAJ, Pandey A, Soccol CR (2019) New method for the extraction of single-cell oils from wet oleaginous microbial biomass: efficiency, oil characterisation and energy assessment. Waste Biomass Valor 11(7):3443–3452. https://doi.org/10.1007/s12649-019-00705-x
38. Vijayakumar S, Kumutha K, Santhana Krishnan P, Gopal H (2010) Effect of carbon sources on lipid and biomass production by oleaginous yeast cultures. MAJ 97(1/3):62–64
39. Ratledge C, Wynn JP (2002) The biochemistry and molecular biology of lipid accumulation in oleaginous microorganisms. Adv Appl Microbiol 51:1–52
40. Tehlivets O, Scheuringer K, Kohlwein SD (2007) Fatty acid synthesis and elongation in yeast. Biochem Biophys Acta 1771:255–270
41. Huang YT, Su CP (2014) High lipid content and productivity of microalgae cultivating under elevated carbon dioxide. Int J Environ Sci Technol 11(3):703–710
42. Syed MA, Singh SK, Pandey A, Kanjilal S, Prasad RBN (2006) Effects of various process parameters on the production of a-Linolenic acid in submerged fermentation. Food Technol Biotechnol 44:282–287
43. Leesing R, Baojungharn R (2011) Microbial oil production by isolated oleaginous yeast *Torulaspora globosa* YU5/2. WASET 76:799–803
44. Papanikolaou S, Aggelis G (2003) Modeling lipid accumulation and degradation in *Yarrowia lipolytica* cultivated on industrial fats. Curr Microbiol 46(6):398–402
45. Papanikolaou S, Aggelis G (2011) Lipids of oleaginous yeasts. Part I: Biochemistry of single cell oil production. Eur J Lipid Sci Tech 113(8):1031–1051
46. Leesing R, Baojungharn R (2011) Microbial oil production by isolated oleaginous yeast *Torulaspora globosa* YU5/2. WASET. Int J Biotechnol Bioeng 5(4):246–250
47. Li Y, Zhao Z, Bai F (2007) High cell-density cultivation of oleaginous yeast *Rhodosporidium toruloides* Y4 in fed-batch culture. Enzyme Microbiol Technol 41:312–317
48. Wang R, Wang J, Xu R, Fang Z et al (2014) Oil production by the oleaginous yeast *Lipomyces starkeyi* using diverse carbon sources. BioResources 9(4):7027–7040
49. Papanikolaou S, Aggelis G (2011) Lipids of oleaginous yeasts. Part II: Technology and potential applications. Eur J Lipid Sci Technol 113(8):1052–1073
50. Shah Y, Gardner T (2012) Biomass torrefaction: applications in renewable energy and fuels. In: Lee S (ed) Encyclopedia of chemical processing. Taylor & Francis Group, New York, pp 1–18
51. Hochman G, Rajagopal D, Timilsina GR (2014) Impacts of biofuels on food prices. In: Timilsina GR, Zilberman D (eds) The impacts of biofuels on the economy, environment, and poverty, vol 41. Springer, New York, pp 47–64
52. Shinoj P, Raju SS, Joshi PK (2011) India's biofuels production programme: need for prioritizing the alternative options. Indian J Agric Sci 81:391–397
53. Singh S, Adak A, Saritha M, Sharma S, Tiwari R, Rana S, Nain L (2017) Bioethanol production scenario in India: potential and policy perspective. In: Sustainable biofuels development in India. Springer, New Delhi, pp 21–37. https://doi.org/10.1007/978-3-319-50219-9_2
54. Khatri N, Ojha A (2016) Crony capitalism in India: establishing robust counteractive institutional frameworks. In: Khatri N, Ojha A (eds) Crony capitalism in India: establishing robust counteractive institutional frameworks. Palgrave Macmillan, p 144
55. Government of India (2018) Press information bureau. Retrieved from Press Information Bureau, Government of India, New Delhi
56. Poundrik S (2018) Transport and renewable energy policies in India. Government of India, New Delhi (Retrieved from Transport and renewable energy policies in India)
57. Rangarajan C (2012) Report of the regulation of sugar sector in India: the way forward. PRS Legislative Research, New Delhi

58. McCormick RL, Westbrook SR (2010) Storage stability of biodiesel and biodiesel blends. ACS Energy Fuels 24(1):690–698. https://doi.org/10.1021/ef900878u
59. Saada R (2020) Green transportation in green supply chain management. In: Bányai T, Kaczmar I (eds) Green supply chain. https://doi.org/10.5772/intechopen.93113
60. National Research Council (2011) Renewable fuel standard: potential economic and environmental effects of U.S. biofuel policy. The National Academies Press, Washington, DC. https://doi.org/10.17226/13105
61. Da Silva JG, Serra GE, Moreira JR, Concalves JC, Goldenberg J (1978) Energy balance for ethyl alcohol production from crops. Science 201:903–906
62. Hopkinson CS, Davy JW (1980) Net energy analysis of alcohol production from sugarcane. Science 207:302–304
63. Karlsson H, Ahlgren S, Sandgren M, Passoth V, Wallberg O, Hansson PA (2016) A systems analysis of biodiesel production from wheat straw using oleaginous yeast: process design, mass and energy balances. Biotechnol Biofuels 9(229):2
64. Brandenburg J, Blomqvist J, Pickova J, Bonturi N, Sandgren M, Passoth V (2016) Lipid production from hemicellulose with *Lipomyces starkeyi* in a pH regulated fed batch cultivation. Yeast 33(8):451–462. https://doi.org/10.1002/yea.3160
65. Calvey CH, Su Y-K, Wills LB, McGee MS, Jeffries TW (2016) Nitrogen limitation, oxygen limitation, and lipid accumulation in *Lipomyces starkeyi*. Bioresour Technol 200:780–788
66. Hu C, Zhao X, Zhao J, Wu S, Zhao KZ (2009) Effects of biomass hydrolysis by-products on oleaginous yeast *Rhodosporidium toruloides*. Bioresour Technol 100(20):4843–4847
67. Hensirisak P, Parasukulsatid P, Agblevor FA, Cundiff JS, Velander WH (2002) Scale up of microbubble dispersion generator for aerobic fermentation. Appl Biochem Biotechnol 101(3):211–227
68. Gao M, Zhang S, Ma X, Guan W, Song N, Wang Q, Wu C (2020) Effect of yeast addition on the biogas production performance of a food waste anaerobic digestion system. R Soc Open Sci 7:200443. https://doi.org/10.1098/rsos.200443
69. Cozma P, Ghinea C, Mămăligă I, Wukovits W, Friedl A, Gavrilescu M (2013) Environmental impact assessment of high pressure water scrubbing biogas upgrading technology. Clean: Soil, Air, Water 41(9):917–927
70. Brown A, Waldheim L, Landälv I, Saddler J, Ebadian M, McMillan JD et al (2020) Advanced biofuels—potential for cost reduction. IEA Bioenergy: Task 41.
71. Gladius L (1984) Some aspects of the production of ethanol from sugar cane residues in Zimbabwe. Sol Energy 33:379–382
72. Schmer MR, Vogel KP, Mitchell RB, Perrin RK (2008) Net energy of cellulosic ethanol from switchgrass. Proc Nat Acad Sci USA 105:464–469
73. Peterson R, Nevalainen H (2012) *Trichoderma reesei* RUT-C30—thirty years of strain improvement. Microbiology 158:58–68
74. Seidl V, Gamauf C, Druzhinina IS, Seiboth B, Hartl L, Kubicek CP (2008) The Hypocrea jecorina (*Trichoderma reesei*) hypercellulolytic mutant RUT C30 lacks a 85 kb (29 gene-encoding) region of the wild-type genome. BMC Genomics 9:327
75. Fenel F, Zitting AJ, Kantelinen A (2006) Increased alkali stability in *Trichoderma reesei* endo-1, 4-beta-xylanase II by site directed mutagenesis. J Biotechnol 121:102–107
76. Kovács K, Megyeri L, Szakacsa G, Kubicekc CP, Galbeb M, Zacchi G (2008) *Trichoderma atroviride* mutants with enhanced production of cellulase and β-glucosidase on pretreated willow. Enzyme Microbial Technol 43:48–55
77. Fenel F, Leisola M, Janis J, Turunen O (2004) A de novo designed N-terminal disulphide bridge stabilizes the *Trichoderma reesei* endo-1,4-beta-xylanase II. J Biotechnol 108:137–143
78. Solov'eva IV, Okunev ON, Vel'kov VV, Koshelev AV, Bubnova TV, Kondrat'eva EG, Skomarovskii AA, Sinitsyn AP (2005) The selection and properties of *Penicillium verruculosum* mutants with enhanced production of cellulases and xylanases. Mikrobiologiia 74:172–178
79. Qin Y, Wei X, Song X, Qu Y (2008) Engineering endoglucanase II from *Trichoderma reesei* to improve the catalytic efficiency at a higher pH optimum. J Biotechnol 135:190–195

80. Sandgren M, Stahlberg J, Mitchinson C (2005) Structural and biochemical studies of GH family 12 cellulases: improved thermal stability, and ligand complexes. Prog Biophys Mol Biol 89:246–291
81. Murray P, Aro N, Collins C, Grassick A, Penttila M, Saloheimo M, Tuohy M (2004) Expression in *Trichoderma reesei* and characterisation of a thermostable family 3 beta-glucosidase from the moderately thermophilic fungus *Talaromyces emersonii*. Protein Expr Purif 38:248–257
82. Bower B, Larenas E, Mitchinson C (2005) Exo-endo cellulase fusion protein. Patent WO2005093073
83. Miettinen-Oinonen A, Paloheimo M, Lantto R, Suominen P (2005) Enhanced production of cellobiohydrolases in *Trichoderma reesei* and evaluation of the new preparations in biofinishing of cotton. J Biotechnol 116:305–317
84. Gupte A, Madamwar D (1997) Solid state fermentation of lignocellulosic waste for cellulose and β-glucosidase production by co-cultivation by *Aspergillus ellipticus* and *Aspergillus fumigatus*. Biotechnol Prog 13:166–169
85. Panda T, Bisaria VS, Ghose TK (1983) Studies on mixed fungal culture for cellulase and hemicellulase production. Part 1. Optimization of medium for the mixed culture of *Trichoderma reesei* Dl-6 and *Aspergillus wentii* Pt 2804. Biotechnol Lett 5:767–772
86. Duff SJB (1985) Cellulase and beta-glucosidase production by mixed culture of *Trichoderma reesei* Rut C30 and *Aspergillus phoenicis*. Biotechnol Lett 7:185–190
87. Ahamed A, Vermette P (2008) Enhanced enzyme production from mixed cultures of *Trichoderma reesei* RUT-C30 and *Aspergillus niger* LMA grown as fed batch in a stirred tank bioreactor. Biochem Eng J 42:41–46
88. Dueñas R, Tengerdy RP, Gutierrez-Correa M (1995) Cellulase production by mixed fungi in solid-substrate fermentation of bagasse. World J Microbiol Biotechnol 11:333–337
89. Qin L, Zhao X, Li W, Zhu JQ, Liu L, Li BZ, Yuan YJ (2018) Process analysis and optimization of simultaneous saccharification and co-fermentation of ethylenediamine-pretreated corn stover for ethanol production. Biotechnol Biofuels 11:118. https://doi.org/10.1186/s13068-018-1118-8
90. Yang B, Wyman CE (2008) Pretreatment: the key to unlocking low-cost cellulosic ethanol. Biofuel Bioprod Bioref 2:26–40
91. Qin L, Li WC, Zhu JQ, Liang JN, Li BZ, Yuan YJ (2015) Ethylenediamine pretreatment changes cellulose allomorph and lignin structure of lignocellulose at ambient pressure. Biotechnol Biofuels 8:174
92. Qin L, Li WC, Liu L, Zhu JQ, Li X, Li BZ, Yuan YJ (2016) Inhibition of lignin-derived phenolic compounds to cellulase. Biotechnol Biofuels 9:70
93. Xue S, Uppugundla N, Bowman MJ, Cavalier D, da Costa SL, Dale BE, Balan V (2015) Sugar loss and enzyme inhibition due to oligosaccharide accumulation during high solids-loading enzymatic hydrolysis. Biotechnol Biofuels 8:195
94. Ojeda K, Sánchez E, El-Halwagi M, Kafarov V (2011) Exergy analysis and process integration of bioethanol production from acid pre-treated biomass: comparison of SHF, SSF and SSCF pathways. Chem Eng J 176:195–201
95. Koppram R, Nielsen F, Albers E, Lambert A, Wännström S, Welin L, Zacchi G, Olsson L (2013) Simultaneous saccharification and co-fermentation for bioethanol production using corncobs at lab, PDU and demo scales. Biotechnol Biofuels 6:2
96. Liu ZH, Chen HZ (2016) Simultaneous saccharification and co-fermentation for improving the xylose utilization of steam exploded corn stover at high solid loading. Bioresour Technol 201:15–26
97. Liu G, Zhang Q, Li H, Qureshi AS, Zhang J, Bao X, Bao J (2018) Dry biorefining maximizes the potentials of simultaneous saccharification and co-fermentation for cellulosic ethanol production. Biotechnol Bioeng 115(1):60–69
98. Nguyen TY, Cai CM, Kumar R, Wyman CE (2017) Overcoming factors limiting high-solids fermentation of lignocellulosic biomass to ethanol. Proc Natl Acad Sci USA 114(44):11673
99. Qi B, Chen X, Su Y, Wan Y (2011) Enzyme adsorption and recycling during hydrolysis of wheat straw lignocellulose. Bioresour Technol 102:2881–2889

100. Jin M, Gunawan C, Uppugundla N, Balan V, Dale BE (2012) A novel integrated biological process for cellulosic ethanol production featuring high ethanol productivity, enzyme recycling and yeast cells reuse. Energy Environ Sci 5:7168–7175
101. Visser EM, Leal TF, de Almeida MN, Guimarães VM (2015) Increased enzymatic hydrolysis of sugarcane bagasse from enzyme recycling. Biotechnol Biofuels 8:5
102. Weiss N, Börjesson J, Pedersen LS, Meyer AS (2013) Enzymatic lignocellulose hydrolysis: improved cellulase productivity by insoluble solids recycling. Biotechnol Biofuels 6:5
103. Eriksson T, Börjesson J, Tjerneld F (2002) Mechanism of surfactant effect in enzymatic hydrolysis of lignocellulose. Enzyme Microb Technol 31:353–364
104. Brethauer S, Studer MH, Yang B, Wyman CE (2011) The effect of bovine serum albumin on batch and continuous enzymatic cellulose hydrolysis mixed by stirring or shaking. Bioresour Technol 102:6295–6298
105. Zhang TY, Hu HY, Wu YH, Zhuang LL, Xu XQ, Wang XX, Dao GH (2016) Promising solutions to solve the bottlenecks in the large-scale cultivation of microalgae for biomass/bioenergy production. Renew Sustain Energy Rev 60:1602–1614
106. Xiang LJ, Zhang F, Jiang DD, Li J, Wang FL, Zhang Z, Wang W, Zhao XQ (2020) Diversity of cellulase-producing filamentous fungi from Tibet and transcriptomic analysis of a superior cellulase producer Trichoderma harzianum LZ117. Front Microbiol Fungi Their Interact 11:1617. https://doi.org/10.3389/fmicb.2020.01617
107. Li J, Zhang F, Li J, Zhang Z, Bai F, Chen J et al (2019) Rapid production of lignocellulolytic enzymes by *Trichoderma harzianum* LZ117 isolated from Tibet for biomass degradation. Bioresour Technol 292:122063. https://doi.org/10.1016/j.biortech.2019.122063
108. Teugjas H, Väljamäe P (2013) Product inhibition of cellulases studied with 14C-labeled cellulose substrates. Biotechnol Biofuels 6:104. https://doi.org/10.1186/1754-6834-6-104
109. Verma J, Kumar D, Singh N, Katti S, Shah YT (2021) Electricigens and microbial fuel cells for bioremediation and bioenergy production: a review. Environ Chem Lett. https://doi.org/10.1007/s10311-021-01199-7
110. https://www.teriin.org/dbt-teri-centre-excellence-advanced-biofuels-and-bio-commodities ; https://www.teriin.org/press-release/department-biotechnology-awards-teri-innovation-methane-generation-coal-beds
111. Patel AK, Debroy A, Sharma S, Saini R, Mathur A, Gupta R, Tuli DK (2015) Biohydrogen production from a novel alkalophilic isolate Clostridium sp. IODB-O3. Bioresour Technol 175:291–297. https://doi.org/10.1016/j.biortech.2014.10.110. Epub 2014 Oct 28. PMID: 25459835
112. Kumari S, Das D (2017) Improvement of bio-hydrogen production using acidogenic culture. Int J Hydrogen Energy 42(7):4083–4094. https://doi.org/10.1016/j.ijhydene.2016.09.021
113. Yadav S, Singh V, Mahata C, Das D (2021) Optimization for simultaneous enhancement of biobutanol and biohydrogen production. Int J Hydrogen Energy 46(5):3726–3741
114. Ratafia-Brown J, Haslbeck J, Skone TJ, Rutkowski M (2007) Assessment of technologies for co-converting coal and biomass to clean syngas-task 2 report (RDS), NETL report 2007
115. McDaniel J (2002) Biomass gasification at polk power station-final technical report, DOE award DE-FG26-01NT41365
116. https://www.bharatpetroleum.com/blog/Blog-Detail.aspx?id=21
117. Valdivia M, Galan JL, Laffarga J, Ramos JL (2016) Biofuels 2020: biorefineries based on lignocellulosic materials. Microb Biotechnol 9(5):585–594. https://doi.org/10.1111/1751-7915.12387
118. Ganguli S, Somani A, Motkuri RK, Bloyd C (2018) India alternative fuel infrastructure: the potential for second-generation biofuel technology. Pacific Northwest National Laboratory Prepared for the U.S. Department of Energy under Contract DE-AC05-76RL01830. PNNL 28283
119. Stephen J, Mabee WE, Saddler JN (2012) Will second-generation ethanol be able to compete with first-generation ethanol? Opportunities for cost reduction. Biofuels Bioprod Bioref 6:159–176

120. Balan V (2014) Current challenges in commercially producing biofuels from lignocellulosic biomass. Hindawi Publishing Corporation, ISRN Biotechnology, 31, Article ID 463074
121. Lee S, Shah YT (2012) Biofuels and bioenergy processes and technologies. In: Green chemistry and chemical engineering. CRC Press, Taylor and Francis Group. ISBN-13: 978-1-4200-8956-1
122. Shah YT (2012) Chapter 12—Biomass to liquid fuel via Fischer-Tropsch and related syntheses. In: Lee JW (ed) Advanced biofuels and bioproducts. Springer Book Project, Springer Publ. Co., New York, NY, pp 185–207
123. Bergman PCA, Kiel JHA (2005) Torrefaction for biomass upgrading, published at 14 European biomass conference and exhibition, Paris, France, pp 17–21; report ECNRX-05-1 80, 1-08
124. Boerrigter H, Calis HP, Slort DJ, Bodenstaff H, Kaandorp AJ, den Uil H, Rabou LPLM (2004) Gas cleaning for integrated biomass gasification (BG) and Fischer-Tropsch (FT) systems. Energy Research Center of the Netherland (ECN), Petten, The Netherlands, report C-04-056, p 59
125. Boerrigter H, van der Drift A (2004) Biosyngas: description of R&D trajectory necessary to reach large-scale implementation of renewable syngas from biomass. Energy Research Center of the Netherland (ECN), Petten, The Netherlands, report C-04-112, p 29
126. Park CS, Hackett C, Norbeck JM (2005) Synthetic diesel fuel production from carbonaceous feed stocks. Presentation to ISAF XV. In: International symposia on alcohols and fuels, CERT, University of California, Riverside
127. Bergman PCA, Boersma AR, Kiel JHA (2005) Torrefaction for entrained flow gasification of biomass. Energy Research Center of the Netherland (ECN), Petten, The Netherlands, report C-04-067
128. Shah YT (2014) Chapter 5—Hydrothermal processes in subcritical water. In: Lee S (ed) Water for energy and fuel production. Green chemistry and chemical engineering. Ohio University, Athens, Ohio, USA; CRC Press, Taylor & Francis group, pp 113–148. ISBN-13: 978-1-4822-1619-6 (eBook—PDF)
129. Tao J, Shishi C, Fahai C (2012) Advances in H_2 production from the aqueous-phase reforming of biomass-derived polyols: a review. Chem Ind Eng Progr 31(5):1010–1017
130. Alonso DM, Bond JQ, Dumesic JA (2010) Catalytic conversion of biomass to biofuels. Green Chem 12:1493–1513
131. Blommel PG, Cortright RD (2012) Production of conventional liquid fuels from sugars. A White Paper for European Platform on Biofuels, UK
132. Davda R, Shabaker J, Huber G, Cortright R, Dumesic J (2005) A review of catalytic issues and process conditions for renewable hydrogen and alkanes by aqueous-phase reforming of oxygenated hydrocarbons over supported metal catalysts. Appl Catal B 56:171–186
133. Davda R, Dumesic J (2003) Catalytic reforming of oxygenated hydrocarbons for hydrogen with low levels of carbon monoxide. Angew Chem Int Ed 42:4068
134. D'Angelo M, Ordomsky V, van der Schaaf J, Schouten J, Nijhuis T (2013) Aqueous phase reforming in a microchannel reactor: the effect of mass transfer on hydrogen selectivity. Catal Sci Technol 3:2834–2842
135. D'Angelo MFN, Ordomsky V, Paunovic V, van der Schaaf J, Schouten JC, Nijhuis TA (2013) Hydrogen production through aqueous-phase reforming of ethylene glycol in a washcoated microchannel. Chem Sus Chem 6:1708–1716
136. Shah YT (2014) Chapter 6—Aqueous-phase reforming and bio forming process. In: Lee S (ed) Water for energy and fuel production. Green Chemistry and Chemical Engineering. Ohio University, Athens, Ohio, USA; CRC Press Taylor & Francis group, pp 157–178. ISBN-13: 978-1-822-1619-6
137. Shah YT (2016) Chapter 6—Methane production by anaerobic co-digestion of biomass and waste. In: Lee S (ed) Energy and fuel systems integration. Green chemistry and chemical engineering. Ohio University, Athens, Ohio, USA; CRC Press, Taylor & Francis group, pp 225–273. ISBN-13: 978-1-4822-5312-2 (eBook—PDF)

138. Braun R, Weiland P, Wellinger A (2009) Biogas from energy crop digestion. IEA Bioenergy, Task 37—Energy from biogas and landfill gas, Denmark
139. Nges IA, Escobar F, Fu X, Björnsson L (2012) Benefits of supplementing an industrial waste anaerobic digester with energy crops for increased biogas production. Waste Manage 32(1):53
140. Jepsen S (2011) Co-digestion of animal manure and organic household waste—The Danish Experience, Ministry of Environment and Energy, Danish EPA, Copenhagen, Denmark
141. Murphy J, Braun R, Weiland P, Wellinger A (2011) Biogas from crop digestion. IEA bioenergy, Task 37—Energy from biogas, Denmark
142. Braun R, Wellinger A (2004) Potential of co-digestion. IEA bioenergy, Task 37—Energy from biogas and landfill gas. IEA Bioenergy, Petten, The Netherlands
143. Wallis MJ, Ambrose MR, Chan CC (2008) Climate change: charting a water course in an uncertain future. J Am Water Works Assoc 100:70–79
144. You Z, Wei T, Cheng J (2014) Improving anaerobic co-digestion of corn stover using sodium hydroxide pretreatment. Energy Fuels 28(1):549–554
145. Redman G (2010) A detailed economic assessment of anaerobic digestion technology and its sustainability to UK farming and waste systems, 2nd edn. The Andersons Center, A project funded by DECC and managed by the NNFCC, Melton, Mowbray, Leicestershire, UK
146. Bayr S, Rantanen M, Kaparaju P, Rintala J (2012) Mesophilic and thermophilic anaerobic co-digestion of rendering plant and slaughterhouse wastes. Bioresour Technol 104:28–36
147. Møller HB, Sommer SG, Ahring BK (2004) Methane productivity of manure, straw and solid fractions of manure. Biomass Bioenergy 26:485–495
148. Shah YT (2016) Chapter 3—Synthetic gas production by co-gasification. In: Lee S (ed) Energy and fuel system integrations. Green chemistry and chemical engineering. Ohio University, Athens, Ohio, USA. CRC Press, Taylor & Francis group, pp. 79–129. ISBN-13: 978-1-4822-5312-2 (eBook—PDF)
149. Wright MM, Brown RC (2007) Comparative economics of biorefineries based on the biochemical and thermochemical platforms. In: Biofuels, bioproducts and biorefining 1(1):49–56
150. Rismiller C, Tyner W (2009) Cellulosic biofuels analysis: economic analysis of alternative technologies; Working Paper # 09-06, Department of Agricultural Economics, Purdue University
151. Lali A (2018) DBT-ICT technology platforms for advanced biofuels. In: EU-India conference on advanced biofuels. https://ec.europa.eu/energy/sites/ener/files/documents/13_arvind_lali-dbt-ict.pdf
152. Sorda G, Banse M, Kemfert C (2010) An overview of biofuel policies across the world. Energy Pol 6977–6988
153. Wolinetz M, Hein M (2017) Biofuels in Canada. Navius Research, Vancouver
154. Brown TR (2019) Why the cellulosic biofuels mandate fell short: a markets and policy perspective. Biofuels Bioprod Biorefining 13(4):889–898
155. U.S. Energy Information Administration (EIA). International Energy Outlook 2011, Technical Report. DOE/EIA-0484, September 19

Economic Impact of Biomass Energy Use in the Service Sector: A Comparative Study on Various Crops and Technologies

Sunny Kumar and Nilambar Bariha

Abstract Biomass energy's integration into the service sector involves challenges and opportunities. The economic impact of biomass energy is influenced by several determinants, including the availability and cost of biomass feedstocks, the efficiency and cost of conversion technologies, and the market demand for biomass-derived energy. When biomass energy is used in the service sector, it could also create new job opportunities within the biomass energy domain. This chapter delves into the economic dynamics surrounding biofuel production from agricultural crops. Furthermore, it explores the cost effective and sustainable technologies associated with this sector. When combined, these factors could positively impact the local or regional economy availability and cost of biomass feedstocks, the efficiency and cost of conversion technologies, and the market demand for biomass-derived energy.

Keywords Agricultural crops · Economics · Service sector · Technology

1 Introduction

Biomass energy is derived from organic materials, such as wood, crops, and agricultural and forestry residues [1]. These energy sources can be harnessed to generate electricity, heat, and transportation fuels. The economic impact of biomass energy depends on several factors, including the availability and cost of biomass feedstocks, the efficiency and cost of conversion technologies, and the market demand for biomass-derived energy [2]. One potential economic benefit of biomass energy is the creation of new jobs and economic opportunities in rural areas, where biomass feedstocks are often grown or harvested. The production, transportation, and processing

S. Kumar (✉)
ChBE, Georgia Institute of Technology, Atlanta, GA, USA
e-mail: skumar710@gatech.edu

N. Bariha
Assam Energy Institute, Centre of Rajiv Gandhi Institute of Petroleum Technology, Sivasagar, Assam, India
e-mail: nbariha@rgipt.ac.in

S. Kumar et al. (eds.), *Clean Energy Transition-via-Biomass Resource Utilization*, Green Energy and Technology, https://doi.org/10.1007/978-981-97-6321-4_12

of biomass can also create jobs in urban areas. Using biomass as an energy source can also help reduce reliance on fossil fuels, which can have long-term economic benefits in terms of energy security and reduced greenhouse gas emissions. However, the economic impact of biomass energy may be limited by the high upfront costs of setting up biomass energy systems and the ongoing costs of sourcing and processing biomass feedstocks. It is essential to consider the potential economic benefits and costs of biomass energy in order to make informed decisions about its use. The economic impact of biomass energy use in the service sector can be significant. For example, if a hospital or university switches to using biomass energy to meet its energy needs, it could reduce its energy costs, which could potentially lead to cost savings for the organization and its customers. Biomass energy use in the service sector could also create new job opportunities in the biomass energy industry, such as in the production, transportation, and processing of biomass feedstocks. This could have a positive economic impact on the local or regional economy.

Using biomass energy in the service sector could also help to reduce greenhouse gas emissions, which could have long-term economic benefits in terms of mitigating the adverse effects of climate change [3]. However, it is important to consider the potential costs and challenges of using biomass energy in the service sector. For example, there may be high upfront costs associated with switching to biomass energy, and there may be ongoing costs related to sourcing and processing biomass feedstocks. Additionally, there may be logistical challenges associated with implementing biomass energy systems in the service sector. Overall, the economic impact of biomass energy use in the service sector will depend on various factors, including the availability and cost of biomass feedstocks, the efficiency and cost of conversion technologies, and the market demand for biomass-derived energy.

2 Biomass Production Effect on Economics

Agricultural crops provide economic advantages to various industries and can be categorized as sugar, cellulose, and starch crops. Sugar crops include sugarcane, sugar beet, sweet beet, sweet sorghum, oats, barley, and rye [4]. Starch crops include corn and sorghum, while oilseed crops consist of soybean and canola. Additionally, waste materials, such as household food waste and other byproducts, hold significance. These crops and wastes can be processed into bio-alcohols such as ethanol, butanol, propanol, and other biofuels [5] which can boost up the overall economics of the countries.

3 Ethanol Production from Different Biomass Crop

Sugar cane is the primary source of molasses, which is often used for ethanol and sugar production. The surplus ethanol can be harnessed for energy demand used as fuel. Several countries produce an abundance of ethanol, deriving it from various crops such as beetroot, corn, rice, wheat, and many other crops [6]. For instance, India and Brazil have a surplus of sugarcane. In the US, corn is the primary crop that is utilized for the production of ethanol as a fuel. Therefore, different countries adopt unique strategies for ethanol production on their agricultural resources. A 2006 report by the United States Department of Agriculture report 2006, the cost of ethanol production from feedstocks used such as \$3.48 per gallon for raw sugar, \$3.97 per gallon for refined sugar, \$2.65 per gallon for cellulosic ethanol, and \$0.90–\$1.65 per gallon for corn-based ethanol [7].

4 Biomass Used for Methanol Production

Typically, methanol is derived from coal, naphtha, and natural gas, which are naturally occurring reserved sources in the earth. However, biomass and its byproduct waste can also serve as sources for methanol production [8]. Materials such as rice husk, wheat husk, and sugar bagasse are potential candidates for the production of methanol. Methanol production often depends on the region and the crop predominant there. Utilization of this biomass could be more cost-effective compared to producing methanol directly from natural gas. The economics of methanol production are dependent upon different technologies [9]. According to the National Renewable Energy Laboratory (NREL) report in 2020, bio-methanol costs between \$700 and \$900 per metric ton, E-methanol costs between \$1200 and \$1600 per metric ton, and fossil fuel-based methanol costs between \$100 and \$250 per metric ton [10]. There are various types of methanol natural gases that cost between \$0.5 and \$1.7 per kilogram (kg). Using Carbon capture, utilization, and storage (CCUS) technologies to reduce CO_2 emissions increases the cost to around \$1 to \$2 per kg.

5 Biomass Used for Hydrogen Production

We recognize the need for environmentally sustainable fuels. While we currently rely on petroleum-based fuel, which emits hazardous gases, hydrogen offers a cleaner alternative with zero gas emissions. When hydrogen reacts with oxygen, it only produces water and delivers a high energy output. Given its benefits, hydrogen is a promising alternative fuel, though its production is limited. Biomass, with its various sources and waste products, can serve as an alternative means to produce hydrogen, making it one of the cleanest and most sustainable options. Green hydrogen is produced through electrolysis using renewable power; it costs between \$10 and \$15 per kg. Gray hydrogen is produced with cheap fracked natural gas; it costs \$2 per kg in US and \$5–6 per kg in Europe, Australia, and Asia. Blue hydrogen is produced with fossil fuels but is subject to carbon capture; it costs \$1.8–\$4.7 per kg [11–13].

6 Economic Effects on Biofuel Production Technology for the Different Service Sectors

There are various technologies for biofuel production from biomass, such as enzymatic, fermentation, hydrolysis, pyrolysis, and gasification. Affordable technology is required for economical biofuel production from biomass which should be sustainable and economically viable.

7 Enzyme Technology

Enzymatic hydrolysis of pre-treated lignocellulosic biomass is a suitable substitute for hydrolysis with acids for the manufacture of bio-ethanol, but it is principally constrained by the need for pre-treatment and economic factors resulting from the prices of enzyme manufacturing and specialized activities [14]. Within the next few years, it is reasonable to anticipate significant improvements in enzyme technology for the cost-effective conversion of lignocellulosic biomass as systematic enzyme research follows conventional reaction engineering practices [15].

8 Pyrolysis Technology

The worldwide surge in crop residue generation has necessitated developing strategies for their sustainable disposal. Pyrolysis has been widely adopted to convert crop residue into biochar with bio-oil and gas being two co-products [16]. The use of pyrolysis products has significant potential to reduce greenhouse gas emissions relative to the use of fossil fuels and conventional chemicals. For most agriculture crop residue, feedstocks containing fixed carbon, volatile matter, ash, and moisture content are in the ranges of 3–26%, 65–90%, 1–15%, and 0–10%, respectively. So, pyrolysis is a promising technology for converting biomass into renewable biofuels. Recently, with the advances in environmental and economic aspects of this pyrolysis technology, bio-oil production has been enhanced from lignocellulosic biomass. There are various types of pyrolysis processes, and their reaction-effective parameters are discussed in Table 1.

Table 1 Recent reaction technology and its operating parameters in pyrolysis [16]

Reaction technology	Operating parameters					Reference
	Pressure (Mpa)	Residence time (s)	Heating rate (°C/s)	Pyrolysis temperature (°C)	Gaseous environment	
Fast	0.1	0.5–10	10–200	500–1200	Inert	Tripathi et al. [17], Ghysels et al. [18], Liu et al. [19]
Slow	0.1	300–7200	0.1–1	300–600		Tripathi et al. [17], Biswas et al. [20]
	0.1	300–1800	0.03–0.08	400–700		Tippayawong et al. [21]
	–	<120	–	400–600		Digman et al. [22]
	–	–	0.1–10	400		Kumar and Nanda [23]
Flash	0.1	<1	>1000	900–1300		Li et al. [16], Sekar et al. [24]
Vacuum	0.01–0.20	<1	0.1–1	300–700		Garcìa-Pèrez et al. [25]
MWP	5–20	<30	0.5–2	300–700		Li et al. [26], Nzediegwu et al. [27], Lam et al. [28]
Hydro	10–17	60–120	10–300	350–600	Hydrogen	Kong et al. [29], Oh et al. [30]

Depending on the feedstock utilized, the conversion, the size of the pyrolysis unit, and the application of co-products, these technologies may help lower the life cycle greenhouse emissions from pyrolysis. Since slow pyrolysis yields more biochar, which can be used to improve soil quality and absorb carbon, it may have a bigger positive environmental impact. However, faster pyrolysis produces higher-value products, it offers a better economic return. The cost of the feedstock, product yields, the capacity to generate higher-value products, and the size of the manufacturing process all affect the profitablity of pyrolysis. Expanding the production of renewable energy through pyrolysis requires a combination of technological advancements and renewable energy policies. To reduce environmental impact and economic risks, this technology should take a wider sustainability study in any technology [31].

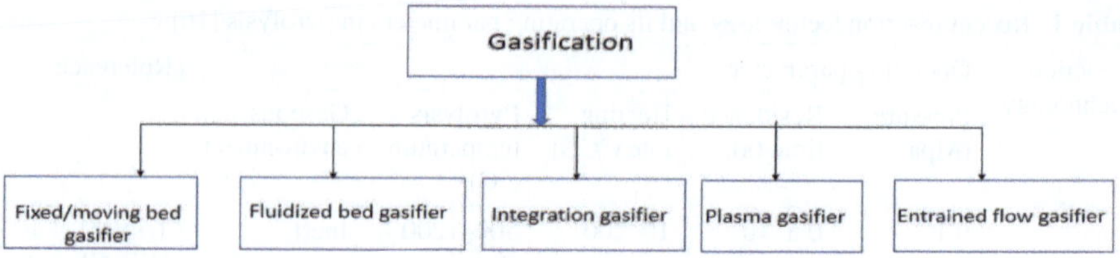

Fig. 1 Different types of gasifier technology

9 Gasification Technology

The gasification technique is an alternative pathway for biofuel production. The process design and integration of biofuel generation have an impact on the socio-environmental impacts and development of eco-friendly technology. The modification in catalyst and reactor of gasification technology can generate various important biofuels such as bio-methanol, bio-ethanol and higher alcohols, bio-dimethyl ether, Fischer–Tropsch fuels, bio-methane, bio-hydrogen, and algae-based fuels [32, 33].

There are different types of gasifiers, such as fixed bed/moving bed gasifiers, fluidized bed gasifiers, entrained flow gasifiers, integration gasification, and plasma gasifiers [34] as shown in Fig. 1. The major advantages of the gasifier are simple and robust design, such as economical at small scale tolerance of a large range of biomass. The disadvantage is not only that plasma gasifiers require high energy and maintenance costs but it is also capable of treatment of hazardous and toxic wastes. The fluidized bed gasifier is uneconomical for small scales, which is based on high particulate matter in syngas, but it reduces less tar and char yield.

10 Catalytic Conversion

The catalyst conversion technology in pyrolysis and gasification can be improved based on the product conversion factor by the reaction [35]. The different catalysts can improve the production yield, which will help to reduce the cost of biofuel. Currently, there are various catalysts such as zeolite, different metal alkali and alkaline earth metals, carbon and metal oxides used for different biomass compositions. It is also used to lower the operating temperature and pressure which saves energy and reduces the cost of wear and tear during the process. There are two types of catalysts such as homogeneous and heterogeneous processes, which are used for biofuel production. The higher yield and lower cost of catalysts in heterogeneous processes reduce the maintenance and operating cost of biofuel (biodiesel) production [36].

11 Economics of Different Energy Sectors

The economics of biomass energy is based on the energy produced and utilized energy and factor effects such as the price of biomass, biomass technologies, and other market factors [37–39]. The share of biomass is more than one-third of primary energy requirements. In rural areas, 90% of energy is obtained by biomass, and in urban areas, it is 40%. Biomass energy contributes to 10–14% of total energy demand worldwide. The agriculture sector provides the largest contribution to total biomass energy production. In total, 140 billion metric tons of biomass are produced by agriculture every year in India [40]. Rural India needs 1927TJ/day (2.75 MJ/capita-day) energy to meet the cooking energy demand. This waste biomass is sufficient to reduce the dependency on petroleum-based fuel and increase the economics of the agriculture sector. The cost of pulverized coal can be replaced by logging residue (wood sawdust), which reduces CO_2 emissions and generates a lower cost of electricity [41]. The quality improvement of biomass will enhance the biofuel production quality which will enhances economic benefits.

12 Conclusion

This chapter summarizes the economic potential of biomass energy in the service sector. Biomass can be considered an alternative to fossil fuels with low emissions of greenhouse gases. Biomass production and transportation can create employment in service sectors. This sector can reduce the exploitation of farmers and increase the per-year income from the agriculture sector. This biofuel production will enhance the carbon–neutral and sustainable environments. Innovation, research, and development can improve bioenergy technology, business opportunities in the service sector and products profits.

References

1. Serrano-Ruiz JC (2020) Biomass: a renewable source of fuels, chemicals and carbon materials. Molecules 25:5217. https://doi.org/10.3390/molecules25215217
2. Reid WV, Ali MK, Field CB (2020) The future of bioenergy. Glob Change Biol 26:274–286. https://doi.org/10.1111/gcb.14883
3. Thornley P, Gilbert P, Shackley S, Hammond J (2015) Maximizing the greenhouse gas reductions from biomass: the role of life cycle assessment. Biomass Bioenergy 81:35–43. https://doi.org/10.1016/j.biombioe.2015.05.002
4. Balan V (2014) Current challenges in commercially producing biofuels from lignocellulosic biomass. Int Sch Res Not 2014:e463074. https://doi.org/10.1155/2014/463074
5. Bryan BA, Ward J, Hobbs T (2008) An assessment of the economic and environmental potential of biomass production in an agricultural region. Land Use Policy 25:533–549. https://doi.org/10.1016/j.landusepol.2007.11.003

6. Geddes CC, Nieves IU, Ingram LO (2011) Advances in ethanol production. Curr Opin Biotechnol 22:312–319. https://doi.org/10.1016/j.copbio.2011.04.012
7. ARS Annual Performance Report for 2006: USDA ARS (n.d.). https://www.ars.usda.gov/docs/ars-annual-performance-report-for-2006/. Accessed 2 Nov 2023
8. Poluzzi A, Guandalini G, Romano MC (2022) Flexible methanol and hydrogen production from biomass gasification with negative emissions, sustain. Energy Fuels 6:3830–3851. https://doi.org/10.1039/D2SE00661H
9. Poluzzi A, Guandalini G, Guffanti S, Martinelli M, Moioli S, Huttenhuis P, Rexwinkel G, Palonen J, Martelli E, Groppi G, Romano MC (2022) Flexible power and biomass-to-methanol plants with different gasification technologies. Front Energy Res 9. https://www.frontiersin.org/articles/10.3389/fenrg.2021.795673. Accessed 2 Nov 2023
10. Kylee K, Harris K, Grim R, Tao L (2021) A comparative techno-economic analysis of renewable methanol synthesis pathways from biomass and CO_2, Volume 11: Sustainable energy solutions for changing the world: Part III. https://doi.org/10.46855/energy-proceedings-7301
11. Ball M, Weeda M (2016) 11—The hydrogen economy—vision or reality?. In: Ball M, Basile A, Veziroğlu TN (eds) Compendium of hydrogen energy. Woodhead Publishing, Oxford, pp 237–266. https://doi.org/10.1016/B978-1-78242-364-5.00011-7
12. Executive summary—Global hydrogen review 2021—Analysis. IEA (n.d.). https://www.iea.org/reports/global-hydrogen-review-2021/executive-summary. Accessed 2 Nov 2023
13. Economics, SGH2 energy (n.d.). https://www.sgh2energy.com/economics. Accessed 2 Nov 2023
14. Ab Rasid NS, Shamjuddin A, Abdul Rahman AZ, Amin NAS (2021) Recent advances in green pre-treatment methods of lignocellulosic biomass for enhanced biofuel production. J Clean Prod 321:129038. https://doi.org/10.1016/j.jclepro.2021.129038
15. Zhang Z, Donaldson AA, Ma X (2012) Advancements and future directions in enzyme technology for biomass conversion. Biotechnol Adv 30:913–919. https://doi.org/10.1016/j.biotechadv.2012.01.020
16. Li Y, Gupta R, Zhang Q, You S (2023) Review of biochar production via crop residue pyrolysis: development and perspectives. Bioresour Technol 369:128423. https://doi.org/10.1016/j.biortech.2022.128423
17. Tripathi M, Sahu JN, Ganesan P (2016) Effect of process parameters on production of biochar from biomass waste through pyrolysis: a review. Renew Sustain Energy Rev 55:467–481. https://doi.org/10.1016/j.rser.2015.10.122
18. Ghysels S, Estrada Léon AE, Pala M, Schoder KA, Van Acker J, Ronsse F (2019) Fast pyrolysis of mannan-rich ivory nut (Phytelephas aequatorialis) to valuable biorefinery products. Chem Eng J 373:446–457. https://doi.org/10.1016/j.cej.2019.05.042
19. Liu R, Sarker M, Rahman MdM, Li C, Chai M, Nishu N, Cotillon R, Scott N (2020) Multi-scale complexities of solid acid catalysts in the catalytic fast pyrolysis of biomass for bio-oil production—a review. Prog Energy Combust Sci 80:100852. https://doi.org/10.1016/j.pecs.2020.100852
20. Biswas B, Pandey N, Bisht Y, Singh R, Kumar J, Bhaskar T (2017) Pyrolysis of agricultural biomass residues: comparative study of corn cob, wheat straw, rice straw and rice husk. Bioresour Technol 237:57–63. https://doi.org/10.1016/j.biortech.2017.02.046
21. Tippayawong N, Kinorn J, Thavornun S (2008) Yields and gaseous composition from slow pyrolysis of refuse-derived fuels. Energy Sour Part Recovery Util Environ Eff 30:1572–1580. https://doi.org/10.1080/15567030701258550
22. Digman B, Joo HS, Kim D-S (2009) Recent progress in gasification/pyrolysis technologies for biomass conversion to energy. Environ Prog Sustain Energy 28:47–51. https://doi.org/10.1002/ep.10336
23. Kumar V, Nanda M (2016) Biomass pyrolysis-current status and future directions. Energy Sour Part Recovery Util Environ Eff 38:2914–2921. https://doi.org/10.1080/15567036.2015.1098751
24. Sekar M, Mathimani T, Alagumalai A, Chi NTL, Duc PA, Bhatia SK, Brindhadevi K, Pugazhendhi A (2021) A review on the pyrolysis of algal biomass for biochar and bio-oil—bottlenecks and scope. Fuel 283:119190. https://doi.org/10.1016/j.fuel.2020.119190

25. Garcìa-Pèrez M, Chaala A, Roy C (2002) Vacuum pyrolysis of sugarcane bagasse. J Anal Appl Pyrolysis 65:111–136. https://doi.org/10.1016/S0165-2370(01)00184-X
26. Li X, Peng B, Liu Q, Zhang H (2022) Microwave pyrolysis coupled with conventional pre-pyrolysis of the stalk for syngas and biochar. Bioresour Technol 348:126745. https://doi.org/10.1016/j.biortech.2022.126745
27. Nzediegwu C, Arshad M, Ulah A, Naeth MA, Chang SX (2021) Fuel, thermal and surface properties of microwave-pyrolyzed biochars depend on feedstock type and pyrolysis temperature. Bioresour Technol 320:124282. https://doi.org/10.1016/j.biortech.2020.124282
28. Lam SS, Lee XY, Nam WL, Phang XY, Liew RK, Yek PN, Ho YL, Ma NL, Rosli MH (2019) Microwave vacuum pyrolysis conversion of waste mushroom substrate into biochar for use as growth medium in mushroom cultivation. J Chem Technol Biotechnol 94:1406–1415. https://doi.org/10.1002/jctb.5897
29. Kong L, Zhang L, Gu J, Gou L, Xie L, Wang Y, Dai L (2020) Catalytic hydrotreatment of kraft lignin into aromatic alcohols over nickel-rhenium supported on niobium oxide catalyst. Bioresour Technol 299:122582. https://doi.org/10.1016/j.biortech.2019.122582
30. Oh S, Lee J, Lam SS, Kwon EE, Ha J-M, Tsang DCW, Ok YS, Chen W-H, Park Y-K (2021) Fast hydropyrolysis of biomass conversion: a comparative review. Bioresour Technol 342:126067. https://doi.org/10.1016/j.biortech.2021.126067
31. Roy P, Dias G (2017) Prospects for pyrolysis technologies in the bioenergy sector: a review. Renew Sustain Energy Rev 77:59–69. https://doi.org/10.1016/j.rser.2017.03.136
32. Sikarwar VS, Zhao M, Fennell PS, Shah N, Anthony EJ (2017) Progress in biofuel production from gasification. Prog Energy Combust Sci 61:189–248. https://doi.org/10.1016/j.pecs.2017.04.001
33. Molino A, Larocca V, Chianese S, Musmarra D (2018) Biofuels production by biomass gasification: a review. Energies 11:811. https://doi.org/10.3390/en11040811
34. Basu P (2010) Chapter 6—Design of biomass gasifiers. In: Basu P (ed) Biomass gasification and pyrolysis. Academic Press, Boston, pp 167–228. https://doi.org/10.1016/B978-0-12-374988-8.00006-4
35. Alonso DM, Bond JQ, Dumesic JA (2010) Catalytic conversion of biomass to biofuels. Green Chem 12:1493–1513. https://doi.org/10.1039/C004654J
36. Kiss FE, Jovanović M, Bošković GC (2010) Economic and ecological aspects of biodiesel production over homogeneous and heterogeneous catalysts. Fuel Process Technol 91:1316–1320. https://doi.org/10.1016/j.fuproc.2010.05.001
37. Stocker A, Großmann A, Madlener R, Wolter MI (2011) Sustainable energy development in Austria until 2020: insights from applying the integrated model "e3.at". Energy Policy 39:6082–6099. https://doi.org/10.1016/j.enpol.2011.07.009
38. Madlener R, Koller M (2007) Economic and CO_2 mitigation impacts of promoting biomass heating systems: an input–output study for Vorarlberg, Austria. Energy Policy 35:6021–6035. https://doi.org/10.1016/j.enpol.2007.08.023
39. Trink T, Schmid C, Schinko T, Steininger KW, Loibnegger T, Kettner C, Pack A, Töglhofer C (2010) Regional economic impacts of biomass based energy service use: a comparison across crops and technologies for East Styria, Austria. Energy Policy 38:5912–5926. https://doi.org/10.1016/j.enpol.2010.05.045
40. Kumar R, Kumar V, Nagpure AS (2023) Bio-energy potential of available livestock waste and surplus agriculture crop residue: an analysis of 602 rural districts of India. Sci Total Environ 889:163974. https://doi.org/10.1016/j.scitotenv.2023.163974
41. Gan J, Smith CT (2006) A comparative analysis of woody biomass and coal for electricity generation under various CO_2 emission reductions and taxes. Biomass Bioenergy 30:296–303. https://doi.org/10.1016/j.biombioe.2005.07.006

MIX
Papier aus verantwortungsvollen Quellen
Paper from responsible sources
FSC® C105338

If you have any concerns about our products,
you can contact us on
ProductSafety@springernature.com

In case Publisher is established outside the EU,
the EU authorized representative is:
Springer Nature Customer Service Center GmbH
Europaplatz 3, 69115 Heidelberg, Germany

Printed by Libri Plureos GmbH
in Hamburg, Germany